BISTATIC SAR DATA PROCESSING ALGORITHMS

BISTATIC SAR DATA PROCESSING ALGORITHMS

Xiaolan Qiu
Chibiao Ding
Donghui Hu
Institute of Electronics, Chinese Academy of Sciences, P.R. China

This edition first published 2013

Published by John Wiley & Sons Singapore Pte. Ltd., 1 Fusionopolis Walk, #07-01 Solaris South Tower, Singapore 138628, under exclusive license by Science Press in all media and all languages throughout the world excluding Mainland China and excluding Simplified and Traditional Chinese languages.

For details of our global editorial offices, for customer services and for information about how to apply for permission to reuse the copyright material in this book please see our website at www.wiley.com.

Library of Congress Cataloging-in-Publication Data

Qiu, Xiaolan.
Bistatic SAR data processing algorithms / Xiaolan Qiu, Chibiao Ding, and Donghui Hu.
pages cm
Includes bibliographical references and index.
ISBN 978-1-118-18808-8 (hardback)
1. Bistatic radar. 2. Signal processing. 3. Synthetic aperture radar. 4. Algorithms. I. Ding, Chibiao. II. Hu, Donghui. III. Title.
TK6592.B57Q58 2013
621.3848′5–dc23

2012050150

Typeset in 11/13pt Times by Aptara Inc., New Delhi, India
Printed and bound in Singapore by Markono Print Media Pte Ltd

Contents

About the Authors

Xiaolan Qiu received her B.S. degree in electronic engineering from the University of Science and Technology of China (USTC) in 2004 and a Ph.D. degree in signal and information processing from the Graduate University of the Chinese Academy of Sciences (GUCAS), Beijing, China, in 2009.

While working towards her Ph.D. degree, she was working as an intern in the Institute of Electronics, Chinese Academy of Sciences (IECAS) (website: www.ie.cas.cn) and worked on the IECAS bistatic SAR program. Her Ph.D. thesis was 'Research on Precise Imaging Algorithms for Bistatic SAR', and she won the Special Prize of President of The Chinese Academy of Sciences (*the highest honor of GUCAS and only 20 winners in China per year*) as a Ph.D. candidate in 2009 on this subject. She was then nominated to be a member of the Youth Innovation Promotion Association of CAS, which gives young researchers financial support for their studying, researching and communicating with other researchers.

After obtaining her Ph.D. degree, she joined IECAS and became a member of the SAR group in the Key Laboratory of Technology in Geo-spatial Information Processing and Application System (GIPAS), Chinese Academy of Sciences (CAS). In IECAS-GIPAS, Xiaolan Qiu was involved in developing processors for IECAS's airborne SARs and also for the Chinese Remote Sensing Satellites. From May to October 2011, she was supported by K.C. Wong Education Foundation, Hong Kong and being a guest scientist in the Center for Sensorsystems (ZESS), University of Siegen. Her current research interests include bistatic SAR, SAR interferometry and SAR simulation.

Chibiao Ding received his B.S. and Ph.D. degrees in electronic engineering from Beihang University, Beijing, China, in 1991 and 1997, respectively.

Since then, he has been working with the Institute of Electronics, Chinese Academy of Sciences "IECAS". In IECAS, he has taken charge of a number of significant projects in China, including the first high-resolution spotlight airborne SAR system and the first airborne InSAR system of China, and the remote sensing satellite data processing and application system. For his excellent work, he won the National Prize for Progress in Science and Technology of China (the first prize in 2006) and the National Youth Science and Technology Award (2007).

Chibiao Ding is currently a Research Fellow and the Vice Director of IECAS. His research interests include advanced SAR (including bistatic SAR, GeoSAR, MIMO SAR) data processing and application.

Donghui Hu received his B.S. degree from Peking University, Beijing, China, in 1992, and the M.S. degree from the Beijing Institute of Technology, Beijing, in 2001.

Since then, he has been with the Institute of Electronics, Chinese Academy of Sciences, where he has been involved in airborne and spaceborne SAR data processing and became a senior expert in this area. As a main contributor, he successfully developed the simulation and calibration software for the first remote sensing satellite of China. As the technical supervisor, he processed the data of IECAS's airborne InSAR and produced DSM with very high precision. Because of his contribution, he won the second prize of Scientific and Technological Progress Awards.

His current research interests include bistatic SAR data processing, SAR interferometry, geostationary orbit SAR data processing and radar moving target identification.

Preface

The synthetic aperture radar (SAR) is a microwave imaging radar that achieves high resolution while taking advantage of pulse compression technology and the Doppler effect. The SAR technology has made progress for over 60 years since the idea of a synthetic aperture was proposed by Wiley in 1951. Nowadays, SAR is widely used in environment protection, disaster detection, ocean observation, resource exploiting, agriculture, forest, space and aerial reconnaissance, and so on. SAR can work in an all-weather condition, day and night, and so is irreplaceable in these areas. As technology progresses, SAR is now desired to achieve improved capabilities, for example, higher resolution combined with a wider swath, multipolarization, three-dimensional resolution and so on. For some of these capabilities, the traditional monostatic SAR system has its limitations. Therefore, bi- and multistatic SAR systems become one of the important trends of SAR development. Therein, bistatic SAR has been paid a lot of attention all over the world since the start of the 21st century.

Data processing is one of the key problems and also a difficulty of bistatic SAR. Quite a lot of work has been done on this topic in the last decade. Especially on the topic of bistatic SAR imaging, many new imaging theories and imaging methods have been proposed. Though bistatic SAR imaging has not achieved its mature stage, the authors think the research on bistatic SAR imaging has been quite fruitful and could be used for reference to the data processing of other advanced SAR systems. To the authors' knowledge, there has not been any book in China that is special to bistatic SAR imaging or that summarizes the research results on bistatic SAR imaging, so they have tried to do this job based on work on bistatic SAR in their laboratories. The authors hope their work could be a help to those working on advanced SAR data processing. Therefore, in April 2010, the authors' book titled *Bistatic SAR Imaging Technology* (in Chinese) was published by the Science Press. After that, they were lucky to get the chance to publish their work in an English version by John Wiley & Sons Singapore Pte. Ltd, so they translated the Chinese book and also added some new achievements on bistatic SAR imaging.

The authors are in two laboratories hosted by the Institute of Electronics, Chinese Academy of Sciences (IECAS): Key Laboratory of Technology in Geo-spatial Information Processing and Application Systems (GIPAS) and The National Key Laboratory of Science and Technology on Microwave Imaging (MITL). In 2003, one of the authors attended the International Geoscience and Remote Sensing Symposium (IGARSS) and was inspired and got the idea to build a bistatic SAR receiver. It was decided that they would carry out bistatic SAR experiments combined with the airborne SAR at MITL, which is called the IECAS SAR, and exploit its potential applications. Then, GIPAS began to do research on bistatic SAR data processing and MITL began to study the bistatic synchronization methods. After about one year of research, MITL succeeded in obtaining funding to build a passive receiver. They started to design and construct the receiver and some indoor experiments were carried out. GIPAS was responsible for processing the experimental data. The bistatic SAR project was suspended for a time as an airborne InSAR project was considered to have priority. As a result, the first airborne–ground bistatic SAR experiment was not carried out until July 2011 and the first bistatic SAR image of IECAS was generated then. Between May 2011 and November 2011, the first author was at the Center for Sensor Systems (ZESS) hosted by the University of Siegen as a visiting scientist by the support of the K.C. Wong Education Foundation of the Chinese Academy of Sciences. In ZESS, she had the chance to process the bistatic SAR experimental data, including airborne bistatic SAR data and satellite–ground one-stationary bistatic SAR data. She verified some of the algorithms the authors had proposed using these data.

The authors wrote this book based on their work on bistatic SAR at MITL and GIPAS for about ten years. This book starts with a brief introduction of bistatic SAR and then provides the resolution theory and classical imaging algorithms for traditional monostatic SAR to be set as a basis. The bistatic SAR resolution is then analyzed in detail and echo simulation methods are discussed. Based on this, we get to the bistatic SAR imaging algorithms. They are categorized into algorithms for a translational invariant bistatic SAR configuration and algorithms for a translational variant configuration. The algorithms are further sorted by their basic idea in each category. Finally, parameter estimation and motion compensation methods closely related to image formation are introduced.

This book can be a reference for researchers and engineers who work on SAR data processing, especially on advanced SAR data processing. Also, it can be helpful to postgraduate students who are beginners on SAR data processing.

As mentioned above, though the authors try to make a summary of bistatic SAR imaging algorithms, this book is mainly based on their own work. Therefore, the authors' work is described in detail, while the work of others is only briefly introduced. The categorizing of the algorithms represents the authors' own opinion and they would be very grateful to have different opinions brought to their attention. The authors apologize in advance for any errors that may be present in the book and would be very grateful to have them brought to their attention.

Acknowledgements

We would like to acknowledge three groups of people who have been very important to this book.

Firstly, we would like to thank those people who have supported and supervised our work. We would like to thank Academician YiRong Wu, who is the director of IECAS, for the kind guidance and the financial support. We also would like to thank Prof. Wen Hong and Prof. XingDong Liang, who are the directors of MITL, and Prof. Kun Fu and Prof. Bin Lei, who are the directors of GIPAS, for their trust and support. In addition, we thank very much the KC Wong Education Foundation of the Chinese Academy of Sciences for supporting the first author to go aboard and work at ZESS, hosted at the University of Siegen. We thank very much Prof. Dr.-Ing. habil. Otmar Loffeld, who is the Speaker and Chairman of ZESS, for accepting the first author as a visiting scientist to work at ZESS half a year and providing her the great chance to work with the SAR group in ZESS and gaining access to real bistatic SAR data. Besides, we would like to thank Prof. Ian G. Cumming and Dr. Yew Lam Neo for their kind encouragement. Without all this support, the authors could not have realized the dream of publishing a book abroad.

Secondly, the authors would like to thank those people who made direct contributions to this book. Dr. LiJia Huang at GIPAS helped to write Section 5.3.3 of the book. Dr. Bin Han and Dr. DaDi Meng at GIPAS provided plenty of the material for Chapters 2 and 7. Liangjiang Zhou was in charge of taking the bistatic SAR experiment. SongYa xiong and Yanlei Li helped to process the IECAS's bistatic experimental data. FangFang Li and Wei Han helped to draw many of the figures in this book. We also thank other members of the SAR group at GIPAS; they are Yuxin Hu, Jiayin Liu, Lihua Zhong, Wenyi Zhang, Hui Zhang, Yueting Zhang and so on. We would like to give our special thanks to Prof. Loffeld at ZESS who helped us to revise the first chapter of this book and gave many valuable comments and made many helpful revisions. We would also like to thank all the SAR group members in ZESS; they are Dr. Holger Nies, Florian Behner, Simon Reuter, JinShan Ding, Wei Yao, Amaya Medrano Ortiz, Qurat Ul-Ann, Ashraf Samarah, Valerij Peters and the former member Robert Wang. The first author has benefited from the discussions with them. The authors would like to thank very much Prof. Jie Chen at Beihang University, Beijing, China, for the kind help with translating the first two chapters of this book from Chinese to English. We also would like to thank DangDang Zhang and Xue Lin at GIPAS for the translation of Chapters 3 and 7. Especially, we would like to thank GuiYing Wang, who is the director of the Information Technology Center hosted at IECAS, for her help with planning this book.

Finally, we would like to thank our families for their patience, support, tolerance, caring and all the things they have done for us.

List of Acronyms

APC	antenna phase center
BIT	Beijing University of Technology
BP	back-projection
BPTRS	bistatic point target reference spectrum
CDE	correlation Doppler estimator
CS	chirp scaling
CT	computed tomography
DEM	digital elevation model
DFG	German Research Foundation
DInSAR	differential interferometric SAR
DLR	German Aerospace Center
DMO	dip-move-out
DSR	double square root
ESA	European Space Agency
EUSAR	European Conference on Synthetic Aperture Radar
FFT	fast Fourier transform
FMCW	frequency modulation continuous wave
HRTI-3	high-resolution terrain information 3
IDW	instantaneous Doppler wavenumber
IECAS	Institute of Electronics, Chinese Academy of Sciences
IFFT	inverse fast Fourier transform
IGARSS	International Geoscience and Remote Sensing Symposium
INS	inertial navigation system
InSAR	interferometric SAR
ISLR	integrated sidelobe ratio
JSTARS	joint surveillance target attack radar system
LBF	Loffeld's bistatic formula
LFM	linear frequency modulated
MTI	moving target indication
NLCS	nonlinear chirp scaling
NuSAR	numerical SAR
PD	phase difference

PGA	phase gradient autofocus
PolInSAR	polarimetric interferometric SAR
PolSAR	polarimetric SAR
POSP	principle of stationary phase
PRF	pulse repetition frequency
PSLR	peak sidelobe ratio
RADAR	radio detection and ranging
RCM	range cell migration
RCMC	range cell migration correction
RCS	radar cross-section
RD	range-Doppler
RMS	root mean square
ROPE	rank one phase estimation
3D SAR	three-dimensional SAR
SAR/FTI	synthetic aperture/fixed target indicator
SIMU	strapdown inertia measuring unit
SNR	signal-to-noise ratio
SRC	second range compression
STFT	short-time Fourier transform
TBP	time bandwidth product
TI configuration	translational invariant configuration
TV configuration	translational variant configuration
UESTC	University of Electronic Science and Technology of China
WAS/MTI	wide-area surveillance/moving target indicator
ZESS	Center for Sensorsystems

1

Introduction

1.1 Overview of SAR Development

Radar is the acronym of "radio detection and ranging". It intends to detect and identify targets through the properties of electromagnetic waves reflected from obstacles (targets). Radar can detect a wide variety of targets, ranging from buildings, roads, bridges, vehicles, aircrafts, ships and other man-made objects to the mountains, rivers, forests, deserts, sea and other natural landscape. Furthermore, radar can detect the existence of long-distance targets without having effects of meteorological factors, such as daylight, clouds and rain conditions. Because of these characteristics of radar, it has been increasingly becoming an important tool in the field of microwave remote sensing and has played an important role in all aspects of remote sensing applications since its appearance in World War II. The history of radar has a close relationship with military applications since its inception. The advantages of radar in the areas of battlefield reconnaissance, target surveillance, weapon guidance and other aspects of performance greatly stimulated the interest of major industrial powers to radar, and this became the driving force of radar technological advances after World War II. Synthetic Aperture Radar (SAR) is a new high-resolution radar system that appeared in this period [1–6].

1.1.1 The History of SAR Development

SAR is an active microwave imaging radar that can achieve two-dimensional high-resolution images. The concept of synthetic aperture was firstly proposed to improve the azimuth resolution of radar. Before the concept was promoted, the traditional airborne radar used real aperture to obtain the images of the ground, that is, to distinguish targets at different locations through the direction of a real antenna beam. There is an inherent defect in real aperture radar, which is that its azimuth resolution is related to the distance between the radar and target. Within the constraints of

Bistatic SAR Data Processing Algorithms, First Edition. Xiaolan Qiu, Chibiao Ding and Donghui Hu.

wavelength and antenna size, as the distance between the radar and target increases, the radar azimuth resolution decreases. In order to obtain a high-resolution image of long-distance targets, the antenna beam must be extremely narrow. However, a narrow beam pattern can only be formed by a large antenna. For an airborne imaging radar, the distance between the radar and target is usually from several tens of kilometers to a hundred kilometers; for a spaceborne imaging radar, the distance between the radar and target is up to several hundred kilometers. In this way, to achieve only tens of meters of the azimuth resolution, the required aperture of the radar antenna would be a few kilometers or even dozens of kilometers, which is impossible to achieve in practice.

People have been looking for new ways to resolve the problem of low azimuth resolution in real aperture radar imaging. Carl A. Wiley, a mathematician at Goodyear Aircraft Corporation in Litchfield Park, Arizona, invented the concept of using azimuth frequency analysis to improve the azimuth resolution, which could provide the basis of SAR theory. In July 1953, Control Systems Laboratory of the University of Illinois obtained the first SAR image based on a nonsynthetic aperture focusing method. In the same year, in the summer seminar organized by the University of Michigan in the United States, many scholars presented the technique to utilize carrier aircraft movement with a real radar antenna to an integrated jumbo size linear antenna array. This indicates the concept of synthetic aperture really entering the domain of radar. On this basis, the SAR was developed in August 1957 for flight experimentation and obtained the first piece of a large area high-resolution SAR image. Since then, SAR has been widely recognized and began to enter the practical stage.

Early, Fourier lens were used by SAR to obtain images. This method is very inconvenient to use, for example, it needs fine adjustment of the lens placed in the optical path; it is difficult to do automatic processing and the images obtained by this method usually have poor quality. In order to overcome the inherent defects in optical methods, people began to study SAR digital signal processing technology [7–12]. Compared with optical processing, digital processing has great flexibility to adapt to the needs of different occasions; in particular, it is able to meet the special requirements of signal processing of spaceborne SAR. The major difficulties of digital technology in SAR signal processing are the requirements for a huge capacity for data storage and high processing speed, which require high performance of both hardware and software; therefore, SAR development has been restricted over a very long period. In the 1970s, with the maturity of computer technology and the development of fast imaging algorithms for SAR, this situation was fundamentally changed [13–18]. In 1976, the digital processing technology was applied, for the first time, to United States marine satellite (SEASAT) imaging, and successfully obtained images of nearly 1 million square kilometers of the earth's surface. After this milestone achievement, SAR began to enter the field of earth observation and a new era of space remote sensing began. Since then, many technological advanced countries have launched their own satellites carrying SAR, such as ALMAZ (Russia), ERS-1/ERS-2 (European Space Agency, ESA), RadarSAT-1 (Canada), and so on, setting off a worldwide interest in SAR research and application.

The major advantages of SAR can be summarized as follows [19]: (1) all-day and all-weather imaging capabilities; (2) high azimuth resolution; (3) independence between image resolution and radar wavelength and range; (4) capability of penetrating a certain degree of shelter by selecting an appropriate wavelength. Because of the advantages of SAR, it has a wide range of applications in resource exploration, disaster prediction, environmental protection, military reconnaissance and other related areas. Today, SAR has become a multidisciplinary research field that is rich in content and continues to open up new research directions. Its important position in the field of space remote sensing is increasingly being reflected.

1.1.2 The Current Status and Trends of SAR Development

The traditional SAR is usually based on a single platform and usually works in a single band and a single polarization mode; however, to meet the application requirements of different resolutions and mapping swaths, SAR usually has a variety of imaging modes. The traditional modes of SAR imaging are as follows (as shown in Figure 1.1):

(1) Stripmap SAR. This is the most common and simplest mode of SAR imaging, in which the antenna keeps pointing in the same direction relative to the platform, so that the antenna beam sweeps evenly on the ground with the platform moving and a continuous image is obtained for a continuous strip. The azimuth resolution in this mode is determined by the antenna length.
(2) Scan SAR. This is an imaging mode that achieves a wide swath at the expense of azimuth resolution. It is suitable for a large area survey. The antenna scans along the range direction, intermittently and periodically, that is, the antenna beam points to the multiswath by turn through an antenna pointing switch while the platform moves. A wide swath is achieved by a multiswath mosaic. To ensure the continuity of observations, the dwell time of each swath is limited, so that the scene target is not completely irradiated by the antenna aperture. Thus the azimuth resolution is determined by the dwell time. Compared with the stripmap mode, the azimuth resolution of a scan SAR is lower.

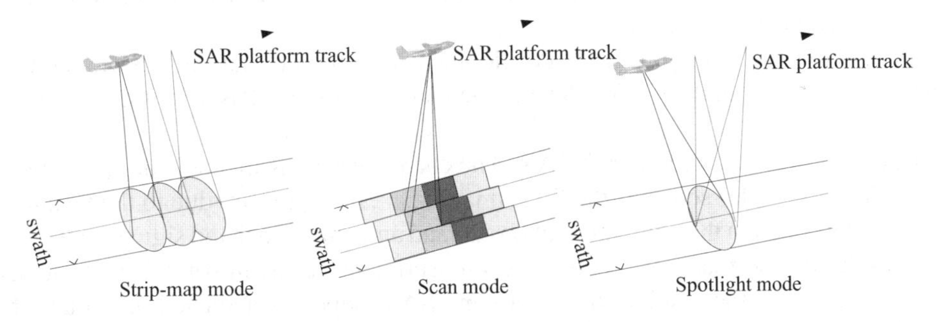

Figure 1.1 Diagram of traditional SAR imaging modes

(3) Spotlight SAR. This is a mode that sacrifices the continuity of the imaging area to obtain a high azimuth resolution. It is suitable for detailed investigation of a limited area. In this mode, the antenna beam direction is gradually adjusted backward while the platform moves forward, so that some circular area on the ground can be covered by the antenna beam for a relatively longer time. By adjustment of the antenna, SAR breaks the constraints of the synthetic aperture length by the antenna length, which results in higher azimuth resolution. However, due to the fact that the antenna beam coverage on the ground is not moving forward with the platform, the spotlight mode can only image a limited circle field on the ground at once and so the obtained images are not continuous.

After the development in the last half century, SAR has become a mature remote sensing tool. With intense awareness of its potential applications and the development of application technology, the traditional SAR systems and imaging modes cannot satisfy the requirements of further applications. Therefore new systems and new concepts of SAR are being put forward to achieve stronger abilities in remote sensing. Table 1.1 summarizes the operating characteristics of some representative and advanced spaceborne and airborne SAR systems.

From the design features of these systems, it can be seen that the current SAR systems have been advanced out of the traditional simple modes (which means single frequency/unipolarization/single-channel-based modes and so on) and evolved into multifrequency/multipolarization/multichannel integrated modes and so on. The transmitted signal is no longer confined to the traditional form of chirp, but also to frequency modulated continuous wave (FMCW), step frequency modulated signal, phase coding signal and other complex waveforms. The application of such new technologies not only further enhances the resolution of SAR systems but also expands the obtained information categories of SAR; for example, other than the geometric features of targets, SAR can obtain more information of targets including the characteristics of polarization, elevation, motion parameters and so on. In addition, with the development of large-scale digital integrated circuits, thin-film antennas, digital beam, pulse labeling and other new technologies, as well as new processes and new materials, the SAR structure is gradually becoming small, lightweight, modular and polymorphic, which greatly broadens SAR applications in the field of remote sensing, both in breadth and in depth.

Integrating the emerging SAR development and the cutting-edge SAR requirements, the major trends of SAR development are summarized as follows:

(1) High-resolution, wide-swath SAR. The changeless SAR theme is of higher resolution and wider scope of the SAR imaging area. At present, due to the technical breakthrough in the realization of broadband and ultra wideband signals, the range resolution can reach decimeter or even centimeter-levels in some SAR systems. For example, the resolution of Germany SAR-Lupe has achieved 0.5 m and the latest series of United States Lacrosse satellites (Lacrosse 5) in fine mode is able

Table 1.1 A summary of advanced spaceborne and airborne SAR systems

	System name	Frequency band	Polarization	Imaging mode	Highest resolution (m)	Launched/ built time	Nations	Others
Spaceborne	RadarSAT-2	C	Full	Stripmap/scan SAR	3	2007	Canada	Dual channel (experimental)
	TerraSAR-X	X	Single/dual/full	Spotlight/stripmap/ scan/sliding spotlight	1	2007	Germany	Dual channel ATI /repeat path PolInSAR
	ALOS/PALSAR	L	Dual/full	Stripmap/scan	7	2006	Japan	/
	SAR-Lupe	X	Multipolarization	Stripmap/spotlight	0.5	2006–2008	Germany	Five satellites (770 kg each)
	Cosmo-skymed	X	Full	Stripmap/scan/spotlight	1	2007–2010	Italy	Four satellites
	Envisat/ASAR	C	Single/dual	Stripmap/scan	28	2002	ESA	/
Airborne	F-SAR	X/C/S/L/P	Full	Stripmap/spotlight	0.3	2006	Germany	ATI/XTIrepeat path PolInSAR, Step frequency signal
	PAMIR	X	Full	Stripmap/spotlight/ sliding spotlight	0.07	2001	Germany	Phased array antenna, InSAR, ISAR, MTI(STAP)
	RAMSES	P/L/S/C/X/ Ku/Ka/W	Full	Stripmap	0.1	—	France	ATI PolInSAR
	PISAR	X/L	Full	—	3	1997	Japan	PolInSAR
	MiSAR	Ka	—	—	0.5	2002	Germany	FMCW, 4 kg
	MiniSAR	Ku/Ka/X	—	Stripmap/spotlight	0.1	2005	USA	12.3 kg

to reach 0.3 m resolution ability; French Airborne SAR RAMSES reaches 0.1 m resolution [20], while German advanced airborne SAR (PAMIR) has achieved centimeter-level ultrahigh resolution. In azimuth, in order to solve the ever-present contradiction between resolution and swath, several new imaging modes (such as the sliding spotlight mode) have been designed, and many new design concepts are proposed, including multiple phase center technology, distributed SAR, high orbit/geosynchronous orbit SAR and so on.

(2) Interferometric SAR, (InSAR). The concept of interference is a revolutionary development in SAR history, which gives SAR the three-dimensional topographic mapping ability for the first time. It arouses widespread concern due to its importance in remote sensing, especially in the military remote sensing area. InSAR receives the ground elevation information through a set of complex SAR image pairs of the same region observed from different angles. There are two different realization methods, single pass and repeat pass. Differential InSAR (DInSAR), which has been further developed based on InSAR technology, can measure the dynamic change of the surface elevation. Thus, it is widely used for surface subsidence, glacial changes and so on. The permanent scatter technique [21–23] is a very active research topic in the DInSAR field.

(3) Polarimetric SAR (PolSAR). In early times, the SAR system usually transmitted and received electromagnetic waves in a single polarization mode (HH or VV, where H is the acronym for horizontal polarization and V is vertical polarization). Using a single polarization means that the electromagnetic wave vector is treated as a scalar and thereby the phase information carried by radar echoes will be lost. As the phase information contains the scattering mechanism of ground targets, using single polarization produces a lack of being able to distinguish different scattering mechanisms. Using multipolarization technology, images of different polarizations (HH, VV, HV, VH) can be integrated together and analyzed, and the scattering mechanisms of different surfaces can be distinguished, thereby improving the classification of SAR images [24–26]. Multipolarization SAR has a unique advantage in vegetation classification, forest mapping and so on, so most advanced SAR systems (such as RadarSAT-2, TerraSAR-X, F-SAR) are all equipped with multipolarization or full-polarization features.

(4) Polarimetric interferometric SAR (PolInSAR). This is a new technology that combines polarization and interference [27,28]. It combines the ability of polarimetric SAR to classify different targets and the ability of InSAR to measure the terrain elevation. It can decompose different types of scattering mechanism at different heights, which has very important values for the physical parameter inversion of vegetation and forest biomass estimation and so on. Meanwhile, the military is also very interested in its capability of detecting foliage covered targets. PolInSAR is a cutting-edge branch of the SAR remote sensing field. It has only been over 10 years since its concept was proposed. Currently, there are some advanced airborne and spaceborne SAR systems that have the ability to obtain polarimetric interferometric data, but the full normal operation mode of

polarization interference in data acquisition has not yet been realized. Published information on the current view, Tandem-L system, which is planned to be launched by 2015 to 2020 (developed by the German Aerospace Center (DLR)), will carry the polarimetric interferometric SAR as an operational mode.

(5) Three-dimensional SAR (3D SAR) [29]. Three-dimensional SAR, a new type of microwave remote sensing technology, was developed based on the traditional two-dimensional imaging model of SAR in recent years. Typical 3D SAR imaging modes include multibaseline SAR tomography, curved SAR, down-looking three-dimensional SAR, circular SAR and so on. In theory, whatever the three-dimensional SAR mode is, it must form a three-dimensional space-distributed sample array, so that resolution ability is achieved in all three directions, which are along the horizontal plane and the vertical direction. 3D SAR can achieve the elevation resolution ability rather than the elevation measurement ability, which is different from InSAR. The main difficulty faced by 3D SAR lies in its sparse spatial sampling. To solve this problem, a few new signal processing theories and methods are required, such as modern spectrum estimation technology, compressed sensing (or namely compressive sampling theory) [30] and so on.

(6) Moving target indication (MTI) [31,32] combined with SAR. MTI is a very important military application field in remote sensing. Combining SAR technology and MTI technology can not only improve radar detection performance to moving targets but also can offer near real-time, high-resolution battlefield environmental images. It can play an invaluable role in grasping the battlefield for the commander to make a good military deployment and it is valuable in guiding attack and assessing the attack effect. The most famous system is American E8C-JSTARS (joint surveillance target attack radar system), which is equipped with two kinds of main battle mode: WAS/MTI (wide-area surveillance/moving target indicator) mode and the SAR/FTI (synthetic aperture radar/fixed target indicator) mode. It played an unprecedented role in the Gulf War in 1991.

For the modes and systems of SAR listed above, there are two technical approaches to realize them, which are single-platform technology and multiplatform technology. In single-platform technology the transmission and reception of electromagnetic waves are carried out by the same radar and antenna in the same spatial location. Nowadays, the majority of existing SAR systems uses this approach. In multiplatform technology the transmission and reception of electromagnetic waves are carried out by two or more radars or antennas at different spatial locations. Generally, it is referred to as distributed SAR. Compared to single-platform technology, multiplatform technology provides more flexibility and can achieve a variety of earth observation SAR capabilities, such as high-resolution/wide-swath imaging, interferometry and moving target detection. Meanwhile, the use of multiple satellites for an observation network can shorten the SAR observation period and improve the timeliness of SAR data acquisition. Due to the incomparable advantages of multiplatform technology, which are superior to those of single-platform technology, it has attracted universal attention in the SAR remote sensing area in recent years. Bistatic SAR is the simplest form in multiplatform SAR, which is considered to be the basis of the more complex multiplatform systems. Therefore, it has recently become one of the hotspots in the SAR field. Section 1.2 will give a separate description of bistatic SAR.

1.2 Brief Introduction of Bistatic SAR

1.2.1 Basic Concept of Bistatic SAR

Bistatic SAR is the simplest form of multiplatform SAR system. It refers to the SAR system in which the transmitting and receiving antenna are placed on two different platforms. From a signal processing point of view, bistatic SAR also refers to the SAR system in which the transmitting and receiving antenna phase centers of the same pulse are in different spatial positions. Therefore, strictly, the traditional monostatic SAR should be a kind of bistatic SAR as to the latter definition because a certain time elapses during pulse transmission, forward propagation, scattering and return to the receiving antenna. The SAR moves during this short period, so the antenna phase center is at different positions when the same pulse is transmitted and received. However, the SAR moving velocity is very small compared with the microwave propagation velocity. Thus the movement of the antenna phase center can be ignored and the “stop–go–stop” approximation can be applied to consider the antenna phase center while transmitting and receiving the same pulse to be at the same position. Without special instructions, transmitting and receiving antennas of bistatic SAR in this book are physically separated.

1.2.2 The Advantages and the Prospects of Bistatic SAR

Compared with the traditional monostatic SAR, bistatic SAR has many advantages and broad application prospects, some of which are listed as follows:

(1) Military applications. Because the receiver works “silently”, bistatic SAR is good in hiding, antijamming and is of high security. It can perform in the way of “transmitting faraway, receiving nearby” to increase radar effective range or to improve the antiseized ability by reducing the transmission power. In addition, because the receiver does not contain high-power devices, it is cheap to build and easy to implement.
(2) Interferometric applications. Compared to repeat-pass monostatic InSAR, bistatic SAR interferometry can avoid the temporal decoherence; hence it can achieve a high accuracy of interferometry. Compared to dual-antenna interferometry by monostatic SAR, bistatic SAR can have a longer baseline, which the former finds hard to achieve, so that it can improve the accuracy of terrain height measurement.

(3) Target classification and recognition. Bistatic SAR can acquire the scattering information of a target's radar cross-section (RCS) from different directions. This helps it to measure the surface roughness and dielectric constant (especially when the RCS is not very strong in the monostatic case). This also helps it to study the surface clutter scattering mechanism. Due to the fact that RCS varies with bistatic angle, bistatic SAR helps to improve the capacity of image classification and recognition. In urban areas, bistatic SAR can avoid intense backscattering building tops to reduce the image dynamic range and to improve the signal-to-noise ratio (SNR) of vehicles.

(4) Marine applications. A boarder bandwidth of marine spectrum can be obtained by observation of bistatic SAR, which introduces significant features of sea waves in the SAR imaging model [33, 34].

1.2.3 The Present Status of Bistatic SAR Development

Owing to the above advantages, bistatic SAR technology attracted widespread interest. The research of bistatic SAR technology first appeared in the 1970s. In 1977, the research of United States Xonics Company showed that in bistatic configuration it can realize moving target detection and synthetic aperture imaging. In 1979, Goodyear Company and Xonics Company signed a contract with the US Air Force for the formal implementation of the "tactical bistatic radar verification". The program, conducted in May 1983, demonstrated good bistatic SAR images and successfully detected slowly moving tanks hidden in woods.

After the 1980s, some patents on bistatic SAR image processing, data correction, double-spaceborne SAR imaging system, as well as bistatic SAR synchronization technology and other relevant technology were put forward in the United States.

In the twenty-first century, due to the technological improvements in timing, communication and navigation, bistatic SAR technology development set a new trend. The published literature about bistatic SAR, for example, in system design, synchronous technology and image processing, increases rapidly. Meanwhile, bistatic SAR gains more and more attention in international conferences. In 2002, there were some articles about bistatic SAR shown in the International Geosciences and Remote Sensing Symposium (IGARSS) conference. Since then, there has been a "bi-/multistatic SAR" topic every year in the IGARSS conference. Since 2004, the European Conference on Synthetic Aperture Radar (EUSAR) meeting has also set up a special subject of bistatic SAR, and many invited reports and academic papers on bistatic SAR have been presented in the EUSAR conferences. Furthermore, some technologically advanced countries have carried out bistatic SAR experiments of different configurations, such as airborne bistatic SAR experiments [35–37], ground–airborne bistatic SAR experiments [38,39], satellite–ground bistatic SAR experiments [40–44] and satellite–airborne bistatic SAR experiments [45–47], and obtained bistatic SAR images of good quality.

In 2002, a bistatic SAR experiment was carried out by QinetiQ in the United Kingdom. The frequency of this experiment is in the X-band, both the receiver and the transmitter use the spotlight imaging mode and bistatic angles of 50 and 70° are tested in this campaign. The image results of this experiment were first exhibited in [36]. Because of the different directions of incident and reflected waves, a double-shadow phenomenon of the trees in the bistatic SAR image was shown in [36], which may be able to provide additional information for the object height extracted based on its shadows.

In 2003, a bistatic airborne SAR experiment was carried out with the cooperation of DLR and ONERA. The experiment is described and the image results are exhibited in [35]. The experimental frequency is also in the X-band and the transmitted signal bandwidth is 100 MHz. Three different bistatic SAR configurations are carried out in this experiment, which are: tandem configuration, parallel-track configuration with a large incidence angle and parallel-track configuration with a small incidence angle. Three images are obtained in the three configurations, respectively, which are fused together with different colors. The pseudo-color composite image in [35] verifies that fusion of the images obtained by different bistatic SAR configurations can improve the ability of land feature recognition and classification.

In November 2007, a spaceborne–airborne bistatic SAR experiment was carried out by German Aerospace Centre (DLR). In this experiment, the TerraSAR-X is used to be the transmitter and the advanced airborne SAR (F-SAR) is used as the passive receiver. The resulting images of this experiment are shown in [45]. The satisfied image quality in [45] validates the success of the experiment.

Only a little later in April 2008 and funded by German Research Foundation (DFG) a large collaborative project "Bistatic Exploration" between DFG FHR and ZESS (Center for Sensorsystems) of the University of Siegen was carried out with an even more sophisticated experimental setup [46,47]. While in the DLR TerraSAR-X/F-SAR experiment TerraSAR-X operated in the sliding spotlight mode and F-SAR in the strip map mode, FHR's PAMIR receiver system operated in a new mode of an inverse sliding spotlight mode to prolong the cooperation time of the transmitter and receiver, thus extending the scene extension. Further successful experiments demonstrated bistatic forward-looking imaging [48], bistatic imaging with orthogonal tracks and so on. While the focusing for the DLR experiments was basically performed using a time domain back-projection algorithm, ZESS and FHR developed efficient frequency domain algorithms for the focusing of rather general bistatic configurations including the spaceborne/airborne configuration [49–53].

Since 2009, ZESS has carried out several bistatic SAR experiments using their passive receive system, named "HITCHHIKER" [41–44]. In their experiments, the TerraSAR-X is used as a transmitter of opportunity. The HITCHHIKER receive system was extended to three echo channels in 2010 and so it has the ability to perform multibaseline interferometry or both interferometry and polarimetry during one data acquisition. Bistatic SAR interferometry results acquired by two receive antennas and a fully polarimetric image acquired by two receive channels of different polarizations (H and V) receiving the echo from the dual-polarization mode of TerraSAR-X are shown in [44]. These good results validate the capability of the receive system.

The research on bistatic SAR in China started a little later. It was about 2003 when the research reports on bistatic SAR began to increase. Nowadays, a number of institutions and universities are currently studying bistatic SAR. The Institute of Electronics Chinese Academy of Sciences (IECAS), University of Electronic Science and Technology of China (UESTC), Xidian University, Beihang University, Beijing University of Technology (BIT), National University of Defense Technology and so on have all carried out theoretical research in the fields of synchronization, image processing and so on. Among the research work, Tang Ziyue from IECAS published the first book on bistatic SAR [54], which is mainly about bistatic SAR system theory. In addition, the UESTC conducted the first bistatic SAR experiment in China, which is a vehicle-based bistatic SAR experiment carried out in 2006. The experiment is described and the experimental images are shown in [55] and [56].

From the published literature, we can see that Germany has carried out many bistatic SAR experiments with a sophisticated setup (e.g., by FHR and DLR) and has developed a number of processing approaches to obtain quality images from these experiments of complex configurations (e.g., by FHR and ZESS), which verifies its advanced position in this field at this moment. Other than Germany, published literature shows that countries like France, Britain, Spain, Sweden, Switzerland and China have also carried out bistatic SAR field experiments and reported related data processing methods in recent years. As to the spaceborne bistatic case, the TanDEM-X Mission by DLR and EADS [57] turns out to be the world's first successful operational spaceborne bistatic SAR. The first TerraSAR-X radar satellite was successfully launched in June 2007 and the second satellite was launched in June 2010. The two satellites fly in Helix orbit formation to generate a global digital elevation model (DEM) with an accuracy corresponding to the High-Resolution Terrain Information 3 (HRTI-3) specifications through bistatic InSAR operation. Besides, Italy has promoted a BISSAT plan [58], which aims to launch a small satellite to form an onboard spaceborne bistatic SAR system.

Throughout the history and the current development of bistatic SAR, it can be seen that bistatic SAR is a topic of common interest at present, and is in a booming, thriving and ever-progressing stage.

1.2.4 The Key Problems of Bistatic SAR

The advantages of bistatic SAR mentioned above are gained at the cost of system complexity, in which the first key issue is the synchronization problem between the transmitter and the receiver, including beam synchronization, time synchronization and phase synchronization. Among these three synchronization problems, the most difficult one is phase synchronization. Another key issue is the signal processing problem aroused by the separation of the transmitter and receiver in bistatic SAR systems. Compared with the monostatic SAR, challenges of bistatic SAR signal processing can be concluded as follows:

(1) The analysis of the capability of the bistatic SAR system is more complicated. Because of the separation of the transmitter and the receiver, the echo from a certain target is weighted both by the transmitting antenna pattern and the receiving antenna pattern. In addition, the Doppler history of the echo is determined by the transmitter and the receiver simultaneously. Moreover, the relative position of the transmitter and the receiver is changing with time in some configurations. These make the calculation of the systematic parameters, for example, the synthetic aperture time and the ambiguity ratio, quite difficult, and especially make the analysis of the bistatic SAR's two-dimensional resolution a very difficult job.
(2) Scene raw data simulation of bistatic SAR is more difficult than monostatic SAR. Usually, due to the large computational burden of scene raw data simulation using the target-by-target method, it is required to investigate faster simulation techniques. For monostatic SAR, with the emergence of new complex monostatic SAR modes, the echo simulation method is still further deepened. For bistatic SAR, because the geometry is more complex than that of monostatic SAR, the monostatic SAR echo simulation methods cannot be directly applied to bistatic SAR. Therefore, the scene raw data simulation of bistatic SAR is one of the issues to be studied.
(3) Bistatic SAR imaging is more difficult owing to the different characteristics of each configuration. The range history of bistatic SAR is determined both by the relative movement of the transmitting antenna to the target and by the relative movement of the receiving antenna to the target, so it is different from that of the monostatic SAR. Classical frequency-domain or mixed-domain algorithms of monostatic SAR are not suitable for bistatic SAR imaging. In addition, some bistatic SAR configurations have two-dimensional spatial variant properties, which imposes a great challenge for imaging. Therefore, keeping the characteristics of different configurations in view, imaging algorithms with high performance, high accuracy, able to handle large swath and be adapted to different configurations are required to be explored.
(4) Application potential in special configurations of bistatic SAR is required to be explored. Bistatic SAR has flexible configurations and thus contains the potential for some special applications (such as the forward-looking imaging, three-dimensional resolution, etc.). However, the existence of these special applications is often associated with some unique problems. To exploit the advantages and to improve the performance of bistatic SAR, it is necessary to explore and study the potential applications of bistatic SAR.
(5) Doppler parameter estimation methods of bistatic SAR require improvements. Because of the separation of the transmitter and receiver, the error sources of beam pointing increase and Doppler parameter estimation based on echo data

becomes more important. Existing monostatic SAR Doppler parameter estimation methods usually have some shortcomings, for example, a strong dependence on scenario contents and lack of robustness. So there is still space for improvement. In addition, a bistatic SAR echo has some unique properties; for example, the echo spectrum may be asymmetrical as a result of the modulation of the transmitting antenna and the receiving antenna. Therefore, the Doppler parameter estimation of bistatic SAR requires consideration under specific circumstances.

(6) Motion compensation of bistatic SAR is more complex. Motion compensation is a key problem of airborne SAR. Because there are two carrier aircraft platforms in the airborne bistatic SAR system, which means more motion error sources than monostatic SAR, the measurement and the analysis of the motion error are more difficult. Besides, the design and implementation of motion compensation techniques are complex due to the different effects of transmitter and receiver motion errors to the image.

1.3 Contents of the Book

This book mainly talks about the bistatic SAR signal processing problems, especially focuses on the imaging processing. It explains systematically and comprehensively about the issues of bistatic SAR signal processing, including resolution analysis, echo simulation, imaging algorithms, parameter estimation and motion compensation.

This book is divided into seven chapters.

Chapter 1 describes the SAR development status and trends, which leads to the concept of bistatic SAR and points out the advantages and the difficulties of bistatic SAR.

Chapter 2 briefly describes the principles of SAR and some conventional imaging methods. It provides basic knowledge of traditional monostatic SAR signal processing, which leads to better access to the content of bistatic SAR.

Chapter 3 firstly introduces the classification of bistatic SAR configurations and then analyzes the imaging performance of different configurations, focusing on the two-dimensional resolution. Readers can gain a better understanding of bistatic SAR characteristics through this chapter.

Chapter 4 describes the scene raw data simulation method of bistatic SAR, which provides a simulated data basis for the introduction of the follow-up imaging algorithms.

Chapter 5 presents a quite comprehensive study of various bistatic SAR algorithms for translational invariant configuration and also presents a performance comparison between different algorithms.

Chapter 6 introduces several imaging algorithms for the translational variant configuration.

Chapter 7 briefly describes the Doppler parameter estimation and the motion compensation techniques, which are closely related to imaging.

References

[1] Skolnik, M.I. (1990) *Radar Handbook*, McGraw-Hill, New York.

[2] Revillon, G. (1965) Synthetic aperture antennas and their applications on side-looking radars. *L'Onde Electrique*, **45** (458), 561–567.

[3] Brown, W.M. (1967) Synthetic aperture radar. *IEEE Transactions on Aerospace and Electronic Systems*, **3** (2), 217–219.

[4] Brown, W.M. and Porcello, L.J. (1969) An introduction to synthetic aperture radar. *IEEE Spectrum*, **6** (9), 52–62.

[5] Harger, R.O. (1970) *Synthetic Aperture Radar System: Theory and Design*, Academic Press, New York.

[6] Develt, J.A. (1973) Performance of a synthetic aperture mapping radar system. *IEEE Transactions on Aerospace and Navigational Electronics*, **11** (3), 173–179.

[7] Kirk, J.C. (1975) A discussion of digital processing in synthetic aperture radar. *IEEE Transactions on Aerospace and Electronic Systems*, **11** (3), 326–337.

[8] Sondergard, F. (1977) A dual mode digital processor for medium resolution synthetic aperture radars. IEEE International Radar Conference, pp. 384–390.

[9] Kelly, P.M. (1964) Data processing for synthetic aperture radar. *Proceedings of the IEEE*, **52** (2), 194–200.

[10] Kirk, J.C. (1975) Digital synthetic aperture radar technology. IEEE International Radar Conference, pp. 482–487.

[11] Perry, R.P. and Smith, R.J. (1973) Pulse compression techniques. Report AFAL-TR-73-143, Northrop Corporation Aircraft Division.

[12] Sack, M., Ito, M.R. and Cumming, I.G. (1985) Application of efficient linear FM matched filtering algorithms to synthetic aperture radar processing. *Proceedings of the IEEE*, **132** (1), 45–47.

[13] Wu, C. (1976) A digital system to produce imagery from SAR data. In: *Proceedings of the AIAA Systems Design Driven by Sensors Conference*.

[14] Cumming, I.G. and Bennett, J.R. (1979) Digital processing of SEASATSAR data. International Conference on Acoustics, Speech and Signal Processing, Washington DC.

[15] Jin, M.Y. and Wu, C. (1984) A SAR correlation algorithm which accommodates large range migration. *IEEE Transactions on Geosciences and Remote Sensing*, **22**(6), 592–597.

[16] Smith, A.M. (1991) A new approach to range-Doppler SAR processing. *International Journal of Remote Sensing*, **12** (2), 235–251.

[17] Raney, K. and Vachon, P. (1989) A phase preserving SAR processor. In: *Proceedings of the IEEE International Geoscience and Remote Sensing Symposium (IGARSS'89)*, vol. 4, pp. 2588–2591.

[18] Runge, H. and Bamler, R. (1992) A novel high precision SAR focusing algorithm based on chirp scaling. In: *Proceedings of the IEEE International Geoscience and Remote Sensing Symposium (IGARSS'92)*, Houston, pp. 372–375.

[19] Zhang, C. (1989) *Synthetic Aperture Radar: Principles, Analysis and Utilization*, Science Press, Beijing.

[20] Cantalloube, H.M.J., Dubois-Fernandez, P. and Dupuis, X. (2005) Very high resolution sar images over dense urban area. In: *Proceedings of the IEEE International Geoscience and Remote Sensing Symposium (IGARSS'05)*, vol. 4, pp. 2799–2802

[21] Ferretti, A., Prati, C. and Rocca, F. (2000) Non-linear subsidence rate estimation using permanent scatterers in differential SAR interferometry. *IEEE Transactions on Geoscience and Remote Sensing*, **38** (5), 2202–2212.

[22] Kampes, B. and Hassen, R.F. (2004) Ambiguity resolution for permanent scatterer interferometry. *IEEE Transactions on Geoscience and Remote Sensing*, **42** (11), 2446–2453.

[23] Tang, Y.X., Zhang, H. and Wang, C. (2007) Long term monitoring of urban subsidence by permanent scatterer DIn-SAR. *Progress in Natural Science*, **17** (1), 107–111.

[24] Cloude, S.R. and Pottier, E. (1997) An entropy based classification scheme for land applications of polarimetric SAR. *IEEE Transactions on Geoscience and Remote Sensing*, **35**(1), 68–78.
[25] Doulgeris, A.P., Anfinsen, S.N. and Eltoft, T. (2008) Classification with a non-Gaussian model for PolSAR data. *IEEE Transactions on Geoscience and Remote Sensing*, **46** (10), 2999–3009.
[26] Lee, J.S., Ainsworth, T.L, Kelly, J.P., *et al.* (2008) Evaluation and bias removal of multilook effect on entropy/ alpha/ anisotropy in polarimetric SAR decomposition. *IEEE Transactions on Geoscience and Remote Sensing*, **46** (10), 3039–3052.
[27] Raimadoya, M., Trisasongko, B., Zakharova, L., *et al.* (2007) ALOS-Indonesia PolinSAR experiment (AIPEX): a preliminary result. Microwave Conference.
[28] Pipia, L., Fabregas, X., Aguasca, A., *et al.* (2009) Polarimetric differential SAR interferometry: first results with ground-based measurements. *IEEE Transactions on Geoscience and Remote Sensing*, **6**(1), 167–171.
[29] Tan, W. (2009) Synthetic Aperture Radar Theory and Method of Three-dimensional Imaging. PhD thesis, Institute of Electronics, Chinese Academy of Sciences, Beijing.
[30] Candes, E.J. and Wakin, M.B. (2008) An introduction to compressive sampling. *IEEE Signal Processing Magazine*, **25** (2), 21–30.
[31] Keith, R.R. (1971) Synthetic aperture imaging radar and moving targets. *IEEE Transactions on Aeospace and Electronic Systems*, **17** (3), 499–505.
[32] Chapin, E. and Chen, C.W. (2009) Airborne along-track interferometry for GMTI. *IEEE Aerospace and Electronic Systems Magazine*, **24** (5), 13–18.
[33] Moccia, A., Rufino, G. and Luca, M.D. (2003) Oceanographic applications of space borne bistatic SAR. In: *Proceedings of the IEEE International Geoscience and Remote Sensing Symposium (IGARSS'03)*, pp. 1452–1454.
[34] Alpers, W. and Bruning, C. (1986) On the relative importance of motion-related contributions to the SAR imaging mechanism of ocean surface waves. *IEEE Transactions on Geoscience and Remote Sensing*, **GE-24** (6), 873–885.
[35] Dubois-Fernandez, P., Cantalloue, H., Vaizan, B., *et al.* (2006) ONERA-DLR bistatic SA R campaign: planning, data acquisition, and first analysis of bistatic scattering behavior of natural and urban targets. In: *IEE Proceedings: Radar, Sonar and Navigation*, pp. 214–223.
[36] Yates, G., Horne, A.M., Blake, A.P., *et al.* (2004) Bistatic SAR image formation. EUSAR Conference, pp. 581–584.
[37] Barmettler, A., Zuberbuhler, L., Meier, E., *et al.* (2008) Swiss airborne monostatic and bistatic dual-Pol SAR experiment at the VHF-band. EUSAR Conference, pp. 1–4.
[38] Balke, J. (2005) Field test of bistatic forward-looking synthetic aperture radar. IEEE Conference on Radar, pp. 424–429.
[39] Ulander, L.M.H., Flood, B., Frolind, P.O., *et al.* (2008) Bistatic experiment with ultra-wideband VHF-b and synthetic aperture radar. EUSAR Conference, pp. 1–4.
[40] Sanz-Marcos, J., Prats, P. and Mallorqui, J.J. (2007) SABRINA: a SAR bistatic receiver for interferometric applications. *IEEE Geoscience and Remote Sensing Letters*, **4** (2), 307–311.
[41] Reuter, S., Behner, F., Nies, H., *et al.* (2010) Development and experiments of a passive SAR receiver system in a bistatic spaceborne/stationary configuration. In: *Proceedings of the IEEE International Geoscience and Remote Sensing Symposium (IGARSS2010)*, pp. 118–121.
[42] Nies, H., Behner, F., Reuter, S., *et al.* (2010) SAR experiments in a bistatic hybrid configuration for generating PolInSAR data with TerraSAR-X illumination. 8th European Conference on SAR, pp. 1–4.
[43] Behner, F. and Reuter, S. (2010) HITCHHIKER – hybrid bistatic high resolution SAR experiment using a stationary receiver and TerraSAR-X transmitter. 8th European Conference on SAR, pp. 1–4.
[44] Nies, H., Behner, F. and Reuter, S. (2010) Polarimetric and interferometric applications in a bistatic hybrid SAR mode using Terrasar-X. In: *Proceedings of the IEEE International Geoscience and Remote Sensing Symposium (IGARSS2010)*, pp. 110–113.

[45] Baumgartner, S.V., Rodriguez-Cassola, M., Nottensteiner, A., *et al.* (2008) Bistatic experiment using TerraSAR-X and DLR's new F-SAR system. EUSAR Conference, pp. 57–60.

[46] Loffeld, O. and Ender, J. (2011) Focusing of Bistatic SAR Raw Data. Project Report submitted to DFG, in review, to be published.

[47] Walterscheid, I., Espeter, T., Brenner, A.R., *et al.* (2010) Bistatic SAR experiments with PAMIR and TerraSAR-X – setup, processing, and image results. *IEEE Transactions on Geoscience and Remote Sensing*, **48** (8), 3268–3279.

[48] Espeter, T., Walterscheid, I., Klare, J., *et al.* (2011) Bistatic forward-looking SAR: results of a spaceborne–airborne experiment. *IEEE Geoscience and Remote Sensing Letters*, **8** (4), 765–768.

[49] Wang, R., Loffeld, O., Ul-Ann, Q., *et al.* (2008) A bistatic point target reference spectrum for general bistatic SAR processing. *IEEE Geoscience and Remote Sensing Letters*, **5** (3), 517–521.

[50] Wang, R., Loffeld, O., Neo, Y.L., *et al.* (2010) Focusing bistatic SAR data in airborne/stationary configuration. *IEEE Transactions on Geoscience and Remote Sensing*, **48** (1), 452–465.

[51] Wang, R., Loffeld, O., Neo, Y.L., *et al.* (2010) Extending Loffeld's bistatic formula for general bistatic SAR configurations. *IET Radar, Sonar and Navigation*, **4** (1), 74–84.

[52] Wang, R., Loffeld, O., Nies, H., *et al.* (2010) Frequency-domain bistatic SAR processing for spaceborne/airborne configuration. *IEEE Transactions on Aerospace and Electronic Systems*, **46** (3), 1329–1345.

[53] Wang, R., Loffeld, O., Nies, H., and Ender, J. (2009) Focusing hybrid spaceborne/airborne bistatic SAR data using wavenumber domain algorithm. *IEEE Transactions on Geoscience and Remote Sensing*, **47** (7), 2275–2283.

[54] Tang, Z. and Zhang, S. (2003) *Bistatic Synthetic Aperture Radar System Principle*, Science Press, Beijing.

[55] Xian, L., Xiong, J., Huang, Y., *et al.* (2007) Research on airborne bistatic SAR squint imaging mode algorithm and experiment data processing. APSAR Conference, pp. 618–621.

[56] Huang, Y., Yang, J., Xian, L., *et al.* (2007) Vehicleborne bistatic synthetic aperture radar imaging. In: *Proceedings of the IEEE International Geoscience and Remote Sensing Symposium (IGARSS'07)*, pp. 2164–2166

[57] Krieger, G., Fiedler, H., Hajnsek, I., *et al.* (2005) Tandem-X: mission concept and performance – an analysis. In: *Proceedings of the IEEE International Geoscience and Remote Sensing Symposium (IGARSS'05)*, pp. 4890–4893.

[58] Moccia, A., Rufino, G., D'Errico, M., *et al.* (2002) Bissat: a bistatic SAR for earth observation. In: *Proceedings of the IEEE International Geoscience and Remote Sensing Symposium (IGARSS'02)*, pp. 2628–2630.

2

Signal Processing Basis of SAR

SAR is a kind of radar that has high resolution in both range and azimuth direction. Its high-resolution ability is gained by a special signal design and a special working strategy. In this chapter, basic disciplines of SAR resolution and classic algorithms in SAR imaging processing are introduced, which provides the basic knowledge of SAR signal processing.

2.1 Range Resolution of SAR

2.1.1 Basic Concept of Range Resolution

As mentioned in Chapter 1, the main task of early radar was to measure the range of targets. Based on such a purpose, range resolution is one of the most important parameters to assess the performance of radar systems. Through long-term experiences, people conclude that the range resolution of radar is decided by the pulse width of transmitted signal, which is

$$\rho_r = \frac{cT}{2} \tag{2.1}$$

where ρ_r is the range resolution of radar, c is the speed of a microwave and T is the width of the transmitted pulse in the time domain.

It is very easy to comprehend such a definition of range resolution. Assuming that the width of the transmitted pulse is very narrow so that the back edge of the echo pulse from one target can be received earlier than the front edge of the echo pulse from a nearby target which is a little farther, then these two different targets can be directly distinguished from each other (as shown in Figure 2.1). Therefore, the minimum interval between the front edges of two echo pulses that can be distinguished is the width of the transmitted pulse. This makes the range resolution like that in Equation (2.1).

If the transmitted signal of a radar is an impulse $\delta(t)$, then the best range resolution can be obtained. However, $\delta(t)$ is just a kind of mathematics abstraction and can only be approximately substituted by a narrow pulse. Moreover, even narrow pulses that

---

Bistatic SAR Data Processing Algorithms, First Edition. Xiaolan Qiu, Chibiao Ding and Donghui Hu.

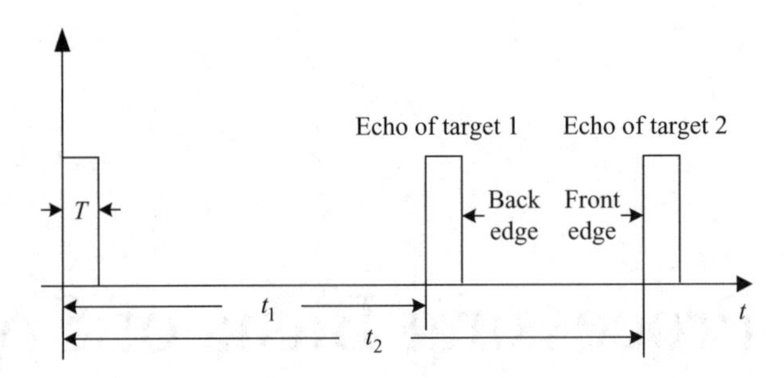

Figure 2.1 Illustration of the resolution principle in a narrow-pulse radar

can be used to detect a target are hard to generate in practice, as we know that the echo received by a receiver is stained by noises. In this case, the purpose of detecting targets can be concluded as distinguishing a signal from noises. Signals with low power will be submerged in noises and cannot be detected. According to the discipline of microwave transmitting and scattering and the characteristics of radar receivers, the signal-to-noise ratio (SNR) of a radar system can be depicted as follows [1]:

$$SNR = \frac{P_t G^2 \lambda^2 \sigma}{(4\pi)^3 k \bar{T} B F_n R^4} \tag{2.2}$$

where *SNR* is the signal-to-noise ratio at the receiver port, P_t is the peak value of the transmitting power, G is the antenna gain, λ is the wavelength, σ is the RCS in m^2, R is the range from the radar to the target, k is the Boltzmann constant, $\bar{T}$ is the noise temperature of the receiver, B is the bandwidth of the receiver and F_n is the noise factor of the receiver. Through Equation (2.2) we can see that the *SNR* of a radar will attenuate with the fourth power of the range. For a radar that can detect targets within several hundred miles, if a strong capability in target detection is required, the transmitting power needs to be extremely high. For a narrow-pulse radar, this means that it needs to reach an extremely high power within an extremely short time, which is very hard to realize.

2.1.2 Classical Theory of SAR Resolution

Obviously, the above definition about SAR resolution is experiential and rough, which cannot be regarded as strict radar resolution theory. After Woodward introduced Probability Theory and Information Theory to the radar domain, he and other later scholars systematically built an entire serious theory of ambiguous function [2]. According to this theory, and assuming the transmitted signal to be *s*(*t*), the time ambiguous function is

$$\chi(\tau) = \int_{-\infty}^{+\infty} s(t+\tau)s^*(t)dt \tag{2.3}$$

From signal processing theory [3], $|\chi(\tau)|$ is symmetrical to the origin and has a peak value at $\tau = 0$, which results in a wave lobe around the origin. This wave lobe is usually called the main lobe. According to the theory of ambiguous function, radar resolution is usually defined as the half-power width of the main lobe, because roughly speaking the half-power width is the threshold of the interval between the main lobes of two targets with the same power but which can be distinguished from each other. With this definition, we solve equation $|\chi(\tau)| = 1/\sqrt{2}$ to find a suitable solution of τ (written as τ_0), and due to the symmetry of the ambiguous function, we find $2\tau_0$ to be the nominal resolution of radar. The reason here to call it a "nominal" resolution is that this result is obtained under the ideal condition of no noise and no specific characters of the targets, so it may not be suitable for real cases. In practice, where noise often exists, targets with an interval larger than $2\tau_0$ may not be separated well. However, nominal resolution is still an effective criterion to measure the resolution ability of a radar system. In this book, radar resolution always refers to the nominal resolution if no special note is added.

From formula (2.3), the form of ambiguous function is related to the signal formation. Therefore an expression of nominal resolution varies with different transmitted signal waveforms. However, according to the computation of several classic waveforms in [4], when ignoring some constant factors, nominal resolution can be unified as

$$\Delta\tau \approx \frac{1}{B} \tag{2.4}$$

where B is the frequency bandwidth of the transmitted signal. If expressing the resolution in range, then we have

$$\rho_r \approx \frac{c}{2B} \tag{2.5}$$

Here, factor 2 in the denominator is owed from the double-propagate attribution of the signal.

The radar resolution theory described above tells us that range resolution is determined by the frequency bandwidth of the transmitted signal rather than the transmitted pulse width. Range resolution has an inverse ratio with the bandwidth. Hence, we are inspired to look for a kind of waveform that has both board frequency bandwidth B and long duration time T. Therefore, we are able to get not only fine range resolution but also achieve high transmitted power to extend the radar effective range. From research, many waveforms of such a kind have been found, such as a linear frequency modulated (LFM) signal, a phase coding signal, a step-frequency signal and so on. The most classic and commonly used signal is the LFM signal. With the improvement in SAR resolution and antijamming ability and with the innovation of SAR systems, the design of an SAR signal waveform becomes more and more complex and flexible;

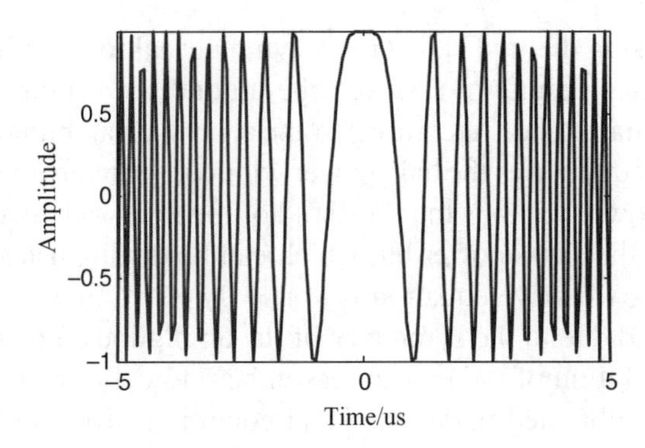

Figure 2.2 Waveform of a chip in the time domain

however, the basic discipline remains the same and the LFM signal is still the most important and classic signal for SAR. As a result, the LFM signal is regarded as a foundation of the following discussion in this book.

2.1.3 Linear Frequency Modulated Signal (Chirp Signal)

A linear frequency modulated signal is a kind of signal whose frequency changes linearly within the pulse duration time. This kind of signal is also called a chirp signal, which has a common form as follows:

$$s_p(t) = \mathrm{rect}(t/T)\cos(2\pi f_0 t + \pi k_r t^2) \tag{2.6}$$

where f_0 is the center frequency, k_r is the chirp rate, T is the pulse duration time and rect() is a rectangular function. It can be seen that the instant frequency of the signal is $f_0 + k_r t$, which has a linear relationship with time. The time-domain waveform of a chirp pulse is shown in Figure 2.2, where it can clearly be seen that the frequency of the chirp signal changes with time. It is this change that results in a frequency bandwidth much larger than *1/T*[1] for a pulse with duration time T and so results in better resolution (as shown in Figure 2.3). This point will also be explained later by using the property of chirp in the frequency domain.

Usually, after receiving a signal, the radar tends to get two channel signals with a 90° phase difference through orthogonal demodulation, and then combine them to get complex signals. The purpose of this is to reduce the frequency bandwidth to a half and so to cut down the sample rate to a half. Hence, the complex form (see formula

[1] As known from the signal processing theory, the spectrum of a rectangular in frequency domain appears as a sinc function, whose first zero point lies at ±1/T, i.e. the mainlobe width is 2/T. Because the half-mainlobe-width is usually roughly used to indicate the frequency bandwidth, here 1/T is used in comparison.

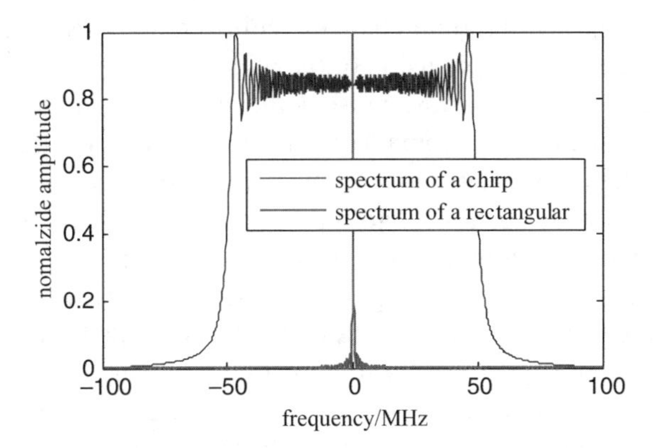

Figure 2.3 Spectral amplitude of a chirp in the frequency domain

(2.7)) of a chirp signal when demodulated to a low frequency is considered here, and the frequency characteristic of the chirp signal is analyzed based on this form:

$$s(t) = \mathrm{rect}(t/T)\exp(\pi k_r t^2), \quad -T/2 \leq t \leq T/2 \tag{2.7}$$

The principle of the stationary phase (POSP) is an effective method used to analyze chirp signals. Detailed descriptions of POSP can be found in [4], but here only some rough introductions are presented.

For a frequency modulated signal with a slowly changing phase $x(t) = a(t)\exp[j\varphi(t)]$, when the change of $\varphi(t)$ exceeds many times 2π, the integral of $x(t)$ is significantly different from zero only around $d\varphi(t)/dt = 0$ (note that the suitable time is t_0). This is the POSP and t_0 is called the point of stationary phase. Using the POSP, the spectrum of $x(t)$ in the frequency domain is

$$X(f) \approx \frac{\sqrt{2\pi}a(t_0)}{\sqrt{|\varphi''(t_0)|}}\exp\left\{-j\left[2\pi f t_0 - \varphi(t_0) - \frac{\pi}{4}\right]\right\} \tag{2.8}$$

where φ'' indicates the second derivative of φ. Therefore the frequency spectrum of a chirp signal can depicted as

$$S(f) \approx \frac{1}{\sqrt{|k_r|}}\mathrm{rect}(f/B)\exp(-j\pi f^2/k_r + j\pi/4) \tag{2.9}$$

where $B = |k_r T|$, which is the frequency bandwidth of the chirp. Figure 2.3 shows the spectrum of a chirp in the frequency domain. It can be seen that the spectrum amplitude is approximately rectangular, except for some tiny fluctuations (usually called a Fresnel fluctuation), and the width of this spectrum is much larger than that of a non-frequency-modulated rectangular pulse. From Equation (2.9) it can be found that the spectrum of a chirp signal can also be approximately regarded as a linear modulated signal. From a frequency domain aspect, it is the time that is modulated (the time is f/k_r), that is, the time changes with frequency. So the LFM pulse can be regarded as a time domain extension of the wide-bandwidth narrow-time-duration pulse, which is to extend a narrow pulse with bandwidth B (usually the pulse is as narrow as $1/B$ in the time domain) to a wide pulse with pulse duration time T, $T = B/|k_r|$. After this extension, the pulse duration time and signal frequency bandwidth can no longer be limited to an inverse ratio. Therefore, we can have a signal with both a wide bandwidth and long duration time, and hence the problem of generating the high-power narrow pulse can be solved. However, compared with regular wide band narrow pulses, considering the fact that the LFM signal is extended in the time domain, some post-processing must be performed to recover its attribute of "narrow", in order to realize the high resolution and high SNR. This process is called pulse compression, which is usually realized by a matched filter.

2.1.4 Matched Filter

A matched filter is a kind of filter that has the largest output SNR on a stationary Gaussian white noise background [5]. According to this definition, the formation of a matched filter both in the time domain and frequency domain can be derived as follows:

$$h(t) = Ks(t_0 - t) \tag{2.10}$$

$$H(w) = KS^*(w)\exp(-jwt_0) \tag{2.11}$$

and the SNR reaches a peak when $t = t_0$. For the sake of simplicity, the constant K is set to 1 in the follow discussion.

From the formula of a matched filter in the frequency domain, it can be found that the amplitude–frequency characteristic of a matched filter is consistent with that of the input signal while the phase–frequency characteristic is not. These characteristics ensure that the phases of the entire components are revised to be the same when they go through the matched filter. As a result, the entire frequency components with the same phase can accumulate to obtain a maximum SNR.

If t_0= 0, the time-domain matched filter of an LFM pulse can written as

$$h(t) = s^*(-t) = \text{rect}\left(\frac{-t}{T}\right)\exp(-j\pi k_r t^2) \tag{2.12}$$

If this filter is used in matched filtering an LFM pulse, the output signal should be

$$\begin{aligned} s_0(t) &= \int_{-\infty}^{+\infty} h(u)s(t-u) \\ &= \int_{-\infty}^{+\infty} \text{rect}\left(\frac{-u}{T}\right)\exp\left(-j\pi k_r u^2\right)\text{rect}\left(\frac{t-u}{T}\right)\exp\left[-j\pi k_r (t-u)^2\right]du \end{aligned}$$

when $0 \leq t \leq T$,

$$s_0(t) = \exp(j\pi k_r t^2) \int_{t-\frac{T}{2}}^{\frac{T}{2}} \exp(-j2\pi k_r t u)du = \frac{\sin[\pi k_r (T-t)t]}{\pi k_r t}$$

when $-T \leq t < 0$,

$$s_0(t) = \exp(j\pi k_r t^2) \int_{-\frac{T}{2}}^{t+\frac{T}{2}} \exp(-j2\pi k_r t u)du = \frac{\sin[\pi k_r (T+t)t]}{\pi k_r t}$$

To make a summary, the output signal can be written as

$$s_0(t) = T\frac{\sin[\pi k_r T t(1-|t|)/T]}{\pi k_r T t}, -T \leq t \leq T \tag{2.13}$$

It can be found that $s_o(t)$ attenuates fast with an increase of $|t|$. Usually only the signal around the main lobe (which means $|t| \ll T$) is taken into account. Thus, the compressed LFM pulse can be approximated as a sinc function, which is

$$s_0(t) \approx T\text{sinc}(k_r T t) \tag{2.14}$$

Figure 2.4 shows the envelope of the compressed LFM pulse. The nominal resolution defined by the half-power width is $\Delta\tau = 0.886/B$. Neglecting the factor of 0.886 we get $\Delta\tau \approx 1/B$, which is consistent with the result of Equation (2.4).

In addition, we should notice that when the time-domain matched filter shown in Equation (2.12) is used to compress the pulse, the peak value of the compressed signal is T instead of 1 before compressing, which means the compression amplitude gain is T. Actually, the frequency matched filter $H(f) = \text{rect}(f/B)\exp(j\pi f^2/k_r)$, which can be derived from Equations (2.9) and (2.11), is usually used for pulse compression in practice, for it can be realized by FFT and hence is efficient. In this case, the waveform after compression is $s_o(t) \approx \sqrt{k_r T^2}\text{sinc}(k_r Tt)\exp(j\pi/4)$, which means that the amplitude compression gain is $\sqrt{k_r T^2}$ (i.e., $\sqrt{\text{BT}}$). Here BT is the time bandwidth product (TBP), usually noted as D. As a result, the power gain of compressing an LFM pulse is usually regarded as equal to its TBP.

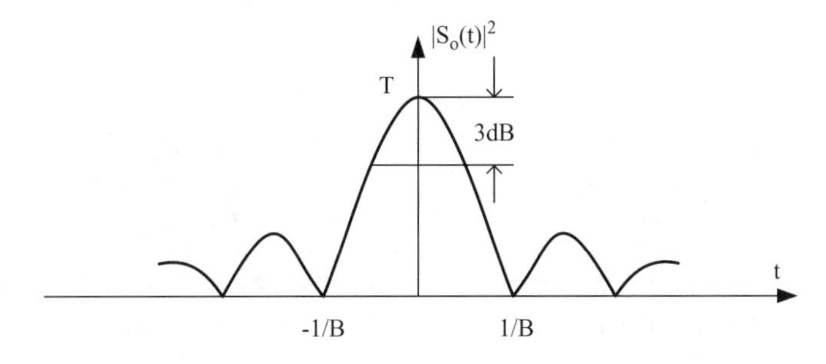

Figure 2.4 Compressed result of an LFM pulse

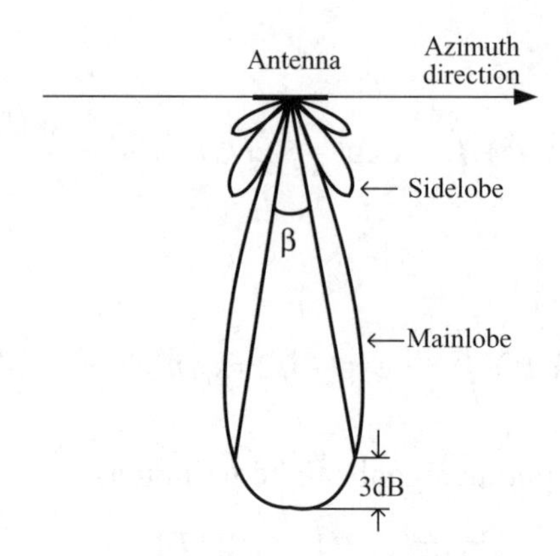

Figure 2.5 A 3 dB beam width of a radar antenna pattern

2.2 Azimuth Resolution of SAR

2.2.1 Basic Concept of Azimuth Resolution

The resolution theory described above is also suitable for azimuth resolution analysis. Classic azimuth resolution is usually defined as 3 dB width of the antenna pattern, as shown in Figure 2.5.

The concept is quite easy to comprehend. If two targets have different positions in the azimuth direction, and the intersection angle to the radar exceeds the beam width of the radar antenna pattern, then the two echo signals will be received in turn and the echo waves will form two peaks, which can be distinguished by the naked eye (shown in Figure 2.6(b)). If the differential angle between two targets to the radar is smaller than the antenna beam width, then the two echo waves will overlie each other and form an indistinguishable echo wave (shown in Figure 2.6(a)).

For a real aperture radar antenna, according to antenna theory, the 3 dB beam width of an antenna pattern can be approximated as follows:

$$\beta \approx \frac{\lambda}{D} \tag{2.15}$$

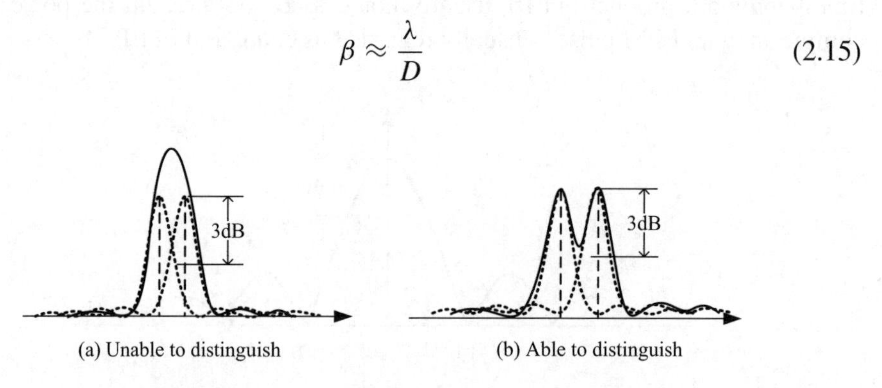

(a) Unable to distinguish (b) Able to distinguish

Figure 2.6 Illustration of azimuth resolution

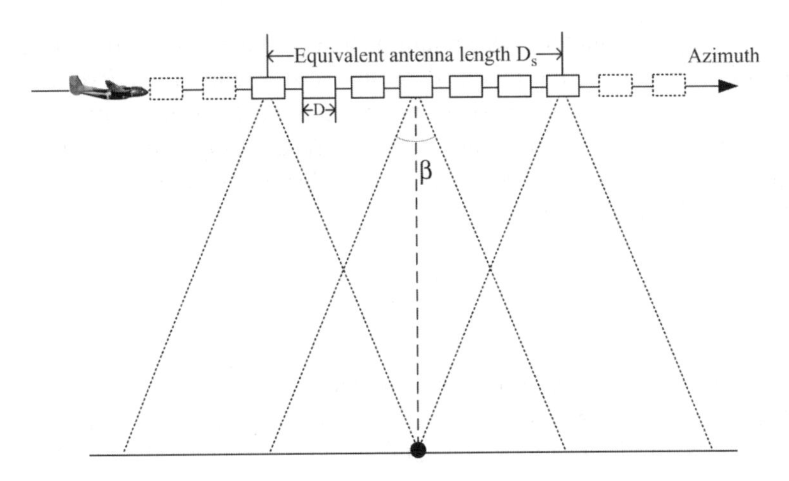

Figure 2.7 Illustration of the synthetic aperture principle

where λ is the wavelength and D is the real aperture length. Azimuth resolution ρ_a in a certain range R is

$$\rho_a \approx R\frac{\lambda}{D} \tag{2.16}$$

2.2.2 *Theory of Synthetic Aperture*

From Equation (2.16) it can be seen that the resolution of a real aperture radar is proportional to the range from the radar to the target (R) and is inversely proportional to the antenna aperture size D. Under the condition of a constant radar wavelength, if high azimuth resolution is desired over a long distance, the only way is to enlarge the size of the radar antenna aperture. However, this is hard to realize in most cases. For instance, if a C band (e.g., $\lambda = 0.56$ m) radar desires to get 1 m resolution in the distance of 800 km (which is typical in the spaceborne case), the antenna aperture must be as large as 448 km. Can the azimuth resolution be improved by a virtual large antenna aperture, which is formed by the movement of the antenna and the synthetic processing of the data collected in different positions? The answer is "yes", and this is exactly what SAR does.

The way to form an SAR antenna is depicted in Figure 2.7. For the convenience of explanation and to simplify some expressions, the discussion of this section is based on the non-squint side-looking mode; that is, the direction of the antenna beam center is orthogonal to the radar movement direction. As shown in Figure 2.7, the radar transmits and receives signals and then stores the received echoes. This process (transmitting, receiving and storing) is conducted at a set of locations along the flying path. The echoes collected at these locations form different units of synthetic aperture called azimuth sample serials. By processing the data collected within the illuminating time of the radar real aperture (which is usually called the synthetic aperture time), an equivalent large radar antenna can be obtained and a high azimuth resolution can be achieved as a result.

It should be noticed that the phase relationship of SAR echoes collected at different locations is different from the phase relationship of a real large aperture radar by a factor of 2. This is because the SAR signal has a double-distance attribute, which is caused by double-direction propagation during transmitting and receiving. This can be regarded as a two-time extension of the synthetic aperture length, so the 3 dB beam width of the synthetic antenna in SAR is

$$\beta \approx \frac{\lambda}{2D_s} \tag{2.17}$$

where D_s is the synthetic aperture equivalent antenna length. For the stripmap mode SAR, D_s is the length covered on the ground by the 3 dB beam width of the real antenna. In a certain distance of R, the value is

$$D_s = R\frac{\lambda}{D} \tag{2.18}$$

Therefore, after processing the entire echoes collected within the synthetic aperture time, the achieved SAR azimuth resolution is

$$\rho_a = R\beta = \frac{R\lambda D}{2R\lambda} = \frac{D}{2} \tag{2.19}$$

It can be found that SAR azimuth resolution is no longer related to the range between the radar and target. Moreover, the shorter the antenna length is, the higher the resolution that can be achieved. This is not hard to understand, as reducing the real antenna aperture size means increasing the synthetic aperture length, and so a narrow synthetic beam can be obtained, resulting in higher azimuth resolution.

Comparing Equations (2.18) and (2.16) it can be found that the synthetic aperture length D_s equals the resolution of a real aperture radar in value. However, if no post-process is added to azimuth echoes, the azimuth resolution of SAR cannot be improved. Therefore, signal processing is the key technique for obtaining a high azimuth resolution. This process is called beam sharpening, which is theoretically equivalent to compressing the signal with a matched filter.

In addition, the theory of synthetic aperture radar can be understood from the aspect of Fresnel diffraction, which is introduced in [4].

2.2.3 *Realizing a Synthetic Aperture Using a Matched Filter*

Figure 2.8 illustrates the geometry of SAR and a target. The radar moves forward with a constant velocity V. At time η_1, the front edge of the radar antenna mainlobe illuminates target P. At time η_2, P is illuminated by the beam center. At time η_3, the back edge of the antenna mainlobe leaves P. The period from η_1 to η_3 is the synthetic

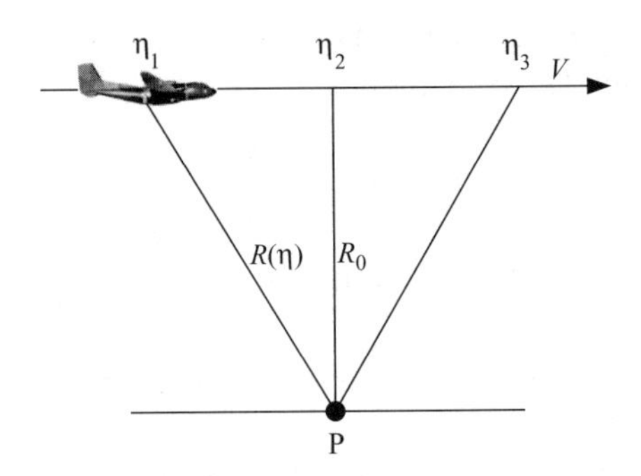

Figure 2.8 Geometry of SAR illuminating a target

aperture time. Supposing η_2 is the time origin; then the range R between the radar and target with respect to η^2 is

$$R(\eta) = \sqrt{R_0^2 + V^2\eta^2}, \qquad -T_{syn}/2 \leq \eta \leq T_{syn}/2 \tag{2.20}$$

where R_0 is the range between the radar and target when the target is illuminated by the beam center and T_{syn} is the synthetic aperture time. In common cases, R_0 is much larger than $|V\eta|$. Therefore R can be approximately expressed by its Taylor expansion at $\eta = 0$ when keeping to the second-order term

$$R(\eta) \approx R + \frac{V^2\eta^2}{2R} \tag{2.21}$$

This approximation is called the Fresnel approximation.

Assuming that the transmitted signal is a continuous wave with a normalized amplitude, and neglecting the effect of the antenna pattern, the transmitted signal is

$$s(\eta) = \exp(j2\pi f_0\eta) \tag{2.22}$$

where f_0 is the signal carrier frequency. Within the synthetic aperture time, the received echo is

$$s_a(\eta) = \exp\left\{j2\pi f_0\left[\eta - \frac{2}{c}R(\eta)\right]\right\} \tag{2.23}$$

It can be found that the phase of echo changes with η. Such a change leads to the change of instant frequency, that is, the Doppler frequency. This is caused by the

[2]Here time t corresponds to the velocity of light, and time η corresponds to the velocity of radar antenna. These two velocities are very different in value. Usually the tiny distance moved by antenna during the signal transmitting to receiving is omitted, which means the "stop-go-stop" approximation is adopted. In other words, time t and η are supposed to be two independent variables. Generally, t is called 'fast time' or 'range time', η is called 'slow time' or 'azimuth time'. To distinguish the difference, two different symbols are used for time in this book.

relative movement between the radar and target. The Doppler frequency can be found from Equations (2.21) and (2.23), which is

$$f_{\eta d}(\eta) \approx -\frac{2V^2}{\lambda R_0}\eta \tag{2.24}$$

Thus, the instant frequency of the azimuth echo is

$$f_\eta(\eta) = f_0 + f_{\eta d}(\eta) \approx f_0 - \frac{2V^2}{\lambda R_0}\eta \tag{2.25}$$

For the existence of the Doppler frequency, the instant frequency of the azimuth echo changes linearly around the carrier frequency, which means the azimuth echo of a target is nearly a chirp signal with a minus chirp rate:

$$f_r = -\frac{2V^2}{\lambda R_0} \tag{2.26}$$

It can be seen that f_r is related to radar movement speed V and range R_0 between the radar and target. If V is a constant, the chirp rates of targets in different ranges are different.

In Section 2.1, high-range resolution can be obtained by compressing the linear frequency modulated pulse using the matched filter. Now that the azimuth echo of SAR can also be regarded as chirp signals, the same strategy of the matched filter can be used to compress the azimuth echo and to get high azimuth resolution. According to the formula of the matched filter, it is necessary to set parameters of the matched filter to be the same as Doppler parameters of the azimuth echo. As a result, in real cases of SAR processing, parameters should be adjusted at different ranges in order to get high azimuth resolution in the entire imaging area. This is the main difference between range processing and azimuth processing.

Similar to the derivation in Section 2.1.4, the normalized output signal after matched filter processing on the azimuth echo can be written as follows:

$$s_{ao}(\eta) = \frac{\sin(\pi f_r T_{syn}\eta)}{\pi f_r T_{syn}\eta} \tag{2.27}$$

The nominal resolution defined by the half-power width is $\Delta\eta = 0.886/\left|f_r T_{syn}\right|$. Neglecting the constant factor, this becomes

$$\Delta\eta \approx \frac{1}{\left|f_r T_{syn}\right|} = \frac{1}{B_a} \tag{2.28}$$

where B_a is the Doppler bandwidth of the signal. Moreover, the synthetic aperture time is

$$T_{syn} = \frac{D_s}{V} = \frac{\beta R}{V} = \frac{\lambda R}{DV} \tag{2.29}$$

so B_a can be written as

$$B_a = |f_r|\,T = \frac{2V^2}{\lambda R}\frac{\lambda R}{DV} = \frac{2V}{D} \tag{2.30}$$

Therefore, by multiplying $\Delta\eta$ by speed V, the SAR azimuth resolution can be expressed by length, which is

$$\rho_a \approx \frac{V}{B_a} = \frac{D}{2} \tag{2.31}$$

This is consistent with the result of Equation (2.19).

It should be noticed that one cannot get the following conclusion through the SAR azimuth resolution formula of (2.31): the SAR azimuth resolution can be improved infinitely by infinitely shortening the antenna size. The reason is that Equation (2.31) is obtained based on the assumption of $R \gg |V\eta|$. Only with this assumption is the Fresnel approximation correct. When the antenna size is shrinking to a very small value, such an approximation is no longer correct. In this case, the azimuth echo cannot be viewed as a chirp signal. Theoretical analysis shows that with an infinite small antenna size, SAR resolution will tend to be a constant value of $\lambda/4$. This is the theoretical limit of the SAR azimuth resolution.

In addition, it is necessary to point out that the foundation of the above analysis is Equation (2.23), which implies that the signal received within the synthetic aperture time should be strictly coherent. Only under this condition can real small antenna be integrated to an equivalent large-size antenna. Coherence is an important precondition for SAR.

2.3 SAR Resolution Cell

Sections 2.1 and 2.2 describe the range and azimuth resolution of SAR. From Equations (2.4) and (2.28), it can be found that the range and azimuth direction correspond to fast time t and slow time η, respectively. Although the range resolution and azimuth resolution are also defined at length (see Equations (2.5) and (2.31), respectively), the physical meanings of these two directions, the so-called "range" and "azimuth", are not clearly explained. Many books on SAR theory have not elaborated this point. For the simple SAR modes, this may not be important, because the two directions can obviously be obtained from the imaging geometries; however, in some complex SAR modes (such as bistatic SAR), the directions of resolutions and the concept of resolution cell are helpful for understanding the two-dimensional resolution ability of SAR. Thus, this book will explain the physical meaning of "range" and "azimuth" direction in SAR and describe the resolution cell of an SAR image. For the convenience of illustration, here monostatic SAR will be taken as an example.

SAR has a narrow pulse in the fast time domain through transmitting a wideband signal and then compressing the echo by a matched filter, and achieves the equivalent narrow beam by the movement of a small antenna that synthesizes a large aperture antenna in the slow time domain. Therefore, we can study the narrow-pulse, narrow-beam imaging radar to illustrate the resolution direction of SAR. After synthetic aperture processing, the synthetic narrow-beam center usually points to the direction at which the radar points to the target at the target's synthetic aperture center. For the SAR modes whose beam direction remains unchanged (such as stripmap SAR), the

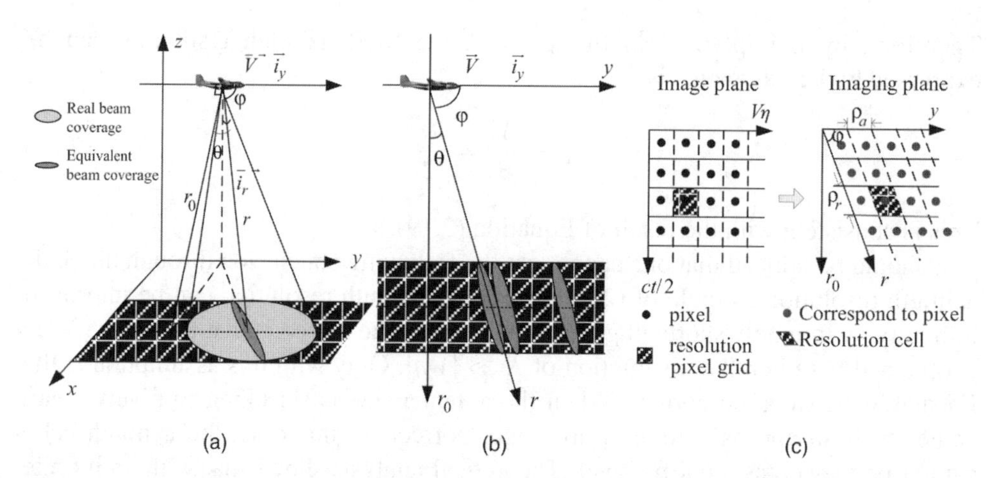

Figure 2.9 Diagram of the resolution cell in a squint SAR: (a) three-dimensional imaging geometry diagram, (b) plane diagram of imaging geometry, (c) resolution cell diagram

direction of the synthetic narrow beam is the direction of the original beam center. Commonly, the complementary angle of the intersection angle between the beam direction and flight direction is called the "squint angle" (see angle θ in Figure 2.9). If this angle is zero, the SAR is in a nonsquint mode; otherwise it is in a squint mode.

Let us now study the equivalent narrow-pulse, narrow-beam imaging radar, which moves along the y axis with constant velocity (velocity vector denoted by $\vec{V}$) and whose antenna pointing direction remains unchanged. For the convenience of illustration, firstly, the monostatic SAR imaging geometry is projected on to a two-dimensional plane (the so-called slant range plane) (as shown in Figure 2.10(b)). At time η, the narrow-beam radar only receives the signal coming from the beam direction at that

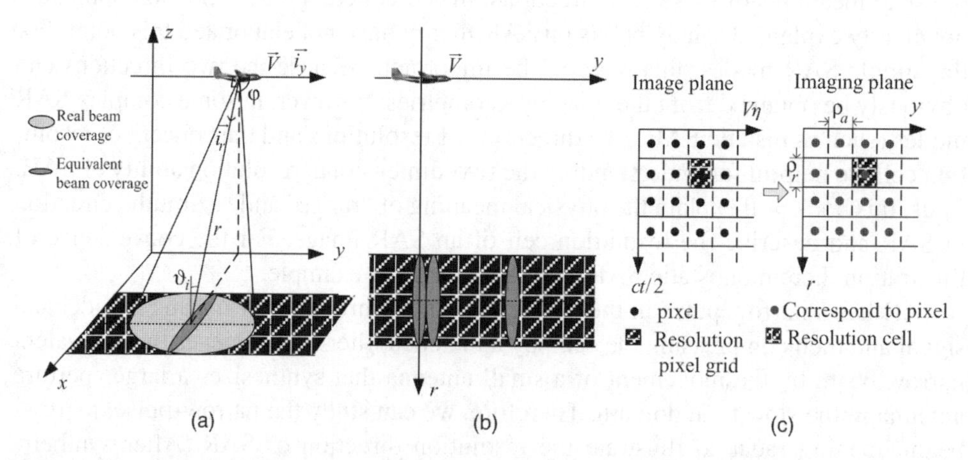

Figure 2.10 Diagram of the resolution cell in nonsquint SAR: (a) three-dimensional imaging geometry diagram, (b) plane diagram of imaging geometry, (c) resolution cell diagram

moment. With the movement of the radar, the antenna beam moves forward on the ground along the direction of $\vec{V}$, so that the radar can receive echoes of targets at different locations gradually in order. Therefore, the two axes η and t of the radar echo obviously correspond to the beam moving direction $\vec{i}_y$ and the narrow-beam point direction $\vec{i}_r$, respectively. These are the range and azimuth direction of the narrow-beam radar, which are also the range and azimuth direction of the monostatic one. It can be seen that $\vec{i}_r$ and $\vec{i}_y$ are orthogonal in a nonsquint SAR, and then the image plane grid (show as a pixel), determined by two-dimensional resolutions ρ_r and $\rho_{a\perp}$ (because the azimuth resolution direction $\vec{i}_y$ is perpendicular to the range resolution direction $\vec{i}_r$, so here adds "$\perp$" to the azimuth resolution) in the data space $(ct/2, V\eta)$, exactly corresponds to a rectangular element of the imaging plane. Its area is $\Delta s = \rho_{a\perp}\rho_r$. When the antenna is squint, the angle between $\vec{i}_r$ and $\vec{i}_y$ is not 90° and the resolution grid in the image plane then corresponds to the parallelogram in the imaging plane, whose area is as follows:

$$\Delta s = \rho_r \rho_a \sin\varphi \tag{2.32}$$

It should be pointed out that from the above equation the following conclusions cannot be deduced: if φ is smaller (i.e., the squint angle is bigger), the area of the parallelogram is smaller and so the SAR two-dimensional resolution is finer. Because under the same conditions the azimuth resolution is related to the squint angle, which is that the greater the squint angle, the narrower the Doppler bandwidth and hence the lower the azimuth resolution. As can be found later in Section 3.4.2, the azimuth resolution ρ_a of squint SAR and the azimuth resolution $\rho_{\mathrm{a}\perp}$ of nonsquint SAR meet the relationship of $\rho_\mathrm{a} = \rho_{\mathrm{a}\perp}/\sin\varphi$. Therefore, in monostatic SAR, the quadrilateral area in the imaging plane, which corresponds to a resolution grid in the image plane, is the same whether it is in the squint mode or the nonsquint mode. The range and azimuth direction are not necessarily orthogonal in monostatic SAR, so sometimes the direction, which is perpendicular to $\vec{i}_r$, is defined as the cross-range direction, and the resolutions in the cross-range direction and the range direction are directly analyzed to measure the resolution ability of SAR.

The resolution cell of an imaging plane is defined as the quadrilateral corresponding to the resolution grid in the image plane, where the resolution grid is determined by the resolution of η and t (as shown in Figure 2.10). Since it is the ground that SAR is imaging, the resolution cell area on the ground can truly indicate the resolution ability of SAR. Therefore, the resolution cell referred to here is taken to be on the ground plane unless special instructions are given. In the case of the nonsquint mode, the ground range resolution and the slant range resolution have the following relationship:

$$\rho_{rg}(x) = \rho_r / \sin\vartheta_i(x) \tag{2.33}$$

Therefore, the ground resolution cell area is

$$\Delta s_g(x) = \frac{\rho_{a\perp}\rho_r}{\sin\vartheta_i(x)} \tag{2.34}$$

in which $\vartheta_i(x)$ is the incidence angle of the electromagnetic waves at x, ρ_{rg} is the ground range resolution and Δs_g is the area of the resolution cell on the ground (here this means the (x, y) plane).

It can be seen that the resolution is changing within the imaging swath from near to far. SAR can get better resolution at the far range than at the near range. In the squint case, the relationship between the slant range plane and the ground plane is more complex. The ground resolution cell area can be calculated by the gradient method, which will be introduced in Chapter 3.

2.4 SAR Processing Model – Single-Point Target Imaging

As stated above, SAR is a kind of microwave imaging radar with a two-dimensional high-resolution capability. Its main purpose is to obtain two-dimensional high-resolution images of desired targets or ground surface by receiving the scattering echo of the scenario. Different from the "point-to-point" imaging mechanism of an optical radar, the SAR received echo must be processed to rebuild the image of the scenario due to its special range and azimuth resolution accessing mechanism. In this section, an ideal analytical model of the "single-point target" received echo is constructed, and the characteristics of the SAR signal and the key issues of SAR signal processing are illustrated based on this model. Furthermore, two simple imaging methods for SAR are given based on this model.

2.4.1 SAR Echo Model of a Single-Point Target

Assuming that the radar only receives echoes from one specific point target on the ground (as shown in Figure 2.7), the radar continuously transmits a chirp signal to the target during its flying along the track and receives the scattering echo of the target. If the radar transmitted signal waveform is

$$s(t) = \text{rect}\left(\frac{t}{T}\right)\exp(2\pi f_0 t + \pi k_r t^2) \tag{2.35}$$

then the received echo of the target is

$$s_r(t,\eta) = A(t-\alpha,\eta)\text{rect}\left(\frac{t-\alpha}{T}\right)\exp[j2\pi f_0(t-\alpha) + j\pi k_r(t-\alpha)^2] \tag{2.36}$$

where α is the time delay of the echo, A is the amplitude envelope of the echo and

$$A(t, \eta) = \sigma_0 \omega_r(t) \omega_a(\eta) \tag{2.37}$$

where σ_0 is the target's backscattering coefficient, W_r is the two-way antenna pattern weighting in the range direction multiplied by the propagation distance decay factor and W_a is the two-way antenna pattern weighting in the azimuth direction. For nonquantitative applications, the constant factors can be ignored and antenna pattern modulation can usually be simply approximated as a rectangular envelope. For simplicity, these simplifications are used below if not specified.

The time delay α can be written as follows:

$$\alpha = \frac{R_T(\eta) + R_R(\eta)}{c} = \frac{2R(\eta)}{c} \tag{2.38}$$

where R_T is the transmitting range and R_R is the receiving range. The two ranges can be considered as the same in monostatic SAR; therefore the unique one-way propagation range $R(\eta)$ is used here to represent them both. Because of the relative motion between the radar and target, the distance between them is changing with time; hence the time delay is changing correspondingly. The effect of this on the radar echo data space is that the received echo from the same target at different azimuth sample times will distribute at different cells along the range direction (i.e., different range samples), as shown in Figure 2.11. This phenomenon is called range cell migration (RCM) in the SAR signal processing field. It is the existence of RCM that causes the problem of two-dimensional coupling, which is range and azimuth coupling, in SAR echo data. The key issue of SAR imaging is the two-dimensional decoupling.

The SAR received echo from the targets is usually called the raw data. It can be seen from Equation (2.36) that the raw data of one target is not the target's backscattering coefficient σ_0 itself, but the σ_0 that is modulated on a two-dimensional frequency modulated signal with two-dimensional coupling. To recover σ_0, signal processing must be done on the raw data. In the following, two simple imaging methods for a single-point target will be described in the frequency domain and the time domain, respectively.

2.4.2 Single-Point Target Imaging

2.4.2.1 Frequency Domain Imaging – Based on a Two-Dimensional Matched Filter

Based on the knowledge obtained from the foregoing content, the imaging process of a single-point target can be achieved by a two-dimensional matched filter. According to

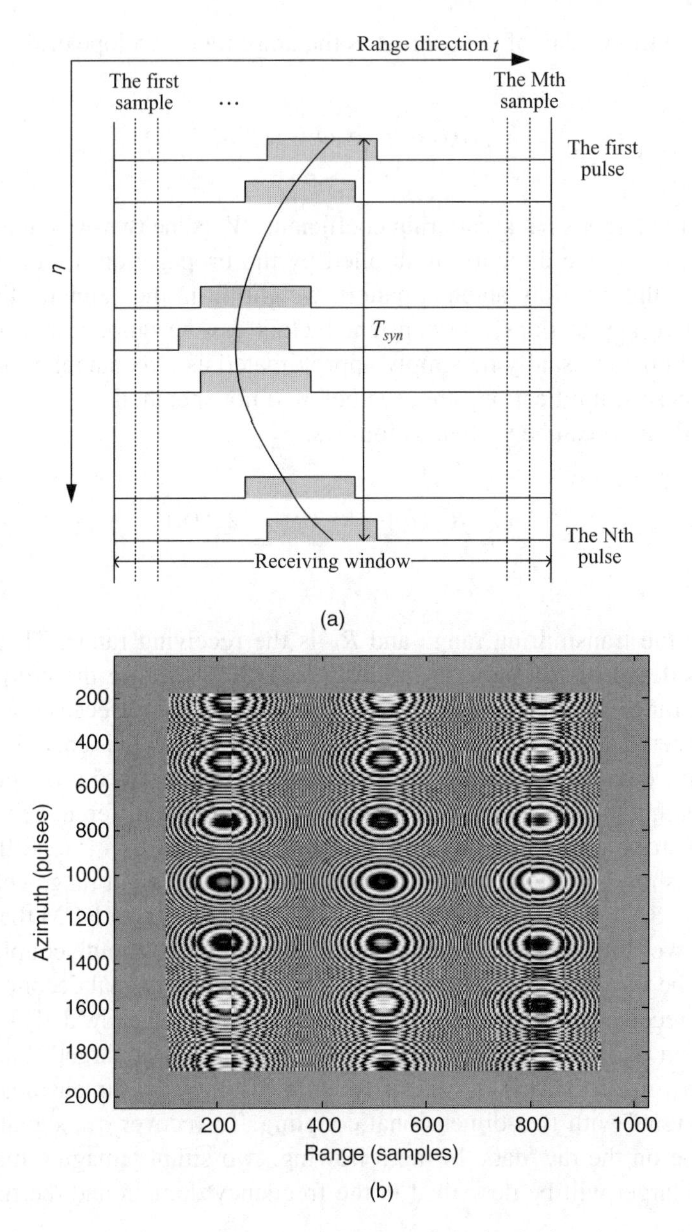

Figure 2.11 Illustration of the two-dimensional SAR raw data of a single-point target: (a) SAR echo schematic diagram, (b) SAR echo

the definition of a matched filter and combining Equation (2.36), the two-dimensional matched filter for the echo of a single target can be expressed as

$$H(f, f_\eta) = S^*(f, f_\eta) \tag{2.39}$$

where $S(f, f_\eta)$ is the frequency domain expression of $s_r(t, \eta)$ and the superscript "$*$" stands for conjugation. It can be found by POSP that

$$S(f, f_\eta) \approx A(f, f_\eta)\exp[j(\phi_1 + \phi_2)] \tag{2.40}$$

where $A(f, f_\eta)$ is the amplitude spectrum of the raw data, ϕ_1 is the phase spectrum of radar transmitted signal and ϕ_2 is the two-dimensional phase spectrum where range is coupled with azimuth:

$$\phi_1 = -\frac{\pi f^2}{k_r} \tag{2.41}$$

$$\phi_2 = -\pi \frac{cR}{2(f_0 + f)V^2} f_\eta^2 \tag{2.42}$$

In order to get the image of the point target, it only needs to multiply $H(f, f_\eta)$ with two-dimensional spectrum of the raw data and then perform the inverse Fourier transform. With the approximation of a rectangular antenna pattern, the matched filter output signal of the point target is

$$s_0(t, \eta) \approx \sigma_0 G \frac{\sin(\pi k_r T \eta)}{\pi k_r T \eta} \frac{\sin(\pi f_r T_{syn} \eta)}{\pi f_r T_{syn} \eta} \tag{2.43}$$

where G is a constant that represents the processing gain of the compression process. This output signal is a two-dimensional sinc function, whose peak value is proportional to the backscattering coefficient σ_0. Thus, the single-point target imaging process has been completed.

2.4.2.2 Time-Domain Imaging – Back-Projection (BP) Algorithm

Similar to the imaging method in the frequency domain based on a two-dimensional matched filter principle, a focused image can also be obtained through coherent processing of a SAR echo in the time domain, taking advantage of the coherence between the echo pulses from one target. Classical imaging methods in the time domain includes the back-projection (BP) algorithm [6–8], which was first derived from computed tomography (CT) imaging technology. The common processing steps are as follows: calculate the range and time delay between the radar and the targets (which are represented by the image pixels) according to the information of image pixel positions and the radar position, then back-project the raw data to the appropriate positions in the image domain based on the time delay and, lastly, coherently accumulate the projected signal in each pixel in the image plane to obtain a two-dimensional image. This algorithm is a kind of universal algorithm because it can be used in any imaging geometry and has no special requirement for radar platform moving tracks. The BP algorithm will be explained from the aspect of time-domain coherent processing as follows.

Assume that the grid of the pixel in the image plane corresponding to the desired imaging area is noted by (x_i, r_j), the radar transmitted signal is $s(t)$ and the location of SAR is $(x(\eta), -R_0)$. Thus, the echo corresponding to pixel (x_i, r_j) can be expressed as follows:

$$s_{rij}(t, \eta) = \sigma_{ij} s \left\{ t - \frac{2\sqrt{(r_i + R_0)^2 + [x_j - x(\eta)]^2}}{c} \right\} \tag{2.44}$$

The received echo signal can be expressed as

$$s_r(t, \eta) = \sum_{i,j} \sigma_{ij} s \left\{ t - \frac{2\sqrt{(r_i + R_0)^2 + [x_j - x(\eta)]^2}}{c} \right\} \tag{2.45}$$

The BP imaging process for pixel (x_i, r_j) can be described as

$$\begin{aligned} f(x_i, r_j) &= \int_\eta \int_t s_r(t, \eta) s^* \left\{ t - \frac{2\sqrt{(r_i + R_0)^2 + [x_j - x(\eta)]^2}}{c} \right\} dt d\eta \\ &= \int_\eta \int_t s_r(t, \eta) s^* \left\{ t - t_{ij}(\eta) \right\} dt d\eta \end{aligned} \tag{2.46}$$

where $s^*[t - t_{ij}(\eta)]$ is the matched filter of pixel (x_i, r_j). It can be seen that the imaging results at pixel (x_i, r_j) is the convolution of the SAR echo and the matched filter of the point target at this pixel.

Because the matched filter in the range direction is fixed, the frequency-domain methods can usually be used for rapid processing of range compression. If the SAR signal after range matched filtering is defined as follows:

$$s_M(t, \eta) = s_r(t, \eta) \otimes s^*(-t) = \int_\tau s_r(\tau, \eta) s^*(\tau - t) d\tau \tag{2.47}$$

then Equation (2.46) can be written as

$$f(x_i, r_j) = \int_\eta s_M[t_{ij}(\eta), \eta] d\eta \tag{2.48}$$

Therefore, to get the output $f(x_i, r_j)$ of one pixel, it only needs to do rangematched filtering at first and then calculate the time delay $t_{ij}(\eta)$ and find the signal time delay $t_{ij}(\eta)$ in the raw data, and finally coherently accumulate these signals at $t_{ij}(\eta)$ along the azimuth direction. On the contrary, if the reconstruction of the image is started from the raw data (not from the image pixel of view), then the BP algorithm process

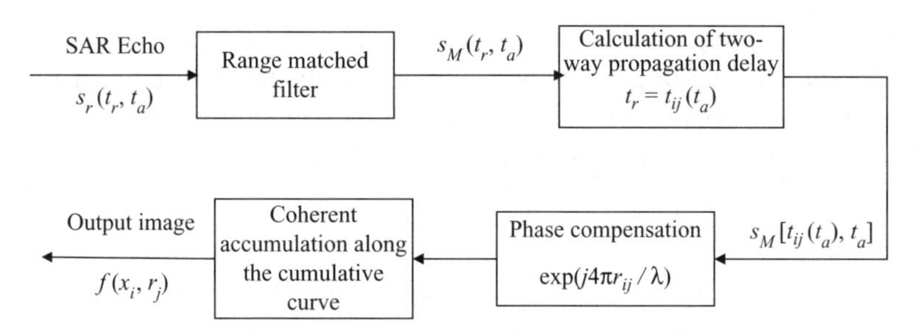

Figure 2.12 Block diagram of the BP algorithm

can be described as follows: firstly, do range matched filtering to the raw echo; then, for each signal at azimuth time η, according to the time delay in each range cell and the position of SAR at this moment, calculate its corresponding pixel position in the image plane and back-project the signal to its corresponding pixel. Thus, we can get an image for each azimuth sample time η. The final SAR image could be obtained by coherently accumulating the corresponding pixel values of all images. In practice, the image grid is usually determined before back-projection and then the appropriate echo location for each image pixel at each time η is calculated and found through interpolation. The processing flow described above is shown below in Figure 2.12.

It should be noted that, due to the fact that the Doppler parameters of the target echoes at different positions may be different, imaging methods both in frequency domain (based on a two-dimensional matching filter) and in the time domain (BP algorithm) need to constantly update the processor parameters according to the location of the target; hence, these methods are known as the pixel-by-pixel approach. These pixel-by-pixel methods can achieve the best focusing result of all targets, so they are the most accurate imaging approaches. However, the disadvantages of these methods are also the large amount of calculating burden and hence low efficiency. These make them unable to be a real-time processor and are not suitable to be applied to practical systems. To solve these problems, methods for rapidly processing the SAR massive data are also proposed during a long-term exploration, among which, the range-Doppler (RD) algorithm, the chirp scaling (CS) algorithm and the wave number domain algorithm (ω-k algorithm), are classic ones that are most representative. These algorithms will be introduced one by one in the following section.

2.5 Brief Introduction to Efficient SAR Imaging Algorithms

In essence, all of the traditional efficient imaging algorithms are based on the following two characteristics of SAR echo: firstly, the azimuth signal is approximately a chirp signal and, secondly, the spatial-variant property of Doppler parameters along the azimuth direction can be ignored (this is reasonable in most spaceborne and airborne systems). The first assumption ensures that the time-domain and the frequency-domain signals of SAR in the azimuth direction have a single-mapping relationship, while the second assumption avoids the parameter updating of the azimuth processor along the azimuth direction. Thus, the SAR processor could take advantage of the high-performance characteristics of the fast Fourier transform (FFT) to perform batch processing on a large number of targets through the frequency domain filters, and hence greatly enhance the efficiency of SAR imaging.

2.5.1 RD Algorithm

The RD algorithm is a quite "old" algorithm. The SEASAT SAR launched in 1978 brought forward an urgent demand for efficient SAR processing algorithms, thereby inducing the birth of the RD algorithm [9–12]. From 1976 to 1978, the development of the RD algorithm solved the bottleneck problem of SAR data processing for the first time, making it possible to process the massive amount of SAR data efficiently. The basic idea of the RD algorithm is: taking advantage of the fact that the range signal and the azimuth signal can be decoupled into two one-dimensional signals under certain conditions, the imaging processing can be simplified as two one-dimensional pulse compressions, that is, the range compression and azimuth compression. Of these, the range compression is relatively simple. It can be realized by matched filtering the SAR echo in the range frequency domain based on the information of a transmitted signal. Meanwhile, the azimuth compression is a little more complex. Based on the fact that the range migration curves of the different targets in the same range cell are approximately identical in the range time, the azimuth frequency domain (i.e., the range-Doppler domain), the RD algorithm fulfills the azimuth compression through two operations in the range-Doppler domain, which are RCM correction (RCMC) and azimuth matched filtering. The main processes of the RD algorithm are shown in Figure 2.13. Because the key operation (RCM correction) for decoupling the range and azimuth signal is accomplished in the range-Doppler domain, this algorithm is named the RD algorithm.

The RD algorithm will be further explained below through formulas. To make the meaning of the variables in the following formulas clear, SAR imaging geometry is shown in Figure 2.14, where the radar flies along the y axis, the position of the target is described by the slant range and azimuth position, that is, by $P(r, y)$ and the azimuth location of the radar at $\eta = 0$ is zero. Supposing that the radar transmits LFM pulses and the azimuth antenna pattern is approximated to be rectangular, the received echo signal of the scenario can be written as

$$\begin{aligned} s_r(t,\eta) = \int_r\int_y drdy \Big\{ \sigma(r,y)\mathrm{rect}\left[\frac{t-2R(\eta;r)/c}{T}\right]\mathrm{rect}\left(\frac{\eta-y/V}{T_{syn}}\right) \\ \times \exp\left\{j\pi k_r[t-2R(\eta;r)/c]^2\right\}\exp[-j4\pi R(\eta;r)/\lambda]\Big\} \end{aligned} \tag{2.49}$$

where

$$R(\eta;r) = \sqrt{r^2+(V\eta-y)^2} \tag{2.50}$$

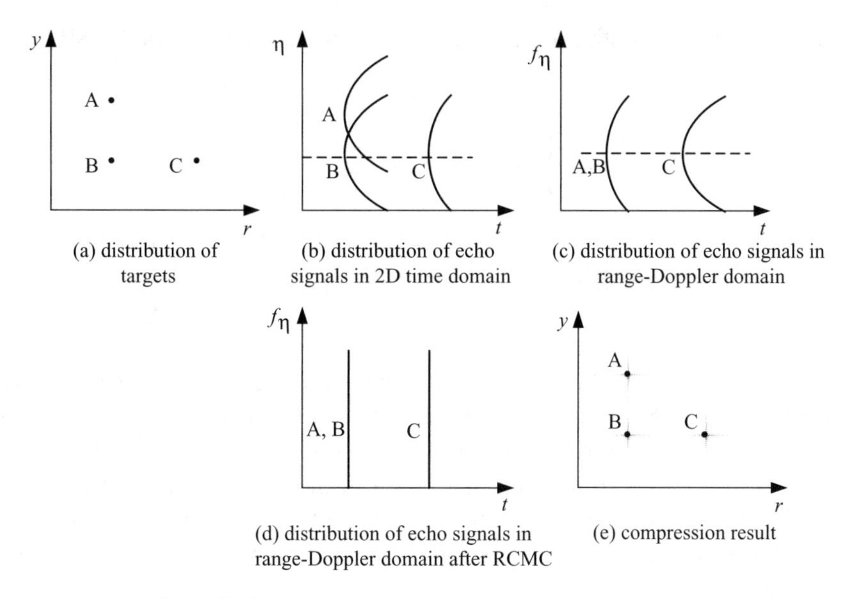

Figure 2.13 Illustration of the process of the RD algorithm

Applying the Fourier transform to Equation (2.49) in the range direction and ignoring the constant terms, we have

$$S_r^t(f, f_\eta) = \int_r\int_y drdy \left\{\sigma(r, y)\text{rect}\left(\frac{\eta - y/V}{T_{syn}}\right) \times \exp(-j\pi f^2/k_r)\exp\left[-j\frac{4\pi(f_0 + f)}{c}R(\eta; r)\right]\right\} \tag{2.51}$$

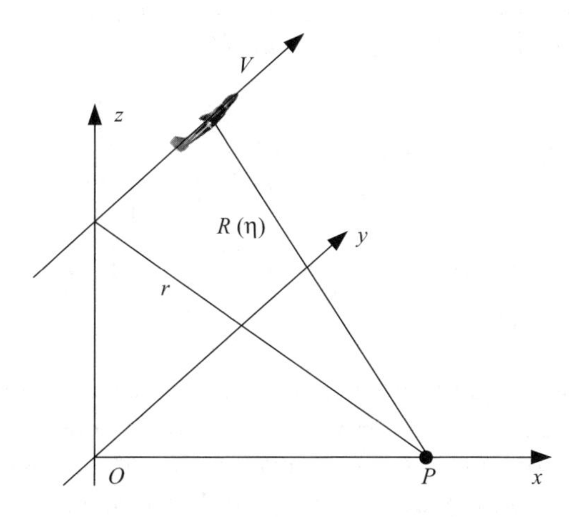

Figure 2.14 Monostatic SAR imaging geometry

Moreover, apply the azimuth Fourier transform to the signal based on POSP. Then we have

$$S_r(f, f_\eta) = \int_r \int_y drdy \left\{ \sigma(r, y) \exp(-j2\pi f_\eta y/V) \exp(-j\pi f^2/k_r) \times \exp\left\{ -jr\sqrt{\left[\frac{2\pi(f_0+f)}{c/2}\right]^2 - \left(\frac{2\pi f_\eta}{V}\right)^2} \right\} \right\} \tag{2.52}$$

This is the expression of the SAR raw data in the two-dimensional frequency domain. Most of the efficient SAR imaging algorithms are based on this expression; thus, this expression or its transformations will be frequently used in this book. Observing Equation (2.52), it can be found that the phase in the last term is a function of two variables (f and f_η), which represents the coupling of the range and azimuth caused by RCM. It is the existence of this coupled phase that makes the SAR imaging processing very complicated.

Typically, the frequency bandwidth of the transmitted signal is quite narrow compared with the radar carrier frequency (except for ultra wideband radar), that is, $|f| \ll f_0$. We can therefore expand the last term of Equation (2.52) at f_0 and only keep the first few terms, giving

$$\begin{aligned} S_r(f, f_\eta) \approx & \int\int_y drdy\{\sigma(r, y)\exp(-j2\pi f_\eta y/V)\exp(-j\pi f^2/k_r) \\ & \times \exp\left[-j\frac{4\pi}{\lambda}r\sqrt{1-(\lambda f_\eta/2V)^2}\right] (\textit{zero-order term}) \\ & \times \exp\left[-j\frac{4\pi f}{c}r\frac{1}{\sqrt{1-(\lambda f_\eta/2V)^2}}\right] (\textit{first-order term}) \\ & \times \exp\left\{-j\frac{2\pi f^2}{cf_0}r\frac{(\lambda f_\eta/2V)^2}{\left[\sqrt{1-(\lambda f_\eta/2V)^2}\right]^3}\right\} (\textit{second-order term}) \\ & \times \exp\left\{-j\frac{2\pi f^3}{cf_0^2}r\frac{(\lambda f_\eta/2V)^2}{[\sqrt{1-(\lambda f_\eta/2V)^2}]^5}\right\} (\textit{third-order term}) \end{aligned} \tag{2.53}$$

In the above equation, the zero-order term only depends on the azimuth frequency, so there is no coupling in this term. However, the two-dimensional coupling exists in the first and all higher-order terms of the equation. The original RD algorithm ignores the second-order term and higher-order terms in Equation (2.53), and only takes the linear term into account. In this case, performing a range inverse Fourier transform to

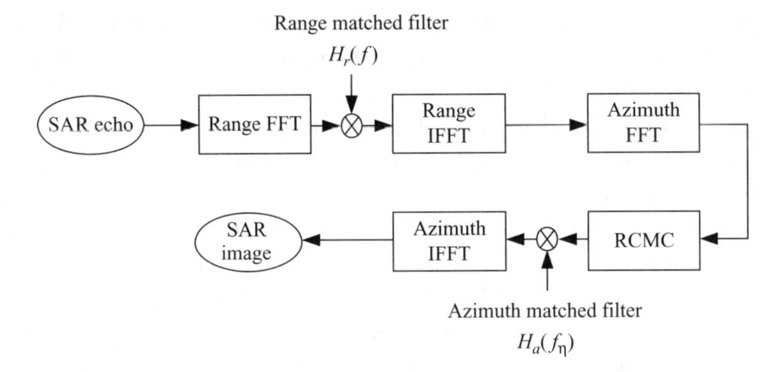

Figure 2.15 Block diagram of the RD algorithm

$S_r(f, f_\eta)$, the resulting target signal position in the range direction will vary with f_η as follows:

$$R(f_\eta) = \frac{r}{\sqrt{1-(\lambda f_\eta/2V)^2}} \tag{2.54}$$

Since it is hoped that the echo signal of a target is in the same range cell (usually at r), the range deviation from its desired position is

$$R_{RCM}(f_\eta) = r\left[\frac{1}{\sqrt{1-(\lambda f_\eta/2V)^2}} - 1\right] \tag{2.55}$$

This is the expression of RCM in the range-Doppler domain. It can be seen that the RCM in the range-Doppler domain is variant with the slant range. However, for those targets with the same slant range, their RCMs will coincide, so they can be corrected simultaneously. It is this property that the RD algorithm uses to achieve batch RCM correction by interpolation in the range-Doppler domain according to Equation (2.55), and hence decouples the imaging process into two separate filtering processes in the range and azimuth directions. Among them, the matched filter of range direction in the frequency domain is

$$H_r(f) = \exp(j\pi f^2/k_r) \tag{2.56}$$

and the matched filter of the azimuth direction in the frequency domain is

$$H_a(f_\eta) = \exp\left\{j\frac{4\pi}{\lambda}r[\sqrt{1-(\lambda f_\eta/2V)^2} - 1]\right\}^3 \tag{2.57}$$

Both filtering processes above could be rapidly performed by FFT.

The block diagram of the RD algorithm is shown in Figure 2.15. Though the traditional RD algorithm has been widely used for its simplicity in implementation, there

[3]The "−1" term in the brackets of azimuth matched filter keeps the range information of the target. It guarantees that the range information could be stored in image's phase for further application such as interferometer processing.

are two drawbacks that limit its application. Firstly, it only considers the first-order coupling term in formula (2.53), so the accuracy of the algorithm is limited. In conditions of severe two-dimensional coupling, such as a large squint angle configuration or a wide bandwidth transmitted signal, a higher-order coupling term will become quite large and cannot be ignored, leading to defocusing and other adverse effects. To solve these problems, many researchers had conducted studies and a number of improved algorithms had been put forward to extend the application scope of the RD algorithm. Readers can refer to literature [13, 14] for further information. Another drawback of the RD algorithm is that the RCM correction needs interpolation, which will bring in interpolation errors. To reach a higher interpolation accuracy, a greater amount of computation is demanded. Thus the interpolation will negatively affect the imaging efficiency or imaging quality. To solve this problem, algorithms without interpolation are explored. The CS algorithm, which will be introduced in the following subsection, is exactly such an algorithm.

2.5.2 CS Algorithm

The CS algorithm is based on scaling theory proposed by Papoulis [15], which is that the position of a compressed LFM signal in the time domain can be shifted by frequency modulation on the LFM signal. Based on this theory, the CS algorithm corrects the RCM by phase multiplication instead of time-domain interpolation. The CS algorithm was first proposed by two research groups independently at the same time, which were Raney and Cumming in Canada [16] and the Bamler team of DLR in Germany [17]. In 1992 the two groups met at an international conference and got to know each other's work, and then published a paper [18] and applied for a patent jointly. The CS algorithm has since become one of the classic algorithms for SAR imaging. As is known from the former subsection, the RCM curve of SAR changes with the slant range. Thus the RCM correction in the CS algorithm is divided into two steps: firstly, eliminate the difference between the different RCM curves in different slant ranges to make all RCM curves identical and, secondly, correct the identical RCM curve to make it straight. The identification of different RCM curves is the key difficulty of SAR imaging and the CS principle is the tool used for handling this difficulty. This is why the CS algorithm got its name.

To make understanding of the CS algorithm a little easier, the CS principle used to shift the signal position is explained first. Suppose that there are two chirp pulses with the same FM rate but different time spans, which can be expressed as

$$x(t) = \sum_{i=1}^{2} P_i \mathrm{rect}\left(\frac{t - t_i}{T}\right) \exp[j\pi k_r (t - t_i)^2] \tag{2.58}$$

where P_i is a complex constant. If the frequency domain filter $H(f) = \exp(j\pi f^2/k_r)$ is used for pulse compression, the output signal can be written as $x_0(t) =$

$\sum_{i=1}^{2} P_i \text{sinc}[k_t T(t - t_i)]$. So the peaks of these two compressed chirp pulses appear at t_1 and t_2, respectively, which correspond to the times when the frequency of the chirp signals are zero. Choose t_1 as the reference time and the modulation signal $x(t)$ by $q(t)$; then

$$y(t) = x(t)q(t) = x(t)\exp[j\pi k_\Delta (t - t_1)^2] \tag{2.59}$$

that is

$$y(t) = P_1 \text{rect}\left(\frac{t - t_1}{T}\right)\exp[j\pi (k_r + k_\Delta)(t - t_1)^2] + P_2 \text{rect}\left(\frac{t - t_2}{T}\right) \times \exp\left[j\pi (k_r + k_\Delta)\left(t - t_2 + \frac{k_\Delta}{k_r + k_\Delta}\Delta t\right)^2\right]\exp\left(j\pi \frac{k_r k_\Delta}{k_r + k_\Delta}\Delta t^2\right) \tag{2.60}$$

where $\Delta t = t_2 - t_1$. Compress $y(t)$ by the frequency domain matched filter $H_\Delta(f) = \exp[j\pi f^2/(k_r + k_\Delta)]$; the output is

$$y_0(t) = P_1 \sin c[(k_r + k_\Delta)T(t - t_1)] + P_2 \sin c\left[(k_r + k_\Delta)T\left(t - t_1 + \frac{k_\Delta}{k_r + k_\Delta}\Delta t\right)\right] \times \exp\left(j\pi \frac{k_r k_\Delta}{k_r + k_\Delta}\Delta t^2\right) \tag{2.61}$$

If these two chirp pulses are considered as the SAR echo signals from two targets, it can be seen from Equation (2.61) that, after frequency modulation and pulse compression, the target at the reference position t_1 remains at its original location while the other target, which initially deviated from the reference target with interval Δt, is shifted to another location with a time shift of

$$\Delta t_{move} = \frac{k_\Delta}{k_r + k_\Delta}\Delta t \tag{2.62}$$

which is a linear function of Δt. In addition, there is a constant phase, which is a function of Δt^2, adding to the signal of the target at t_2. To maintain the target's phase characteristics, this additional phase needs to be removed. The process of shifting the signal using the CS method can be graphically illustrated, as shown in Figure 2.16.

Expressing the RCM function (2.55) as a function of time, we have

$$t_{RCM}(f_\eta) = t\left[\frac{1}{\sqrt{1 - (\lambda f_\eta/2V)^2}} - 1\right] \tag{2.63}$$

It can be seen that the amount of RCM happens to be a linear function of t, which is consistent with the linear characteristic of Equation (2.62). Therefore, the idea of

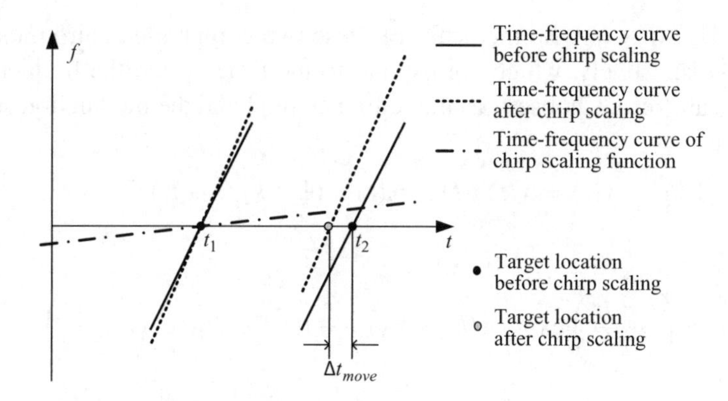

Figure 2.16 Illustration of the CS principle

using CS processing to adjust the RCM curves to identical RCM curves, which are all the same as the RCM curve at the reference range, can be applied in SAR imaging (as shown in Figure 2.17).

Since the CS operation is acting on the linear FM signals, the differential RCM correction by CS processing in the CS algorithm should be executed before range compression. Ignoring the third-order terms in Equation (2.53), after a range inverse Fourier transform the equation can be expressed in the range-Doppler domain as follows:

$$
\begin{aligned}
S_r^{\eta}(t, f_\eta) \approx \int_r\int_y drdy \Bigg\{ & \sigma(r, y)\exp(-j2\pi f_\eta y/V) \\
& \times \text{rect}\left[\frac{t - t_m(f_\eta; r)}{T}\right]\exp\{j\pi b_r(f_\eta; r)[t - t_m(f_\eta; r)]^2\} \\
& \times \exp\left[-j\frac{4\pi}{\lambda}r\sqrt{1 - (\lambda f_\eta/2V)^2}\right]\Bigg\}
\end{aligned}
\qquad (2.64)
$$

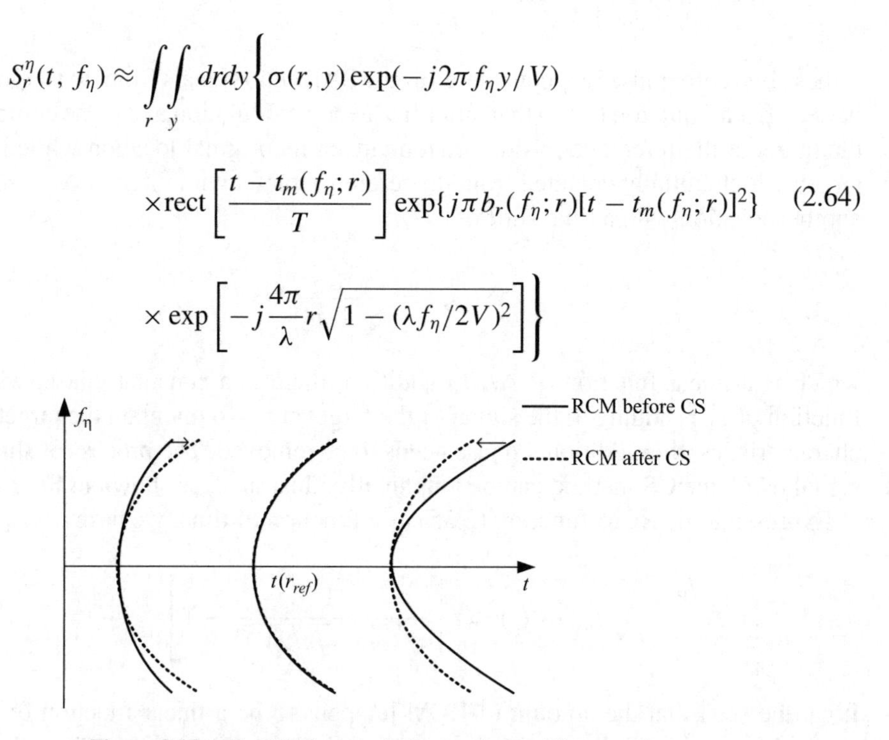

Figure 2.17 The RCM curves before and after CS processing

where

$$b_r(f_\eta; r) = \frac{k_r}{1 - \frac{2k_r}{cf_0} r \frac{(\lambda f_\eta/2V)^2}{[\sqrt{1-(\lambda f_\eta/2V)^2}]^3}} \tag{2.65}$$

$$t_m(f_\eta; r) = 2r[1 + C_s(f_\eta)]/c \tag{2.66}$$

$$C_s(f_\eta) = \frac{1}{\sqrt{1-(\lambda f_\eta/2V)^2}} - 1 \tag{2.67}$$

Ignoring the variance of $b_r(f_n; r)$ with range, that is, supposing $b_r(f_n; r) \approx b_r\left(f_n; r_{ref}\right)$ (R_{ref} is the reference range, generally chosen to be the range of the scene center), the differential RCM correction can be realized by CS processing with the following function:

$$H_{cs}(t, f_\eta) = \exp\{j\pi b_r(f_\eta; r_{ref}) C_s(f_\eta; r_{ref})[t - t_m(f_\eta; r_{ref})]^2\} \tag{2.68}$$

Then the signal is transformed to a two-dimensional frequency domain and both the range compression and the bulk RCM correction are realized by multiplying the signal with the following function:

$$H_r(f, f_\eta) = \exp\left\{ j \frac{\pi f^2}{b_r(f_\eta; r_{ref})[1 + C_s(f_\eta; r_{ref})]} \right\} \exp\left[j \frac{4\pi f}{c} r_{ref}(f_\eta) \right] \tag{2.69}$$

Finally, the signal is transformed to the range time domain and the azimuth compression is conducted by multiplying the signal with the following function in the range-Doppler domain:

$$H_a(f_\eta; r) = \exp\left\{ j \frac{4\pi}{\lambda} r[\sqrt{1-(\lambda f_\eta/2V)^2} - 1] \right\} \exp[j\Theta_\Delta(f_\eta; r)] \tag{2.70}$$

where

$$\Theta_\Delta(f_\eta; r) = -\frac{4\pi}{c^2} b_r(f_\eta; r_{ref}) C_s(f_\eta; r_{ref})[1 + C_s(f_\eta; r_{ref})](r - r_{ref})^2 \tag{2.71}$$

It is used to remove the additional phase brought in by CS processing.

The block diagram of the CS algorithm is shown in Figure 2.18. It should be noted that the traditional CS algorithm described above considers the first and second coupling terms in Equation (2.53), so it is a little more precise than the traditional RD algorithm, which only copes with the first coupling term. However, the traditional CS algorithm neglects the change of $b_r(f_\eta; r)$ with range r while considering the second-order coupling term, and it does not take the third- and higher-order coupling terms into account either. Thus, in the condition of a large squint angle and/or board

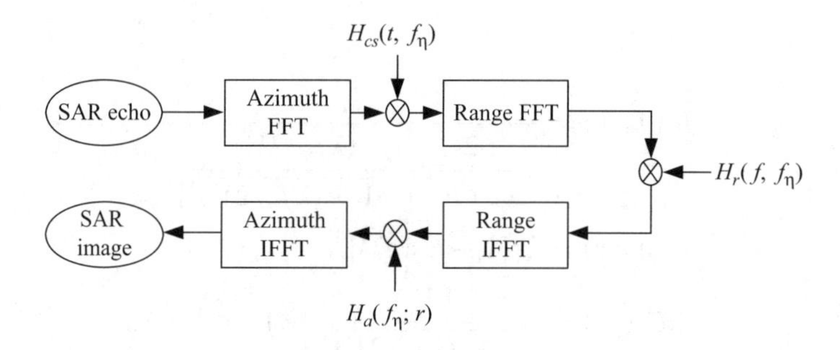

Figure 2.18 Block diagram of the CS algorithm

imaging swath, the desired imaging quality may not be achieved by the traditional CS algorithm. Therefore, improved CS algorithms have been proposed and readers who are interested can refer to the literature [19–26].

2.5.3 ω-k Algorithm

The ω-k algorithm was originally proposed by Cafforio. The principle of the ω-k algorithm is derived from seismic signal processing and is also called the range migration (RM) algorithm [27–31]. The ω-k algorithm is a theoretically accurate imaging algorithm derived from POSP. In this algorithm, the SAR echo is firstly transformed to a two-dimensional frequency domain and is then multiplied by a two-dimensional matched filter corresponding to the reference target, which compensates for the FM effect in the range direction, the FM effect in the azimuth direction and all the range-azimuth coupling terms in the echo phase of the reference target. After this step (often called "bulk focusing"), the targets at the reference range are completely focused while other targets beyond the reference range are not well focused because of the mismatch of the filter. Then, according to the relationship between the variables in the frequency domain and wavenumber domain, a variable substitution (usually called "Stolt conversion" or "Stolt interpolation" [32]) is performed, which causes the RCM curves, the azimuth phase histories and all the range-azimuth coupling terms of targets at different ranges to be exactly the same. This step therefore realizes the "remnant focus" of the targets beyond the reference range. Stolt conversion is the key step of the ω-k algorithm. It is conducted in the two-dimensional frequency domain, which is the significant difference between the ω-k algorithm and the RD/CS algorithm (whose key steps are conducted in the range-Doppler domain).

It can be shown from Equation (2.52) that the range-azimuth coupling is expressed as a complex radical formula and the coupled phase is a function of range. If the matched filter corresponding to the reference range (as shown in Equation (2.72)) is used to focus the raw data, then

$$H_{ref}(f, f_\eta) = \exp(j\pi f^2/k_r)\exp\left\{-jr_{ref}\sqrt{\left[\frac{2\pi(f_0+f)}{c/2}\right]^2-\left(\frac{2\pi f_\eta}{V}\right)^2}\right\} \tag{2.72}$$

and

$$S_r^{\Delta}(t, f_\eta) = \int_r \int_y drdy \left\{\sigma(r, y) \exp(-j2\pi f_\eta y/V)\right. \\ \left. \times \exp\left\{-j(r - r_{ref})\sqrt{\left[\frac{2\pi(f_0 + f)}{c/2}\right]^2 - \left(\frac{2\pi f_\eta}{V}\right)^2}\right\}\right. \tag{2.73}$$

If we define the wave number variables as $k = \frac{2\pi(f_0+f)}{c/2}$, $k_u = \frac{2\pi f_n}{V}$, then Equation (2.73) can be written as

$$S_r^{\Delta}(k, k_u) = \int_r \int_y \left\{\sigma(r, y) \exp(-jk_u y) \exp[-j(r - r_{ref})\sqrt{k^2 - k_u^2}]\right\} drdy \tag{2.74}$$

If the following substitution is adopted,

$$\begin{cases} k_y = k_u \\ k_r = \sqrt{k^2 - k_u^2} \end{cases} \tag{2.75}$$

then

$$S_r^{\Delta}(k_r, k_y) = \int_r \int_y \left\{\sigma(r, y) \exp(-jk_y y) \exp[-jk_r(r - r_{ref})]\right\} drdy \tag{2.76}$$

It can be seen that $s_r^{\Delta}\left(k_r, k_y\right)$ is the two-dimensional spectrum of $\sigma\,(r, y)$. Therefore, if $s_r^{\Delta}\left(k_r, k_y\right)$ is obtained, $\sigma\,(r, y)$ can be obtained directly by a two-dimensional inverse Fourier transform. However, what we have is the spectrum $s_r^{\Delta}\,(k_r, k_u)$ on the uniform grid (k, k_u). To get the desired spectrum $s_r^{\Delta}\left(k_r, k_y\right)$, what is needed is just an interpolation operation according to Equation (2.75), which intends to get the value of $s_r^{\Delta}\left(k_r, k_y\right)$ on the uniform grid $\left(k_r, k_y\right)$ from $s_r^{\Delta}\,(k_r, k_u)$ ($s_r^{\Delta}\,(k_r, k_u)$ is nonuniformly scaled in the $\left(k_r, k_y\right)$ domain). This is the step called Stolt conversion or Stolt interpolation (as shown in Figure 2.19). It should be noted that the reference distance r_{ref} can be chosen arbitrarily in principle. It can even be chosen to be $r_{ref} = 0$, which means the "bulk focusing" step is omitted and the Stolt conversion is responsible for all focusing. However, in practical processing, r_{ref} is usually selected to be the slant range of the scene center. This ensures that the Stolt interpolation performed in a low-frequency area can be carried out directly by the sinc function interpolation method.

The diagram of ω-k algorithm is shown in Figure 2.20. It can be seen from the derivation process above that there is not any approximation in the derivation of the

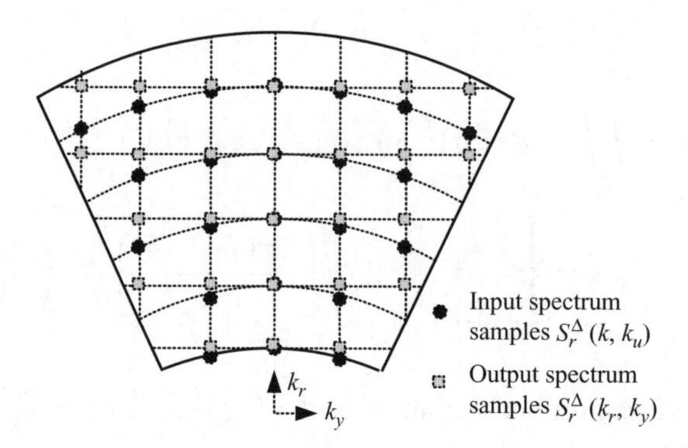

Figure 2.19 Illustration of the Stolt interpolation

ω-k algorithm under the condition that the relative motion between the SAR and different targets in the scenario can be described by an identical uniform velocity. In this case, the ω-k algorithm is more accurate than the RD algorithm and the CS algorithm in principle; thus it is more suitable to cases of high resolution, wide bandwidth transmitted signal and large squint configuration. The ω-k algorithm can be considered as an accurate imaging method for the airborne SAR geometry with constant velocity (except for interpolation errors). However, the ω-k algorithm has limited application in spaceborne SAR imaging, because the equivalent SAR velocity is changing with range due to the surface bending and the earth's rotation. Currently the literature [33] has improved the ω-k algorithm to release this limitation and to extend its scope of application. Besides, another shortcoming of the ω-k algorithm is that Stolt conversion requires interpolation, which affects the accuracy and efficiency of the algorithm.

2.6 Summary

The two-dimensional imaging principle of SAR is described in this chapter with the resolution theory as a start point. The characteristics of the SAR echo are also introduced, based on the point target signal model, and then the pixel-by-pixel imaging methods both in the time domain and frequency domain are introduced. On this basis,

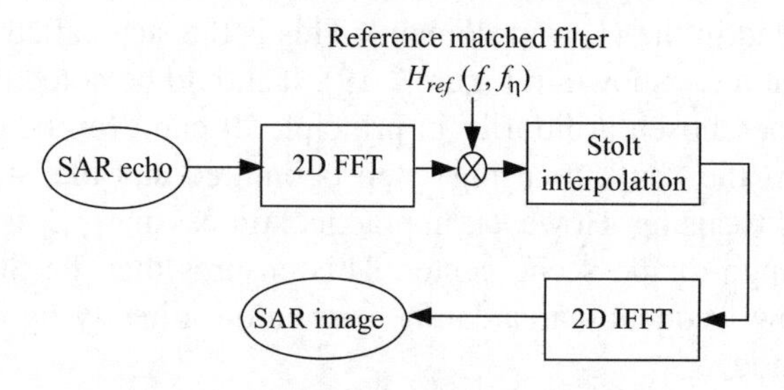

Figure 2.20 Basic flow of the ω-k algorithm

three classic efficient SAR imaging algorithms are introduced, which are the RD algorithm, the CS algorithm and the ω-k algorithm. The advantages and shortcomings of each algorithm are also explained. For a detailed description and comparison of those algorithms, the interested reader can refer to [34]. Since the intention of this chapter is to lay a foundation of SAR signal processing theory, the description of the imaging algorithms here are quite concise and are only based on the simple nonsquint side-looking mode. With regard to the algorithms for complex monostatic SAR modes, the reader can refer to the references listed in this book.

References

[1] Skolnik, M.I. (1990) *Radar Handbook*, McGraw-Hill, New York.

[2] Rihaczek, A.W. (1973) *Principles of High-Resolution Radar*, Science Press, Beijing (Translated by Dong, S.).

[3] Xu, S. (1999) *Signal and System – Theory, Method and Application*, The University of Science and Technology Press, Hefei.

[4] Zhang, C. (1989) *Synthetic Aperture Radar: Principles, Analysis and Utilization*, Science Press, Beijing.

[5] Zheng, J., Ying, Q. and Yang, W. (1999) *Signal and System*, Higher Education Press, Beijing.

[6] Bauck, J.L. and Jenkins, W.K. (1988) Tomographic processing of spotlight-mode synthetic aperture radar signals with compensation for wavefront curvature: acoustics, speech, and signal processing, 1988. *ICASSP-88*, **1** (2), 1192–1195.

[7] Desai, M.D. and Jenkins, W.K. (1992) Convolution backprojection image reconstruction for spotlight mode synthetic aperture radar. *IEEE Transactions on Image Processing*, **1**(4), 505–517.

[8] Jenkins, W.K., Desai, M.D. and Hull, A.W. (1992) Computational strategies for synthetic aperture imaging. IEEE International Conference on Systems Engineering, pp. 523–526.

[9] Wu, C. (1976) A digital system to produce imagery from SAR data. AIAA Conference: System Design by Sensors.

[10] Wu, C. (1977) Processing of SEASAT SAR data. SAR Technology Symo, Las Cruces, NM.

[11] Cumming, I.G. and Bennett, J.R. (1979) Digital processing of SEASAT SAR data. IEEE International Conference on Acoustics, Speech and Signal Processing, Washington, DC.

[12] Bennett, J.R. and Cumming, I.G. (1979) A digital production of SEASAT SAR imagery. In: *Proceedings of the SURGE Workshop*, ESA Publication 154, pp. 16–18.

[13] Schmidt, A.R. (1986) Secondary Range Compression for Improved Range Doppler Processing of SAR Data with High Squint. Master's thesis, The University of British Columbia.

[14] Jin, M.J. and Wu, C. (1984) A SAR correlation algorithm which accommodates large range migrition. *IEEE Transactions on Geoscience and Remote Sensing*, **22** (6), 592–597.

[15] Papoulis, A. (1968) *Systems and Transforms with Applications in Optics*, McGraw-Hill, New York.

[16] Cumming, I.G., Wong, F.H. and Raney, R.K. (1992) A SAR processing algorithm with no interpolation. In: *Proceedings of the IEEE International Geoscience and Remote Sensing Symposium (IGARSS'92)*, Clear Lake, pp. 376–379.

[17] Runge, H. and Bamler, R. (1992) A novel high precision SAR focusing algorithm based on chiro scaling. In: *Proceedings of the IEEE International Geoscience and Remote Sensing Symposium (IGARSS'92)*, Clear Lake, pp. 372–375.

[18] Keith, R., Runge, H., Banler, R., *et al.* (1994) Precision SAR processing using chirp scarling. *IEEE Transactions on Geoscience and Remote Sensing*, **32** (4), 786–799.

[19] Impagnatiello, F. (1995) A precision chirp scaling SAR processor extension to sub-aperture implementation on massively parallel supercomputers. In: *Proceedings of the IEEE International Geoscience and Remote Sensing Symposium (IGARSS'95)*, Florence, Italy, vol. 3, pp. 1819–1821.
[20] Ding, C., Peng, H., Wu, Y. and Jia, H. (1999) Large beamwidth spaceborne SAR processing using chirp scaling. In: *Proceedings of the IEEE International Geoscience and Remote Sensing Symposium (IGARSS'99)*, Hamburg, vol. 1, pp. 527–529.
[21] Wong, F.H. and Yeo, T.S. (2001) New application of non-linear chirp scaling in SAR data processing. *IEEE Transactions on Geoscience and Remote Sensing*, **39** (5), 946–953.
[22] Hawkins, D.W. and Gough, P.T. (1997) An advanced chirp scaling algorithem for synthetic aperture imaging. In: *Proceedings of the IEEE International Geoscience and Remote Sensing Symposium (IGARSS'97)*, Florence, Singapore, 1, pp. 471–473.
[23] Davidson, G.W., Cumming, I.G. and Ito, M.R. (1996) A chirp scaling approach for processing squint mode SAR data. *IEEE Transactions on Aerospace and Electronics Systems*, **32** (1), 121–133.
[24] Hong, W., Mittermayer, J. and Moreira, A. (2000) High squint angle processing of E-SAR stripmap data. In: *Proceedings of the European Conference on Synthetic Aperture Radar (EUSAR'00)*, Munich, Germany, pp. 449–552.
[25] Huang, Y., Li, C., Chen, J., *et al.* (2000) Refined chirp scaling algorithm for high resolution spaceborne imaging. *Acta Electronica Sinica*, **28** (3), 35–38.
[26] Li, C., Huang, Y., Wang, X., *et al.* (1996) A spaceborne SAR imaging processing algorithm using chirp scaling. *Acta Electronica Sinica*, **24** (6), 20–24.
[27] Cafforio, C., Prati, C. and Rocca, F. (1988) Full resolution focusing of SEASAT SAR image in the frequency-wave number domain. In: *Proceedings of the 8th EAR Workshop*, pp. 336–355.
[28] Cafforio, C., Prati, C. and Rocca, F. (1991) SAR data focusing using seismic migration techniques. *IEEE Transactions on Aerospace and Electronic Systems*, **27** (2), 194–207.
[29] Carrara, W.G., Goodman, R.S. and Majewski, R.M. (1995) *Spotlight Synthetic Aperture Radar: Signal Processing Algorithm*, Artech House, Norwood.
[30] Munson, D.C., O'Brian, J.D. and Jenkins, W.K. (1983) A tomographic formulation of spotlight mode synthetic aperture radar. *Proceedings of the IEEE*, **71** (8), 917–925.
[31] Bamler, R. (1991) A comparison of range-doppler and wavenumber domain SAR focusing algorithm. *IEEE Transaction on Geoscience and Remote Sensing*, **12**, 235–251.
[32] Stolt, R.H. (1978) Migration by transform. *Geophysics*, **43**(1), 23–48.
[33] Hu, Y., Ding, C. and Wu, Y. (2005) The wide swath spaceborne SAR imaging based on ω-k algorithm. *Acta Electronica Sinica*, **33** (6), 1044–1047.
[34] Cumming, I.G. and Wong, F.H. (2004) *Digital Processing of Synthetic Aperture Radar Data: Algorithms and Implementation*, Artech House, Norwood.

3

Basic Knowledge of Bistatic SAR Imaging

Chapter 2 has introduced the basic theory of monostatic SAR imaging. From this chapter on, the contents of bistatic SAR signal processing will be introduced. Due to the separation of transmitting and receiving platforms of bistatic SAR, the configurations of bistatic SAR geometry are flexible; the imaging modes can be diverse and the imaging characteristics of different modes are different. Thus, it is necessary to classify bistatic SAR configurations reasonably. The correctly designed configuration and chosen system parameters are preconditions for applications of bistatic SAR. Therefore, it is also very necessary to analyze the system performance of bistatic SAR. SAR system performance includes received SNR, spatial resolution, radiation resolution, system ambiguity, swath width, positioning accuracy and so on. Among them, spatial resolution is an essential performance index for imaging radar like SAR, and is most related to imaging processing. In addition, received SNR is an important parameter related to image quality, and is a factor that must be given priority in SAR system design. Accordingly, as the basis of bistatic SAR imaging processing, this chapter first introduces bistatic SAR configurations and their classification, then describes the bistatic SAR radar equation expressed in the form of the received SNR and finally interprets the bistatic SAR two-dimensional spatial resolution in detail.

3.1 Bistatic SAR Configurations

With regard to bistatic SAR configurations, at the beginning there were some classifications that were not systematic and some appellations that were not very appropriate, such as "along-track configuration", "parallel-track configuration", "one-stationary

Bistatic SAR Data Processing Algorithms, First Edition. Xiaolan Qiu, Chibiao Ding and Donghui Hu.

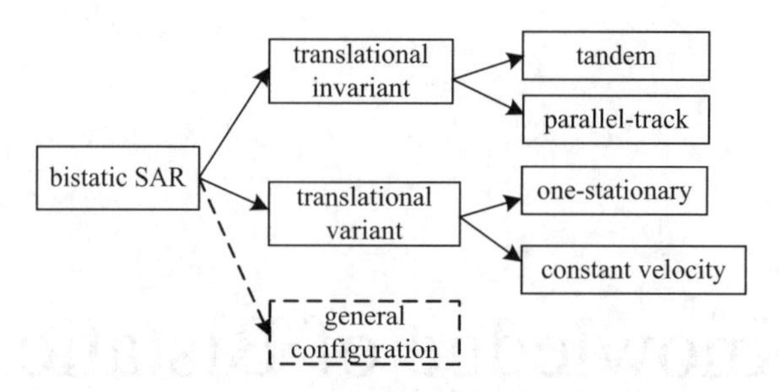

Figure 3.1 Classifications of bistatic SAR configurations

configuration". In 2004, Ender [1] classified bistatic SAR configurations according to their complexity and ranked bistatic SAR in the following stages:

(1) Stage 0. Monostatic configuration.
(2) Stage 1. Tandem configuration, that is, receive and transmit antennas are traveling along the same track with equal and constant velocity vectors.
(3) Stage 2. Translational invariant (TI) configuration, that is, receive and transmit antennas are traveling along parallel tracks with equal and constant velocity vectors.
(4) Stage 3. Constant velocity configuration, that is, receive and transmit antennas are traveling with constant but different velocity vectors.
(5) Stage 4. General configuration, that is, receive and transmit antennas are traveling along arbitrary flight paths.

This classification wass followed afterwards by a number of different ones by other researchers. Referring to the classification above and according to the preference of the authors, bistatic SAR configurations are classified in this book as shown in Figure 3.1.

In Figure 3.1, the TI configuration includes Stage 1 and Stage 2 mentioned above. Under this configuration, receive and transmit antennas travel with equal and constant velocity vectors, that is, the relative position of receive and transmit antenna phase centers remain invariant during SAR illumination. The TI configuration can be divided into the "tandem mode" (Stage 1), in which receive and transmit antennas travel along the same track, and the "parallel-track mode" (Stage 2), in which receive and transmit antennas travel along parallel tracks. The translational variant (TV) configuration is Stage 3 mentioned above, in which receive and transmit antenna phase centers travel with constant velocity vectors, but the relative position changes with time. This configuration includes one stationary configuration, in which one platform remains stationary and the other travels with constant velocity vectors, and a constant velocity configuration, in which the receiver and transmitter travel with constant but different velocity vectors (different magnitude, different direction or both different). In addition, there is a general configuration (Stage 4), in which the receiver, or transmitter, or both

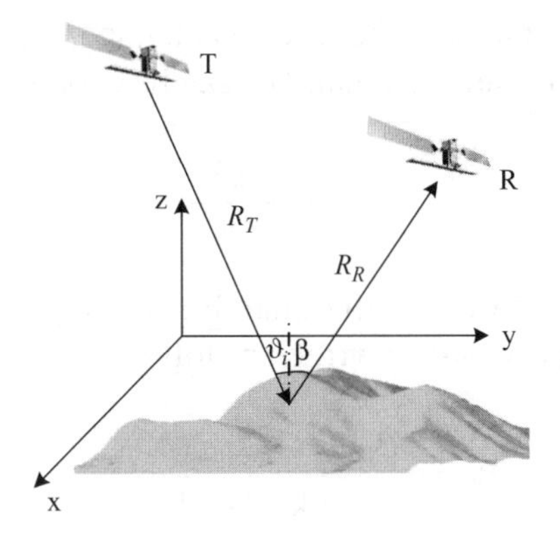

Figure 3.2 Bistatic SAR geometry

of them travel with nonconstant velocity vectors. Because the general configuration is too arbitrary and complicated, this book will not discuss it.

3.2 Radar Equation of Bistatic SAR

The radar equation is an important tool used to measure radar system performance. It plays an important role in radar system designing. Based on the radar equation, the constraint relationship between the parameters of different subsystems can be analyzed and understood. Also, the radar equation can be helpful in correctly choosing and assigning the parameters of each subsystem.

The geometry of bistatic SAR is shown in Figure 3.2. If the transmitter output power is P_T, the transmit antenna gain (antenna gain means power gain in this book) is G_T and the distance between the transmit antenna and the target is R_T, the power density at the target is

$$S_T = \frac{P_T G_T}{4\pi R_T^2} \tag{3.1}$$

The bistatic RCS (denoted by σ here) is a function of the incidence angle ϑ_i and the bistatic angle (the angle between the receive and transmit distance vectors) ξ_{bi}. Then, the power density of the target's scattering energy at the receive antenna can be written as

$$S_R = \frac{S_T \sigma(\vartheta_i, \xi_{bi})}{4\pi R_R^2} \tag{3.2}$$

where R_R is the distance between the target and receive antenna. The signal power received by the receiver is

$$P_R = S_R A_R \tag{3.3}$$

where A_R is the equivalent area of the receive antenna. According to antenna theory, its relationship with receive antenna gain G_R can be expressed as

$$A_R = \frac{G_R \lambda^2}{4\pi} \tag{3.4}$$

Substitute Equations (3.1), (3.2) and (3.4) into Equation (3.3); then the radar equation of bistatic SAR in free space can be written as follows:

$$P_R = \frac{P_T G_T G_R \lambda^2 \sigma(\vartheta_i, \xi_{bi})}{(4\pi)^3 R_T^2 R_R^2} \tag{3.5}$$

Moreover, the relationship between the receive SNR (noted as $(S/N)_r$) and the receive power P_R can be expressed as follows:

$$P_R = kT_R B_n F_n (S/N)_r \tag{3.6}$$

where k is Boltzmann's constant, T_R is the receiver noise temperature, B_n is the noise bandwidth of the receiver, which is the same as the signal bandwidth on the condition of matched filter processing, and F_n is the receiver noise. From Equations (3.5) and (3.6), the radar equation in the form of SNR can be derived, which is

$$(S/N)_r = \frac{P_T G_T G_R \lambda^2 \sigma(\vartheta_i, \xi_{bi})}{kT_R B_n F_n (4\pi)^3 R_T^2 R_R^2} \tag{3.7}$$

Actually, the output SNR of the signal processor will increase because of the two-dimensional compression on SAR raw data during imaging processing. Suppose that

$$(S/N)_o = I_c K_w (S/N)_r \tag{3.8}$$

where I_c is the two-dimensional compression gain; K_w is a constant smaller than 1, which represents the gain attenuation due to weighting and other operations during the processing. Again, suppose the transmitted signal is an LFM pulse with bandwidth B and pulse duration time T. If processed with the matched filter in the frequency domain, the power gain of the signal after range compression is

$$I_{c;rg} = BT \tag{3.9}$$

As SAR achieves high azimuth resolution though coherent integration of azimuth echoes collected during synthetic aperture time T_s, if the processing Doppler bandwidth is B_a, the azimuth compression gain is

$$I_{c;az} = B_a T_s \tag{3.10}$$

To sum up:

$$(S/N)_o = \frac{P_T G_T G_R B_a T_s B T K_W \lambda^2 \sigma(\vartheta_i, \xi_{bi})}{k T_R B_n F_n (4\pi)^3 R_T^2 R_R^2} \tag{3.11}$$

In practice, there are other factors to consider, such as antenna pattern modulation, power loss while transmitting and receiving, and propagation attenuation. Therefore, the definition of the bistatic SAR equation, which is stricter, is as follows:

$$(S/N)_o = \frac{P_T G_T F_T^2(\theta, \varphi) G_R F_R^2(\theta, \varphi) B_a T_s B T K_W \lambda^2 \sigma(\vartheta_i, \xi_{bi})}{k T_R B_n F_n (4\pi)^3 R_T^2 R_R^2 L} \tag{3.12}$$

where $F_T(\theta, \varphi)$ and $F_R(\theta, \varphi)$ are single-pass antenna patterns of the transmitter and receiver (here the antenna pattern refers to the normalized amplitude modulation function) and L represents the energy loss during transmitting, receiving and propagating.

In the case of monostatic SAR, $R_T = R_R = R_m$ (the suffix "m" represents "monostatic"), and usually the receiver and transmitter use the same antenna. Thus, the synthetic aperture time is directly proportional to the distance between the antenna and target, and can be expressed as

$$T_{s,m} = \frac{\lambda R_m}{D_{a,m} V_m} \tag{3.13}$$

where $D_{a,m}$ is the azimuth antenna length of monostatic SAR. Hence, $(S/N)_{o,m} \propto 1/R_m^3$. However, in the case of bistatic SAR, the synthetic aperture time is decided by many factors jointly, such as the transmit and receive antenna beam widths, transmitting and receiving distances and the velocities of the transmitter and receiver. As a result, the relationship between SNR and transmitting and receiving distances needs to be analyzed and designed specifically according to Equation (3.12) for specific configurations.

3.3 Spatial Resolution of Bistatic SAR

Spatial resolution, which is most closely related to SAR imaging, is one of the most important performance indexes for SAR. Due to the complexity of the bistatic geometry configurations, the ground resolution ability of bistatic SAR is more complex than that of monostatic SAR. As known from Section 2.3, in the case where the beam pointing direction of monostatic SAR remains unchanged, the azimuth direction usually denotes the moving direction of the radar, while the range direction denotes the beam pointing direction. However, in the case of bistatic SAR, thanks to the two antennas being separated from each other, the physical meaning of "range direction" is no longer explicit. Moreover, when the flight directions of the receiver and transmitter are not consistent with each other, the physical meaning of the "azimuth direction" also becomes ambiguous. Hence, the understanding of bistatic SAR resolution is more complex than that of monostatic SAR and the resolution calculation is more complicated.

Since the concept of bistatic SAR was put forward, the complexity of its resolution ability has been noticed and quite a lot of work has been done on the two-dimensional resolution ability [2–12]. Some of them are concerned with special cases where the models are simplified, for example, [3] briefly analyzes the airborne cases where the transmitter is far away and the receiver is near the scenario while the receive antenna plays a dominant role; [4] simplifies the bistatic imaging geometry into a two-dimensional plane, that is, the transmitter, receiver and the targets are all assumed to be located in the same plane; [6] places emphasis on the study of spatial resolution in the translational invariant configuration. Also, some literature focuses on the complex configurations of bistatic SAR, in which some comparatively universal resolution calculation methods have been put forward. For example, Cardillo [5] calculates the ground resolution of bistatic SAR via resolving the isogradients of the bistatic range and Doppler frequency, which is a method that can be generally applied and is suitable for computer implementation; Nico and Tesauro [12] introduce a scheme for the computation of the synthetic aperture time in the translational variant configuration under the approximation that receive and transmit beam patterns are rectangular functions. This provides a foundation for an important parameter, that is, the synthetic aperture time, in resolution computations.

In this section, the magnitude and direction of the two-dimensional resolution in translational invariant and translational variant bistatic SAR are analyzed, using the gradient method in combination with the authors' research findings [11]. Hence, one can analyze and calculate bistatic SAR resolution according to this section.

3.3.1 Range Resolution

Similar to monostatic SAR, the fast time t corresponds to the range direction. Suppose the LFM signal with bandwidth B is transmitted. Then the range time resolution is $1/B$, equivalent to a range resolution of c/B in length, that is, as long as the difference between the bistatic ranges (i.e., the sum of the range from the target to the transmit antenna phase center and the range from the target to the receive antenna phase center) of two targets is larger than c/B, the two targets can be distinguished from each other. However, unlike monostatic SAR, because of the separation of the receiver and transmitter, the bistatic range includes distances in two directions, and the slant range defined by the bistatic range cannot be projected into the ground range though simple linear transformation (such as the incidence angle correction in the monostatic case, as shown by Equation (2.33)). Consequently, as the physical meaning of "range direction" in bistatic SAR is not intuitive, the resolution ability of bistatic SAR on the ground cannot be depicted by a simple consideration of the resolution ability of the bistatic range. Therefore, the ground resolution of bistatic SAR will be considered directly in the following. For simplicity, the following analysis is based on the hypothesis that the imaging scenario is flat. In the spaceborne case, if the curvature of the earth cannot be ignored, the analysis can be made by dividing the whole data into blocks along the range.

The geometry of bistatic SAR is shown in Figure 3.2. Suppose the z coordinate of the ground is zero. At azimuth time η, the transmitter and receiver are located at $(x_{T\eta}, y_{T\eta}, h_{T\eta})$ and $(x_{R\eta}, y_{R\eta}, h_{R\eta})$, respectively. It can be seen from bistatic geometry that the surface where the bistatic range is R_{bi} is an ellipsoid surface. When azimuth time is η, the intersection curve of this ellipsoid surface with the ground can be expressed as

$$\sqrt{(x - x_{T\eta})^2 + (y - y_{T\eta})^2 + h_{T\eta}^2} + \sqrt{(x - x_{R\eta})^2 + (y - y_{R\eta})^2 + h_{R\eta}^2} = R_{bi} \quad (3.14)$$

The gradient of an arbitrary point (x, y) on this equidistant line is

$$\nabla R_{bi} = \frac{\partial R_{bi}}{\partial x}\vec{i}_x + \frac{\partial R_{bi}}{\partial y}\vec{i}_y = \left(\frac{x - x_{T\eta}}{R_{T\eta}} + \frac{x - x_{R\eta}}{R_{R\eta}}\right)\vec{i}_x + \left(\frac{y - y_{T\eta}}{R_{T\eta}} + \frac{y - y_{R\eta}}{R_{R\eta}}\right)\vec{i}_y \quad (3.15)$$

where

$$R_{T\eta} = \sqrt{(x - x_{T\eta})^2 + (y - y_{T\eta})^2 + h_{T\eta}^2} \quad (3.16)$$

$$R_{R\eta} = \sqrt{(x - x_{R\eta})^2 + (y - y_{R\eta})^2 + h_{R\eta}^2} \quad (3.17)$$

Define

$$\alpha_{T\eta} = \arcsin\left[\frac{x - x_{T\eta}}{\sqrt{(x - x_{T\eta})^2 + h_{T\eta}^2}}\right] \quad (3.18)$$

$$\alpha_{R\eta} = \arcsin\left[\frac{x - x_{R\eta}}{\sqrt{(x - x_{R\eta})^2 + h_{R\eta}^2}}\right] \quad (3.19)$$

$$\theta_{T\eta} = \arcsin\left[\frac{y - y_{T\eta}}{R_{T\eta}}\right] \quad (3.20)$$

$$\theta_{R\eta} = \arcsin\left[\frac{y - y_{R\eta}}{R_{R\eta}}\right] \quad (3.21)$$

Then the gradient of the bistatic range can be written as

$$\nabla R_{bi} = [\sin\alpha_{T\eta}\cos\theta_{T\eta} + \sin\alpha_{R\eta}\cos\theta_{R\eta}]\vec{i}_x + [\sin\theta_{T\eta} + \sin\theta_{R\eta}]\vec{i}_y \tag{3.22}$$

The magnitude of the gradient is

$$|\nabla R_{bi}| = \sqrt{[\sin\alpha_{T\eta}\cos\theta_{T\eta} + \sin\alpha_{R\eta}\cos\theta_{R\eta}]^2 + [\sin\theta_{T\eta} + \sin\theta_{R\eta}]^2} \tag{3.23}$$

The angle between the direction of the gradient and the x axis is

$$\theta_l = \operatorname{atan}\left(\frac{\sin\theta_{T\eta} + \sin\theta_{R\eta}}{\sin\alpha_{T\eta}\cos\theta_{T\eta} + \sin\alpha_{R\eta}\cos\theta_{R\eta}}\right) \tag{3.24}$$

Since the resolution magnitude of R_{bi} is c/B, the magnitude of ground range resolution at (x, y) along the gradient direction is

$$\rho_{rl} = \frac{c/B}{|\nabla R_{bi}|} = \frac{c/B}{\sqrt{[\sin\alpha_{T\eta}\cos\theta_{T\eta} + \sin\alpha_{R\eta}\cos\theta_{R\eta}]^2 + [\sin\theta_{T\eta} + \sin\theta_{R\eta}]^2}} \tag{3.25}$$

The ground range resolution along an arbitrary direction γ (the angle between γ and the x axis is defined by θ_γ) is

$$\rho_{r\gamma} = \rho_{rl}/\sin(|\theta_l - \theta_\gamma|) \tag{3.26}$$

In the translational invariant configuration, if the receiver and transmitter both fly along the y axis, then $\alpha_{R\eta}$, $\alpha_{T\eta}$ and $\theta_{R\eta}$, $\theta_{T\eta}$ are look angles and squint angles of the receiver and transmitter to the target at (x, y), respectively. Under this situation, to analyze range resolution, the resolution ability in the x-axis direction is usually considered, which is

$$\rho_{rx} = \frac{c/B}{|\sin\alpha_{T\eta}\cos\theta_{T\eta} + \sin\alpha_{R\eta}\cos\theta_{R\eta}|} \tag{3.27}$$

The analysis above provides the ground range resolution of the target (x, y) at a certain time η. For SAR, the target (x, y) can be illuminated by the antenna during the whole synthetic aperture time. Then, which azimuth time should be chosen to calculate the ground range resolution at (x, y)? Similar to monostatic SAR, in bistatic SAR the wideband echo signal in the slow time domain is obtained through the Doppler effect caused by the relative motion between the platform and target and high azimuth resolution is obtained by matched filter processing to this wideband signal. After matched filtering, the target is theoretically supposed to appear at the center of the synthetic aperture time, that is, the time corresponding to the Doppler centroid (if there is no artificial position shift during the imaging process). Thus, the ground range resolution that can be achieved at (x, y) in bistatic SAR should reasonably be calculated at the center of the target's synthetic aperture time.

However, as mentioned in Section 2.4, the SAR imaging process usually does not adopt the pixel-by-pixel computation. Instead, batch processing is often used. For the batch imaging process, a certain length of frequency spectrum (usually referred to as the azimuth processed bandwidth) is chosen for imaging through a window centered on a certain reference Doppler centroid frequency f_{dc_ref}. This is equivalent to changing the window of the target's synthetic aperture time (also changing the width and direction of the combined narrow beam). As a result, the center of the target's synthetic aperture time is altered to the azimuth time corresponding to f_{dc_ref}. Hence, if the imaging processing mentioned above is adopted, the calculation steps of ground range resolution can be described as follows:

(1) The time $\eta_c(x, y)$ corresponding to the reference Doppler frequency f_{dc_ref} of the target at (x, y) of the scene is calculated via the numerical method.
(2) The locations of receive and transmit antenna phase centers at $\eta_c(x, y)$ are obtained based on the original locations and the velocity information of receive and transmit antennas.
(3) The magnitude and direction of the ground range resolution are calculated according to Equations (3.25) and (3.24).
(4) The ground range resolutions of all the locations in the scene are calculated. Accordingly, the ground range resolution field (the ground range resolution of bistatic SAR has magnitude and direction, which is in accord with the definition of field) is described for the convenience of intuitive judgment and parameter determination.

In the following, taking the one-stationary case as an example, simulation results of bistatic SAR range resolution are shown. Suppose that the transmitter is located at (–8 km, 0, 4.5 km) when the azimuth time is zero and travels along the y direction; the transmitted signal bandwidth is 500 MHz; the receiver is located at (–1.3 km, 0, 1 km), which is stationary. The results of Figures 3.3 and 3.4 are obtained through investigating the ground range resolutions within the area of [–1 km, 2 km] in the x direction and [–1.5 km, 1.5 km] in the y direction. In Figure 3.3, the equidistant lines on the ground and the range gradient directions of bistatic SAR are shown. It can be seen that the equidistant lines are sparse at the near range and dense at the far range along the x direction, and is sparse on both sides while dense in the middle along the y direction. This means that the range resolution under this simulation condition is better at the far range than at the near range, and better in the middle of the scene than on both sides (in the y direction). This can also be seen from Figure 3.4. In addition, under these simulation parameters, the range gradient direction also varies with location, which indicates that the magnitude and direction of bistatic SAR range resolution are all space-variant. Usually, for consistency of range resolutions in the scenario, it is better to set the receive and transmit antennas far away from the scenario.

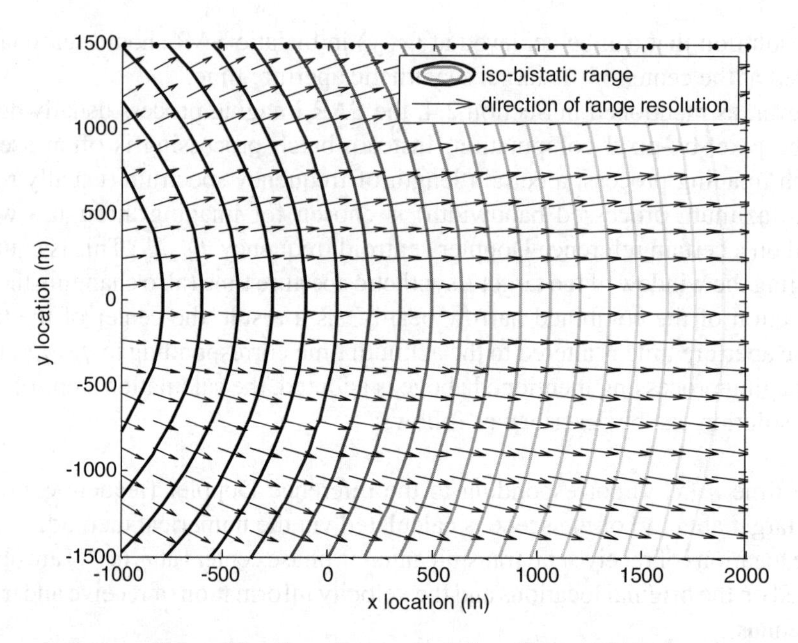

Figure 3.3 Ground range resolution of bistatic SAR

3.3.2 Azimuth Resolution

As mentioned in Chapter 2, supposing the Doppler bandwidth caused by the relative motion between the radar and target during synthetic aperture time T_{syn} is B_a, the resolution in the slow time (usually referred to as the azimuth time) domain is $1/B_a$. As depicted in Section 3.3.1, the target will appear at the azimuth time corresponding to

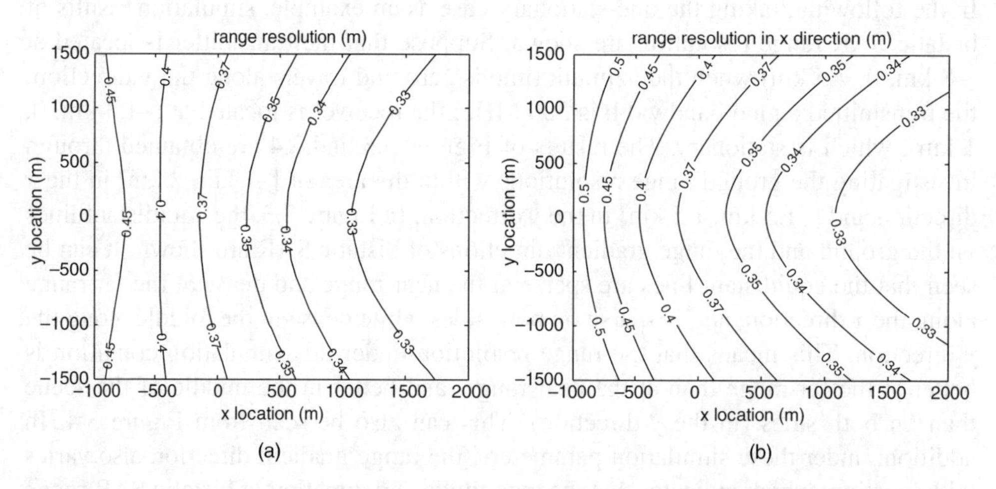

Figure 3.4 Ground equidistant lines and range gradient directions of bistatic SAR: (a) the magnitude of range resolution along the range, (b) the magnitude of range resolution along the gradient in the x direction

the reference Doppler centroid frequency f_{dc_ref} after azimuth compression. Hence, if the difference between the azimuth time of two targets corresponding to f_{dc_ref} is larger than $1/B_a$, the two targets can be distinguished from each other (actually, due to the fact that the azimuth echo signal can be approximated to a linear FM signal, the different choices of f_{dc_ref} will not influence the calculation results of the difference between the azimuth times corresponding to f_{dc_ref} of two targets, and neither will they affect the calculation results of resolution). Suppose that the Doppler FM rate is f_r and the azimuth time of a target corresponding to f_{dc_ref} is η_1. If the Doppler instant frequency of another target is $f_{dc\_ref} + \Delta f_\eta$ at the same time, the Doppler frequency f_{dc_ref} of this target corresponds to the time $\eta_1 + \Delta f_\eta / f_r$. Thus, when $|\Delta f_\eta / f_r| > 1/B_a$, that is, $|\Delta f_\eta| > 1/T_{syn}$, the two targets can be distinguished from each other.

3.3.2.1 Retrospection of the Azimuth Resolution in Different Imaging Modes of Monostatic SAR

According to the discussions above, the azimuth resolution of monostatic SAR is retrospective here and the azimuth resolution formulas in different imaging modes of airborne/spaceborne monostatic SAR are deduced. Then, the gradient method for a resolution calculation is introduced, which can at the same time offer a deeper understanding of the azimuth resolution of monostatic SAR.

In the case of airborne monostatic SAR where the curvature of the earth can be ignored, it can be supposed that the airplane flies along a straight line (suppose it is along the y axis). The imaging geometries of airborne SAR in the strip-map, spotlight and sliding spotlight mode [13, 14] are illustrated in Figure 3.5. Here, suppose the Doppler centroid f_{dc_ref} adopted in the imaging process is zero. If the airplane velocity is denoted as V_p, then $R(\eta) = \sqrt{R_0^2 + (V_p\eta - y)^2} \approx R_0 + \dfrac{(V_p\eta - y)^2}{2R_0}$ and thus the instant Doppler frequency is $f_\eta \approx \dfrac{2V_p(V_p\eta - y)}{\lambda R_0}$. As can be seen from Figure 3.5, no matter which imaging mode it is in, the difference between the time of targets corresponding to f_{dc_ref} can be expressed as

$$\Delta\eta = \Delta y / V_p \tag{3.28}$$

where Δy is the azimuth location difference. Under the approximation that the azimuth signal is a linear FM signal, this formula is independent of the choices of f_{dc_ref}. These three imaging modes differ in the length of synthetic aperture time and the target Doppler bandwidth B_a. As the minimum distinguishable difference between the time corresponding to f_{dc_ref} is $1/B_a$, based on Equation (3.28), the resolution

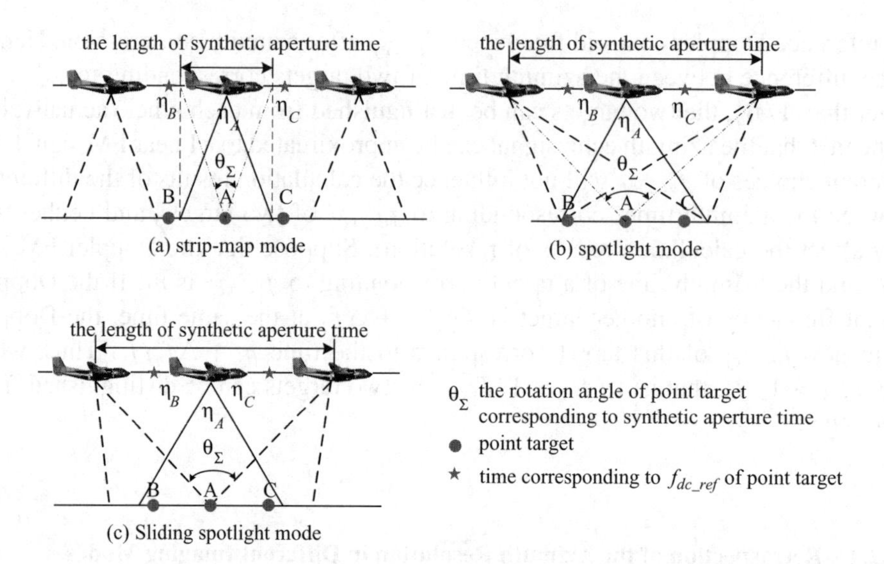

Figure 3.5 Imaging geometries of airborne monostatic SAR

along the y direction in these three modes of airborne monostatic SAR can be denoted uniformly as in the following equation:

$$\rho_{yA} = \Delta y_{\min} = V_p / B_a \tag{3.29}$$

The formula above can also be deduced using the method of an isogradient of instant Doppler frequency. Supposing SAR is located at (x_η, y_η, h_η) at the time η and the target location on the ground illuminated by the beam center is (x, y), then the target Doppler frequency can be written as

$$f_\eta(\eta; x, y) = \frac{2V_p}{\lambda} \frac{y - y_\eta}{\sqrt{(x - x_\eta)^2 + (y - y_\eta)^2 + h_\eta^2}} \tag{3.30}$$

Let $r = \sqrt{(x - x_\eta)^2 + h_\eta^2}$. Project the target to the slant range plane. Then, the Doppler gradient in the slant range plane is

$$\nabla f_\eta = \frac{\partial f_\eta}{\partial r}\vec{i}_r + \frac{\partial f_\eta}{\partial y}\vec{i}_y = -\frac{2V_p}{\lambda R_\eta}\cos\theta_\eta \sin\theta_\eta \vec{i}_r + \frac{2V_p}{\lambda R_\eta}\cos^2\theta_\eta \vec{i}_y \tag{3.31}$$

where

$$R_\eta = \sqrt{r^2 + (y - y_\eta)^2} \tag{3.32}$$

$$\cos\theta_\eta = \frac{r}{\sqrt{r^2 + (y - y_\eta)^2}} \tag{3.33}$$

$$\sin\theta_\eta = \frac{y - y_\eta}{\sqrt{r^2 + (y - y_\eta)^2}} \tag{3.34}$$

Obviously, the magnitude of the Doppler gradient is $|\nabla f_\eta| = \left|\frac{2V_p}{\lambda R_\eta}\cos\theta_\eta\right|$ and the intersection angle between the Doppler gradient direction and negative direction of y axis is θ_η. It can be seen that the Doppler gradient direction happens to be the cross range direction. Suppose a tiny interval between the locations along the direction of Doppler gradient to be Δd; then $\Delta f_\eta = (\nabla f_\eta)\Delta d$. Therefore, the magnitude of the resolution (which can be called the Doppler resolution) in this direction is

$$\rho_{dA} = |\Delta d_{\min}| = \frac{|\Delta f_{\eta\,\min}|}{|\nabla f_\eta|} = \frac{1/T_{syn}}{\left|\frac{2V_p}{\lambda R_\eta}\cos\theta_\eta\right|} = \frac{V_p\cos\theta_\eta}{|f_r T_{syn}|} = \frac{V_p\cos\theta_\eta}{B_a} \tag{3.35}$$

where $f_r = -\frac{2V_p^2\cos^2\theta_\eta}{\lambda R_\eta}$ is the FM rate of monostatic SAR. As shown by the above analysis, in the case of monostatic SAR, the direction of the Doppler resolution is the cross range direction, that is, monostatic SAR has the best Doppler resolution along the cross range direction. One should notice that this direction of the best Doppler resolution is probably not consistent with the platform moving direction. However, in monostatic SAR, the azimuth direction is usually defined as the platform moving direction for the sake of intuitiveness. Hence, in the following, the Doppler frequency gradient along the y direction is considered:

$$|\nabla_y f_\eta| = \frac{2V_p\cos^2\theta_\eta}{\lambda R_\eta} \tag{3.36}$$

Accordingly, the resolution along the y direction is

$$\rho_{yA} = \frac{1/T_s}{|\nabla_y f_\eta|} = \frac{V_p}{B_a} \tag{3.37}$$

Suppose θ_Σ is the rotation angle of the target to the airplane during the synthetic aperture time (as illustrated in Figure 3.5):

$$\theta_\Sigma \approx \frac{V_p T_s}{R_\eta}\cos\theta_\eta \tag{3.38}$$

By combining Equations (3.35), (3.37), (3.38) and the FM rate formula, the azimuth resolution of airborne monostatic SAR can be approximated as

$$\rho_{yA} \approx \frac{\lambda}{2\theta_\Sigma \cos\theta_\eta} \tag{3.39}$$

while the Doppler resolution is

$$\rho_{dA} \approx \frac{\lambda}{2\theta_\Sigma} \tag{3.40}$$

where θ_η is the target's squint angle at the center of the synthetic aperture time. As can be seen from the equations above, in the imaging plane, the Doppler resolution of the target is determined by the wavelength as well as the rotation angle of the target to the airplane during the synthetic aperture time. As to the azimuth resolution, it is also inversely proportional to the cosine of the squint angle. The relationship between the azimuth resolution in the squint case and that in the nonsquint case is $\rho_a = \rho_{a\perp}/\cos\theta_\eta$. Thus, the larger the squint angle, the lower is the azimuth resolution. This can be validated by the extreme case where the squint angle is 90°. When the squint angle is 90°, the targets are along the trajectory of the radar, that is, the distance between the radar and target change linearly. Consequently, the azimuth echo is monochromatic, with no characteristic broadband. As a result, the azimuth resolution ability cannot be achieved.

In the case of spaceborne monostatic SAR where the curvature of the earth cannot be ignored, the satellite travels along a curved track. Take the sliding spotlight mode (as shown by Figure 3.6) as an example (compared with the strip-map mode and

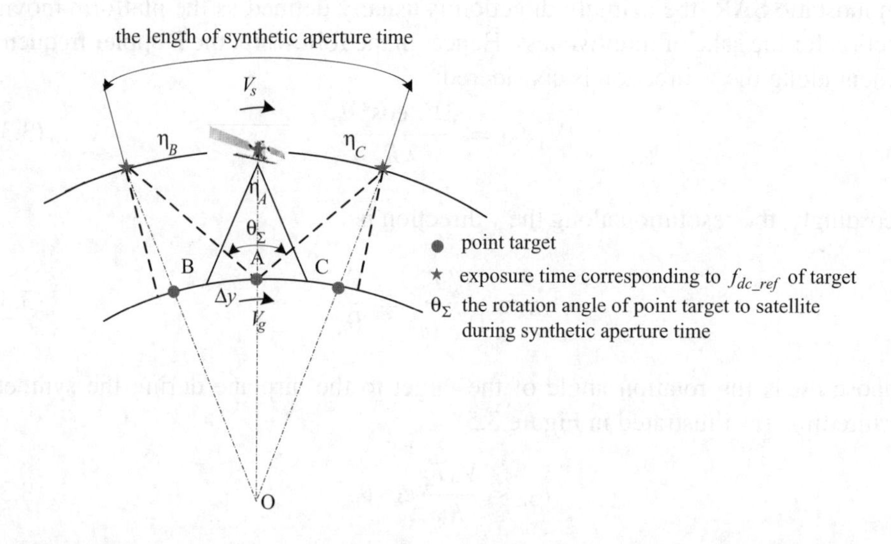

Figure 3.6 Imaging geometry of spaceborne SAR in the hybrid strip-map/spotlight mode

spotlight mode, the sliding spotlight mode is more complex, and the analysis and conclusion about it are also applicable to the strip-map mode and spotlight mode). Suppose the effective relative velocity between the satellite and the ground is V_s; the ground projection velocity of the nadir is V_g. As can be seen from Figure 3.6, the relationship between the difference of the exposure time corresponding to f_{dc_ref} (which is $\Delta\eta$) and the azimuth location difference (which is Δy) is as follows:

$$\Delta\eta = \Delta y / V_g \tag{3.41}$$

Therefore, the azimuth resolution of spaceborne monostatic SAR is

$$\rho_{yS} = V_g / B_a \tag{3.42}$$

It can be seen that it is the ground velocity V_g rather than the relative velocity between the satellite and ground (i.e., V_s) that determines the resolution; that is, the curved geometry of spaceborne SAR improves the resolution.

Assuming θ_Σ is the rotation angle of the target to the satellite during the synthetic aperture time, then

$$\theta_\Sigma \approx \frac{V_s T_s}{R_\eta} \cos\theta_\eta \tag{3.43}$$

Thus, by combining Equations (3.42), (3.43) and the FM rate formula of spaceborne SAR, that is, $f_r = -\frac{2V_s^2 \cos^2\theta_\eta}{\lambda R_\eta}$, the resolution of spaceborne monostatic SAR can be obtained approximately as follows:

$$\rho_{ys} \approx \frac{V_g}{V_s} \frac{\lambda}{2\theta_\Sigma \cos\theta_\eta} \tag{3.44}$$

It can be seen that the wavelength and the rotation angle of the target to the satellite during the synthetic aperture time are the definitive factors for the azimuth resolution of spaceborne SAR, while the ratio of ground velocity to satellite effective velocity improves the resolution.

The analysis above is all done in the slant range plane. As can be discovered, in the slant range plane, the direction of Doppler resolution is perpendicular to the range direction and the magnitude of Doppler resolution is independent of the squint angle (Equation (3.40)). However, as it is the ground that SAR takes images for, the Doppler resolution ability on the ground is considered to be more important. It will therefore be analyzed in the following.

The gradient of Equation (3.30) on the ground can be written as

$$\nabla f_\eta = \frac{\partial f_\eta}{\partial x}\vec{i}_x + \frac{\partial f_\eta}{\partial y}\vec{i}_y = -\frac{2V_p}{\lambda R_\eta}\cos\theta_\eta \sin\theta_\eta \sin\phi \vec{i}_r + \frac{2V_p}{\lambda R_\eta}\cos^2\theta_\eta \vec{i}_y \tag{3.45}$$

where

$$\sin\phi = \frac{x - x_\eta}{\sqrt{(x - x_\eta)^2 + h_\eta^2}} \tag{3.46}$$

and ϕ is the look angle of SAR. Hence, the magnitude of the Doppler resolution on the ground is

$$\rho_{dAg} = \frac{1/T_{syn}}{\left|\frac{2V_p}{\lambda R_\eta}\cos\theta_\eta\right|\sqrt{1 - \sin^2\theta_\eta\cos^2\phi}} \approx \frac{\lambda}{2\theta_\Sigma\sqrt{1 - \sin^2\theta_\eta\cos^2\phi}} \tag{3.47}$$

The angle between the direction of Doppler resolution and the x axis is

$$\theta_{dg} = \text{atan}\left(-\frac{\cos\theta_\eta}{\sin\theta_\eta\sin\phi}\right) \tag{3.48}$$

It can be found that, in the ground plane, the magnitude and direction of the Doppler resolution both relate to the squint angle and the look angle of SAR. When the look angle is changeless, the larger the squint angle, the poorer is the ground Doppler resolution. When the look angle is zero and the beam points to the front of the flight (often referred to as forward looking), the angle between the direction of the Doppler resolution and the x axis is 90°. Under this situation, the direction of the Doppler resolution is the same as that of the range resolution, that is, forward-looking monostatic SAR has resolution ability only in one direction. Thus, two-dimensional images cannot be obtained.

3.3.2.2 Doppler Resolution of Bistatic SAR

The above has analyzed the Doppler resolution and the azimuth resolution of monostatic SAR. In the following, the same methods are used to analyze the Doppler resolution of bistatic SAR. As illustrated by Figure 3.7, assuming that the velocities of the receiver and transmitter are $\boldsymbol{v}_R$ and $\boldsymbol{v}_T$, respectively, the position vectors of the receive and transmit antennas at a certain time η are $\boldsymbol{r}_R$ and $\boldsymbol{r}_T$, respectively, and the position vector of the target $P(x, y)$ that the combined beam center illuminates at time η is $\boldsymbol{r}_P$, then the Doppler phase of this target at the time η is

$$\varphi(\eta; \vec{\boldsymbol{r}}_P) = -\frac{2\pi}{\lambda}(|\vec{\boldsymbol{r}}_T - \vec{\boldsymbol{r}}_P| + |\vec{\boldsymbol{r}}_R - \vec{\boldsymbol{r}}_P|) \tag{3.49}$$

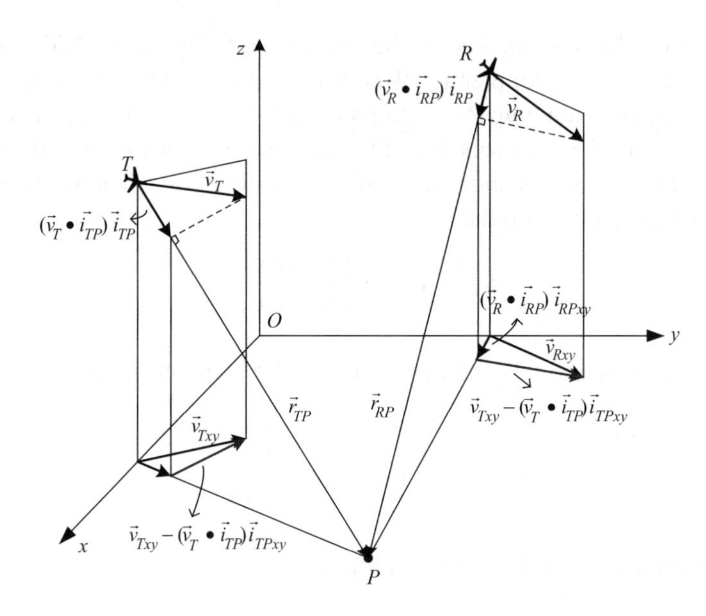

Figure 3.7 Illustration of the vectors related to Doppler resolution of bistatic SAR

Thus, the Doppler frequency is

$$\begin{aligned} f_\eta(\eta;\vec{\boldsymbol{r}}_P) &= \frac{1}{2\pi}\frac{d\varphi(\eta;\vec{\boldsymbol{r}}_P)}{d\eta} \\ &= -\frac{1}{\lambda}\left[\frac{\vec{\boldsymbol{v}}_T\cdot(\vec{\boldsymbol{r}}_T-\vec{\boldsymbol{r}}_P)}{|\vec{\boldsymbol{r}}_T-\vec{\boldsymbol{r}}_P|}+\frac{\vec{\boldsymbol{v}}_R\cdot(\vec{\boldsymbol{r}}_R-\vec{\boldsymbol{r}}_P)}{|\vec{\boldsymbol{r}}_R-\vec{\boldsymbol{r}}_P|}\right] \end{aligned} \tag{3.50}$$

Then, Doppler frequency isolines at that moment can be expressed as

$$f_\eta(\vec{\boldsymbol{r}}_P;\eta) = \text{cons tan t} \tag{3.51}$$

The gradient of Doppler frequency at (x, y) can be obtained:

$$\begin{aligned} \nabla f_\eta &= \frac{\partial f_\eta}{\partial x}\vec{\boldsymbol{i}}_x+\frac{\partial f_\eta}{\partial y}\vec{\boldsymbol{i}}_y \\ &= \frac{1}{\lambda}\left[\frac{\vec{\boldsymbol{v}}_{Txy}-(\vec{\boldsymbol{v}}_T\cdot\vec{\boldsymbol{i}}_{TP})\vec{\boldsymbol{i}}_{TPxy}}{|\vec{\boldsymbol{r}}_T-\vec{\boldsymbol{r}}_p|}+\frac{\vec{\boldsymbol{v}}_{Rxy}-(\vec{\boldsymbol{v}}_R\cdot\vec{\boldsymbol{i}}_{RP})\vec{\boldsymbol{i}}_{RPxy}}{|\vec{\boldsymbol{r}}_R-\vec{\boldsymbol{r}}_p|}\right]\text{(Hz/m)} \end{aligned} \tag{3.52}$$

where $\vec{\boldsymbol{v}}_{Txy}$ and $\vec{\boldsymbol{v}}_{Rxy}$ are the projection vectors of $\vec{\boldsymbol{v}}_T$ and $\vec{\boldsymbol{v}}_R$ in the (x, y) plane, respectively; $\vec{\boldsymbol{i}}_{TP}$ and $\vec{\boldsymbol{i}}_{RP}$ are the unite vectors of the directions of $\vec{\boldsymbol{r}}_P-\vec{\boldsymbol{r}}_T$ and $\vec{\boldsymbol{r}}_P-\vec{\boldsymbol{r}}_R$, respectively; $\vec{\boldsymbol{i}}_{TPxy}$ and $\vec{\boldsymbol{i}}_{RPxy}$ are the projection vectors of $\vec{\boldsymbol{i}}_{TP}$ and $\vec{\boldsymbol{i}}_{RP}$ in the (x, y) plane ($|\vec{\boldsymbol{i}}_{TPxy}| \neq 1$, $|\vec{\boldsymbol{i}}_{RPxy}| \neq 1$), respectively. Figure 3.7 illustrates vividly the meanings of the vectors inside the brackets of Equation (3.52). As can be shown, the direction of Doppler resolution of bistatic SAR relates to the magnitudes and directions of the transmitter and receiver velocities, the receive and transmit beam directions and the distances from the receiver and transmitter to the target. The composition of those factors above determines the direction of the Doppler gradient, whose intersection angle with the x axis is

$$\theta_d = \arctan\left(\frac{\partial f_\eta/\partial y}{\partial f_\eta/\partial x}\right) \tag{3.53}$$

Derived from Equation (3.52), the magnitude of the Doppler gradient is

$$|\nabla f_\eta| = \sqrt{\left(\frac{\partial f_\eta}{\partial x}\right)^2 + \left(\frac{\partial f_\eta}{\partial y}\right)^2} \tag{3.54}$$

Thus, the magnitude of the Doppler resolution is

$$\rho_{dA} = \frac{1/T_{syn}}{|\nabla f_\eta|} \tag{3.55}$$

The magnitude of the Doppler resolution in an arbitrary direction γ (the intersection angle between γ and the x axis is θ_γ) is

$$\rho_{A\gamma} = \frac{\rho_{dA}}{\sin(|\theta_d - \theta_\gamma|)} \tag{3.56}$$

3.3.2.3 Meanings of Azimuth Direction and Azimuth Resolution of Bistatic SAR

The calculation formula for bistatic SAR Doppler resolution has been provided above. However, the magnitude and direction of Doppler resolution described by the formulas are all abstract and it may not be propitious to understand them intuitively. In the case of monostatic SAR, the echo data are recorded (i.e., space sampling) along the SAR moving direction. Hence, the SAR moving direction (not the direction of Doppler resolution) is usually chosen as the azimuth direction and then the factors influencing resolution ability is analyzed through calculating the resolution along this direction. As for the case of bistatic SAR, can such a visualized direction be chosen as the azimuth direction to obtain the azimuth resolution formula, which is easier to understand? When the directions of the velocities of receive and transmit antennas are the same, the answer is simply "yes" and usually the moving direction is chosen to be the azimuth direction. In the following, taking the translational variant mode where the directions of velocities are the same as an example, the formula of the resolution along the azimuth direction (i.e., the moving direction) is deduced and the determinants of azimuth resolution are analyzed.

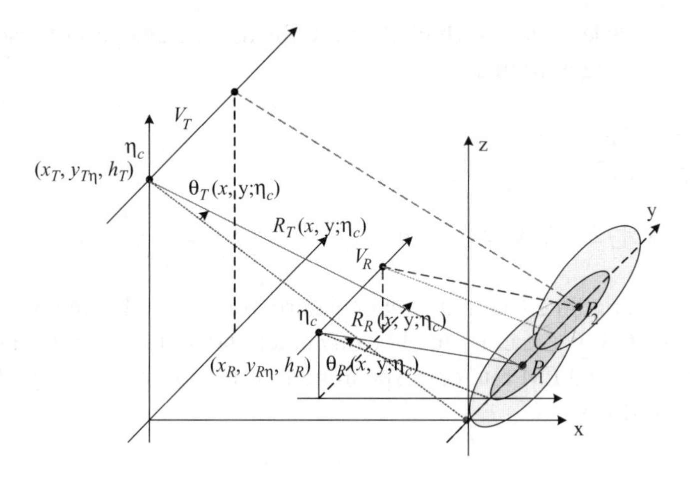

Figure 3.8 Translational variant bistatic SAR where the receiver and transmitter both travel along the y direction

Figure 3.8 illustrates the translational variant configuration of bistatic SAR where the receiver and transmitter both travel along the y axis. Derived from Equation (3.52), the magnitude of the gradient along the y axis is

$$\nabla_y f = \frac{\partial f}{\partial y} = \frac{1}{\lambda}\left[\frac{V_T - (\vec{\boldsymbol{v}}_T \cdot \vec{\boldsymbol{i}}_{TP})|\vec{\boldsymbol{i}}_{TPy}|}{|\vec{\boldsymbol{r}}_T - \vec{\boldsymbol{r}}_p|} + \frac{V_R - (\vec{\boldsymbol{v}}_R \cdot \vec{\boldsymbol{i}}_{RP})|\vec{\boldsymbol{i}}_{RPy}|}{|\vec{\boldsymbol{r}}_R - \vec{\boldsymbol{r}}_p|}\right] \tag{3.57}$$

where $V_T = |\vec{\boldsymbol{v}}_T|$; $V_R = |\vec{\boldsymbol{v}}_R|$; $\vec{\boldsymbol{i}}_{TPy}$ and $\vec{\boldsymbol{i}}_{RPy}$ are projections of $\vec{\boldsymbol{i}}_{TP}$ and $\vec{\boldsymbol{i}}_{RP}$ along the y axis, respectively. Supposing that $\theta_T(x, y; \eta)$ and $\theta_R(x, y; \eta)$ are denoted as the complement of the angle between $\vec{\boldsymbol{r}}_P - \vec{\boldsymbol{r}}_T$ and the y axis and the complement of the angle between $\vec{\boldsymbol{r}}_P - \vec{\boldsymbol{r}}_R$ and the y axis, respectively, the equation above can be written as

$$\nabla_y f = \frac{\partial f}{\partial y} = -\frac{1}{\lambda}\left[\frac{V_T \cos^2\theta_T(x, y; \eta)}{|\vec{\boldsymbol{r}}_T - \vec{\boldsymbol{r}}_p|} + \frac{V_R \cos^2\theta_R(x, y; \eta)}{|\vec{\boldsymbol{r}}_R - \vec{\boldsymbol{r}}_p|}\right] \tag{3.58}$$

By calculating $\nabla_y f$ for the target at (x, y) at the time η_{dc}, which corresponds to f_{dc_ref}, the azimuth resolution can be obtained:

$$\rho_a(x, y) = \frac{1}{T_s(x, y)\left|\dfrac{V_T \cos^2\theta_T(x, y; \eta_{dc})}{|\vec{\boldsymbol{r}}_T - \vec{\boldsymbol{r}}_p|\lambda} + \dfrac{V_R \cos^2\theta_R(x, y; \eta_{dc})}{|\vec{\boldsymbol{r}}_R - \vec{\boldsymbol{r}}_p|\lambda}\right|} \tag{3.59}$$

where $T_s(x, y)$ is the synthetic aperture time of the target at (x, y).

For a deeper understanding of the formula above, it will be deduced again through the matched filtering method from the view of the time domain resolution. Considering the echo signal to be on a certain azimuth line (where the range is all the same) and by ignoring high-order phases (which do not influence the analysis of resolution), the azimuth signal can be written as

$$s(\eta; R_{bi}) = \int_y G\left[\frac{\eta - \eta_c(x, y)}{T_s(x, y)}\right] \times \exp\{j2\pi f_{dc}(x, y)[\eta - \eta_c(x, y)] + j\pi f_r(x, y)[\eta - \eta_c(x, y)]^2\}dy \tag{3.60}$$

where $G(\cdot)$ is the combined azimuth beam pattern, $\eta_c(x, y)$ is the time when the target (x, y) is illuminated by the combined beam center and $f_{dc}(x, y)$ and $f_r(x, y)$ are the Doppler frequency and FM rate corresponding to $\eta_c(x, y)$, respectively, the formulas of them are as follows:

$$f_{dc}(x, y) = \frac{V_T}{\lambda}\sin\theta_T(x, y; \eta_c) + \frac{V_R}{\lambda}\sin\theta_R(x, y; \eta_c) \tag{3.61}$$

$$f_r(x, y) = -\frac{V_T^2\cos^2\theta_T(x, y; \eta_c)}{\lambda R_T(x, y; \eta_c)} - \frac{V_R^2\cos^2\theta_2(x, y; \eta_c)}{\lambda R_R(x, y; \eta_c)} \tag{3.62}$$

Here$R_T(x, y; \eta_c)$, $R_R(x, y; \eta_c)$, $\theta_T(x, y; \eta_c)$ and $\theta_R(x, y; \eta_c)$ are the distances and squint angles of the receive and transmit antennas to the target (x, y) at the time $\eta_c(x, y)$, respectively. The calculation formulas of them are as follows:

$$R_T(x, y; \eta_c) = \sqrt{(x - x_T)^2 + [y - V_T\eta_c(x, y)]^2 + h_T^2} \tag{3.63}$$

$$\theta_T(x, y; \eta_c) = \sin^{-1}\left\{\frac{y - V_T\eta_c(x, y)}{R_T(x, y; \eta_c)}\right\} \tag{3.64}$$

$$R_R(x, y; \eta_c) = \sqrt{(x - x_R)^2 + [y - V_R\eta_c(x, y)]^2 + h_R^2} \tag{3.65}$$

$$\theta_R(x, y; \eta_c) = \sin^{-1}\left\{\frac{y - V_R\eta_c(x, y)}{R_R(x, y; \eta_c)}\right\} \tag{3.66}$$

To analyze the resolution ability determined by the combined beam illumination time, the influence of the beam pattern modulation is ignored here, that is, $G(\cdot)$ in Equation (3.60) is replaced with the rectangular function rect[·]. The effect of beam pattern modulation is regarded as the broadening of resolution.

Assuming that the focusing is performed by the frequency domain matched filtering, Equation (3.60) is transformed into the azimuth frequency domain, which is

$$S(f; R_{bi}) = \int_y \mathrm{rect}\left[\frac{f - f_{dc}(x, y)}{B_a(x, y)}\right] \times \exp\{-j2\pi f\eta_c(x, y) - j\pi[f - f_{dc}(x, y)]^2/f_r(x, y)\}dy \tag{3.67}$$

Due to the fact that $f_r(x, y)$ changes little within the small area where azimuth resolution is analyzed, $f_r(x, y)$ can be supposed to be constant in resolution analysis; therefore it is represented by f_r. As a result, the frequency domain matched filter is

$$H_a(f; R_{bi}) = \exp\{j\pi(f - f_{dc\_ref})^2/f_r\} \tag{3.68}$$

After frequency domain filtering and inverse Fourier transform, the azimuth signal is

$$s_o(\eta; R_{bi}) = \int_y \mathrm{sinc}[\pi B_a(x, y)\eta] * \delta\left[\eta - \eta_c(x, y) - \frac{f_{dc\_ref} - f_{dc}(x, y)}{f_r}\right] dy \tag{3.69}$$

where $*$ denotes convolution. Derived from the sinc function in Equation (3.69), the time resolution of the signal is

$$\Delta\eta_{\min} = 1/B_a(x, y) = 1/\mathrm{abs}[f_r T_s(x, y)] \tag{3.70}$$

Besides, it can be seen according to the $\delta(\cdot)$ function in Equation (3.69) that, after matched filter compression, the target does not appear at its combined beam illumination time η_c. Instead, due to the fact that $f_{dc}(x, y)$ changes with location, the position of the target in azimuth time deviates from η_c after compression. After compression, the relationship between the time interval $\Delta\eta$ and the azimuth location interval Δy of targets satisfies the following equation:

$$\frac{\partial\eta_c(x, y)}{\partial y}\Delta y - \frac{1}{f_r}\frac{\partial f_{dc}(x, y)}{\partial y}\Delta y = \Delta\eta \tag{3.71}$$

Thus, the resolution along the y direction is

$$\rho_a = \Delta y_{\min} = \frac{\Delta\eta_{\min}}{\left|\dfrac{\partial\eta_c(x, y)}{\partial y} - \dfrac{1}{f_r}\dfrac{\partial f_{dc}(x, y)}{\partial y}\right|} \tag{3.72}$$

Substituting Equations (3.61) to (3.66) and Equation (3.70) into Equation (3.72), the azimuth resolution is

$$\rho_a(x, y) = \frac{1}{T_s(x, y)\left[\dfrac{V_T\cos^2\theta_T(x, y; \eta_{dc})}{\lambda R_T(x, y; \eta_{dc})} + \dfrac{V_R\cos^2\theta_R(x, y; \eta_{dc})}{\lambda R_R(x, y; \eta_{dc})}\right]} \tag{3.73}$$

which is completely consistent with Equation (3.59).

If B_{aT} and B_{aR} are defined as the Doppler bandwidth contributed by the transmitter and receiver during the synthetic aperture time, respectively, Equations (3.59) and (3.73) can be simplified as

$$\rho_a(x, y) = \frac{1}{B_{aT}/V_T + B_{aR}/V_R} \tag{3.74}$$

where

$$B_{aT} = \frac{V_T^2 \cos^2 \theta_T(x, y; \eta_{dc})}{\lambda R_T(x, y; \eta_{dc})} T_s(x, y) \tag{3.75}$$

$$B_{aR} = \frac{V_R^2 \cos^2 \theta_R(x, y; \eta_{dc})}{\lambda R_R(x, y; \eta_{dc})} T_s(x, y) \tag{3.76}$$

Suppose $\theta_{R\Sigma}$ and $\theta_{T\Sigma}$ are the rotation angles of the target to receiver and transmitter during the synthetic aperture time, respectively:

$$\theta_{T\Sigma} \approx \frac{V_T T_s}{R_T} \cos \theta_T(x, y; \eta_{dc}), \theta_{R\Sigma} \approx \frac{V_R T_s}{R_R} \cos \theta_R(x, y; \eta_{dc}) \tag{3.77}$$

Therefore, Equation (3.74) can be approximated as

$$\rho_a(x, y) \approx \frac{\lambda}{\theta_{T\Sigma} \cos \theta_T(x, y; \eta_{dc}) + \theta_{R\Sigma} \cos \theta_R(x, y; \eta_{dc})} \tag{3.78}$$

which is similar to Equation (3.39) in the monostatic SAR case. Consequently, for the side-looking case, besides the wavelength, the azimuth resolution of bistatic SAR is determined by the sum of the rotation angles of the target to the receiver and transmitter during the synthetic aperture time.

3.3.2.4 Calculation of the Synthetic Aperture Time for Bistatic SAR

As can be seen from Equations (3.59) and (3.73), one of the most important parameters for calculating Doppler resolution and azimuth resolution of bistatic SAR is the synthetic aperture time of each target. In the translational invariant configuration where the beam direction remains unchangeable, according to receive and transmit beam patterns, the shape of the combined beam projected on the ground can be obtained. Then, the synthetic aperture time can be calculated through the azimuth length covered by the part of the combined antenna beam above –6 dB and the antenna moving velocity. In the case of monostatic SAR, the length of the synthetic aperture time is usually defined according to the boundary of the –3 dB antanna beam width. However, as the receiver and transmitter usually share the same antenna, in fact this antenna modulates the signal both in the process of transmitting and receiving. Consequently, at the boundary of –3 dB of the antenna beam pattern, the signal is actually modulated by –6 dB. However, when the velocities of the receiver and transmitter are different, the length of synthetic aperture time changes with the target locations. In this situation, the beam space–time diagram method [12] can be used to analyze the co-illumination area of the receive and transmit antennas and the length of the synthetic aperture time for the target in this area. In this method, a two-dimensional space is set up, in which time and azimuth location are the two coordinate axes. Then, the motions of the receive beam coverage and the transmit beam coverage can be described in this space–time plane. In the following, the synthetic aperture time and the co-illumination areas under the approximation that receive and transmit beams are both rectangular and are analyzed based on the beam space–time diagram method.

Take the translational variant configuration where the receiver and transmitter both travel along the y axis as an example. Assuming the moving velocities of the two antennas are v_s and v_f (if the antenna moves along the negative y axis, the velocity value will be negative), with $v_f \geq v_s$ (here the subscript "s" denotes "slow", while subscript "f" denotes "fast"), the corresponding azimuth coverage of the part of slow- and fast-moving antenna beams above −3 dB are Y_s and Y_f, respectively. If the time when the two beams start to intersect is set to be the time origin, the beam space–time diagrams under the situations of moving in the same direction and moving in opposite directions can be illustrated by Figures 3.9 and 3.10, respectively. In the figures, $\Delta\eta_{co}$ denotes the maximum co-illuminated time for the targets, during which the targets are illuminated by both of the two parts of antenna beams that are above −3 dB. Meanwhile, Δy_{co} denotes the azimuth width spread by those targets whose co-illumination time is $\Delta\eta_{co}$. Under the approximation that the beam patterns are rectangular, $\Delta\eta_{co}$ is the synthetic aperture time, while Δy_{co} is the effective co-illumination area.

In the following, how the synthetic aperture time and the co-illumination area relate to beam velocity and beam width is analyzed. When the receiver and transmitter

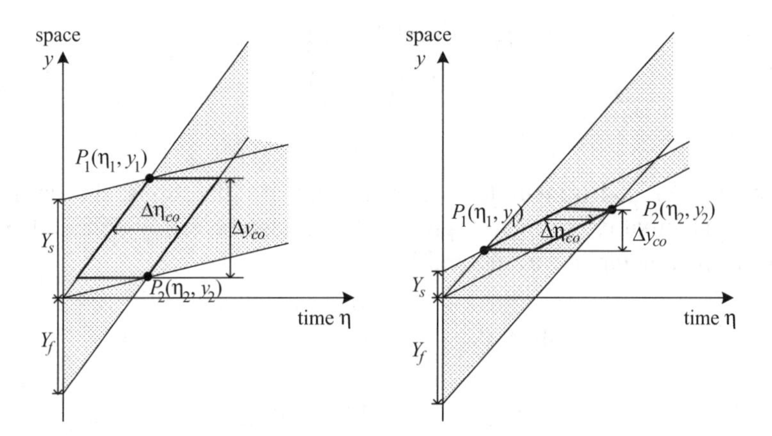

Figure 3.9 Space–time diagrams under the configuration where the receiver and transmitter travel in the same direction

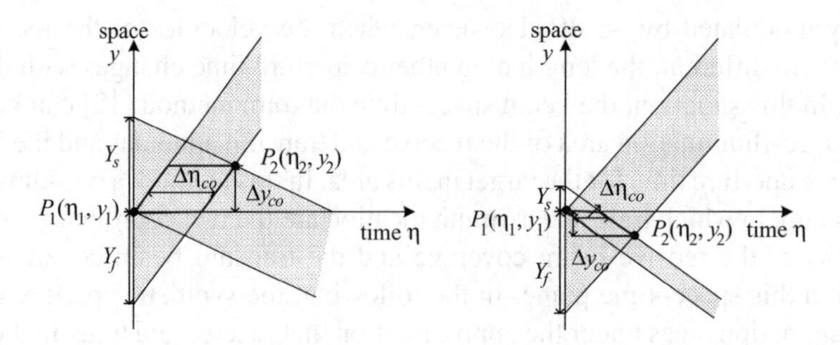

Figure 3.10 Space–time diagrams under the configuration where the receiver and transmitter travel in opposite directions

travel in the same direction, this is a pursuit problem in mathematics, so η_1 and η_2 in Figure 3.9 can be easily derived according to the known parameters v_s, v_f, Y_s and Y_f:

$$\eta_1 = \frac{Y_s}{v_f - v_s}, \quad \eta_2 = \frac{Y_f}{v_f - v_s} \tag{3.79}$$

Thus,

$$y_1 = \frac{v_f Y_s}{v_f - v_s}, \quad y_2 = \frac{v_s Y_f}{v_f - v_s} \tag{3.80}$$

When $y_1 \geq y_2$, that is, $Y_s/v_s \geq Y_f/v_f$, then $\Delta\eta_{co} = Y_f/v_f$ and $\Delta y_{co} = \frac{v_f Y_s - v_s Y_f}{v_f - v_s}$; When $y_1 < y_2$, that is, $Y_s/v_s < Y_f/v_f$, then $\Delta\eta_{co} = Y_s/v_s$ and $\Delta y_{co} = \frac{v_s Y_f - v_f Y_s}{v_f - v_s}$. To sum up, when the receiver and transmitter travel in the same direction,

$$\Delta\eta_{co} = \min\{Y_s/v_s, Y_f/v_f\} \tag{3.81}$$

$$\Delta y_{co} = \frac{|v_s Y_f - v_f Y_s|}{v_f - v_s} \tag{3.82}$$

When the receiver and transmitter travel in opposite directions, this is a meeting problem in mathematics, so η_2 in Figure 3.10 can also be easily obtained according to the known parameters v_s, v_f, Y_s and Y_f:

$$\eta_1 = 0, \quad \eta_2 = \frac{Y_f + Y_s}{v_f + v_s} \tag{3.83}$$

$$y_1 = 0, \quad y_2 = \frac{v_s Y_f + v_f Y_s}{v_f - v_s} \tag{3.84}$$

where $v_s \leq 0$. When $y_2 \geq y_1$, that is, $Y_s/|v_s| \geq Y_f/v_f$, then $\Delta\eta_{co} = Y_f/v_f$ and $\Delta y_{co} = \frac{v_f Y_s + v_s Y_f}{v_f - v_s}$. When $y_2 < y_1$, that is, $Y_s/|v_s| < Y_f/v_f$, then $\Delta\eta_{co} = Y_s/|v_s|$ and $\Delta y_{co} = \frac{-v_s Y_f - v_f Y_s}{v_f - v_s}$. To sum up, when the receiver and transmitter travel in opposite directions,

$$\Delta\eta_{co} = \min\{Y_s/|v_s|, Y_f/v_f\} \tag{3.85}$$

$$\Delta y_{co} = \frac{|v_s Y_f + v_f Y_s|}{v_f - v_s} \tag{3.86}$$

The beam space–time diagram under the approximation of a rectangular beam pattern, which is a concise method used to analyze the synthetic aperture time and the common illumination area, is helpful when designing and analyzing bistatic configurations. However, it is not very accurate for actual applications, because the −6 dB coverage of the combined beam is simply not equal to the intersection of the −3 dB coverage areas of the receive and transmit antennas [15]. If accurate analysis is needed, a third freedom, which is the gain of the receive and transmit beam patterns, should be added when the space–time diagram is drawn (as shown in Figure 3.11). Thereby, more accurate results of the co-illumination area and the length of the synthetic aperture time can be obtained.

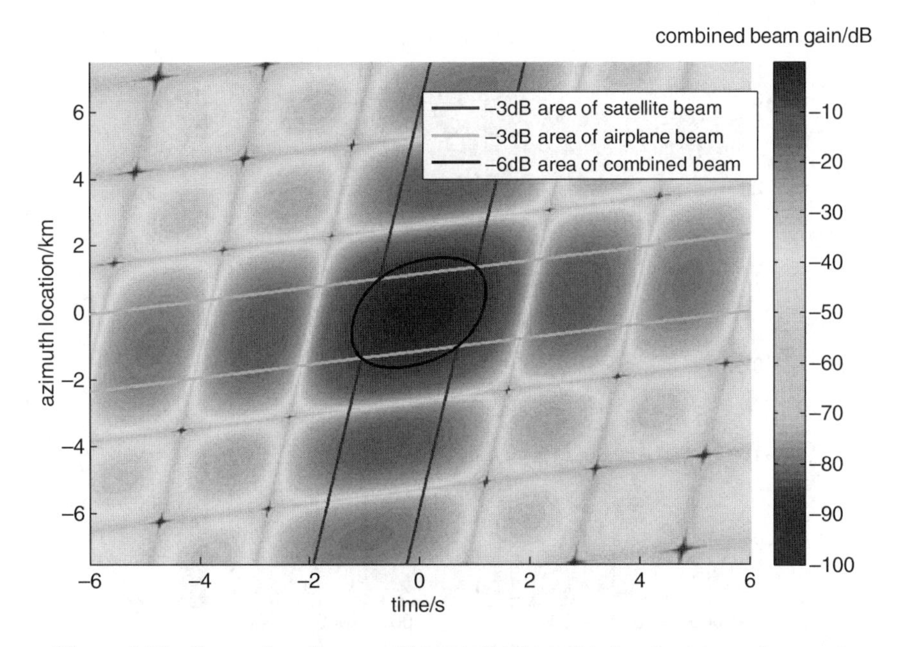

Figure 3.11 Space–time diagram of bistatic SAR considering the antenna beam gain

3.3.2.5 Simulation Examples

In the following, taking the spaceborne–airborne configuration with a translational variant characteristic as an example, the Doppler resolution of bistatic SAR is calculated, and the process of the synthetic aperture time calculation using beam space–time diagrams is illustrated.

Suppose the distance between the L-band spaceborne SAR and the center of the scenario is $R_T = 620$ km, the transmitted bandwidth is 60 MHz, the distance between the airplane and the center of the scenario is $R_R = 30$ km, the velocity of the satellite is $V_T = 7000$ m/s, the one-way beam pattern of the satellite and airplane is in the form of a sinc function, the -3 dB beam width of the satellite is 1.5° and the -3 dB beam width of the airplane is 6.0°. The velocity directions of the airplane and satellite are both along the y axis. Suppose also that the beam centers of the satellite and airplane antennas intersect at the center of the scenario when the time is zero.

If the velocity of the airplane is $V_R = 200$ m/s, the space–time diagram of the airplane and satellite beams projected on the ground is illustrated as shown in Figure 3.12. The -6 dB coverage of the combined beam (not normalized) is the zone within the elliptic curve in the figure. The SNR is relatively poor outside this zone with the peak value of the antenna gain less than -6 dB. If –6 dB is considered to be the lower limit of the peak value of the antenna gain, the co-illumination area is as shown in the figure. The synthetic aperture time of the target within the co-illumination area can be calculated from the space–time diagram. It is considered to be the time during

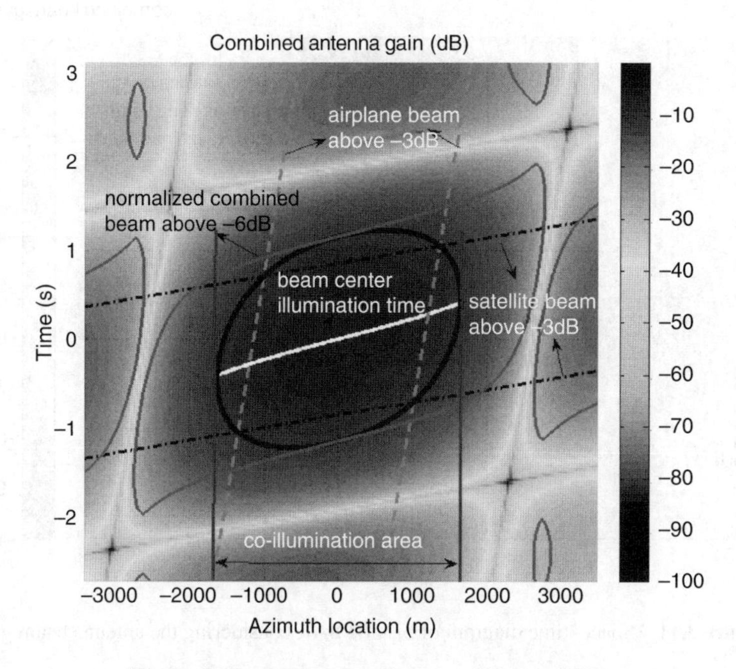

Figure 3.12 Space–time diagram of bistatic SAR

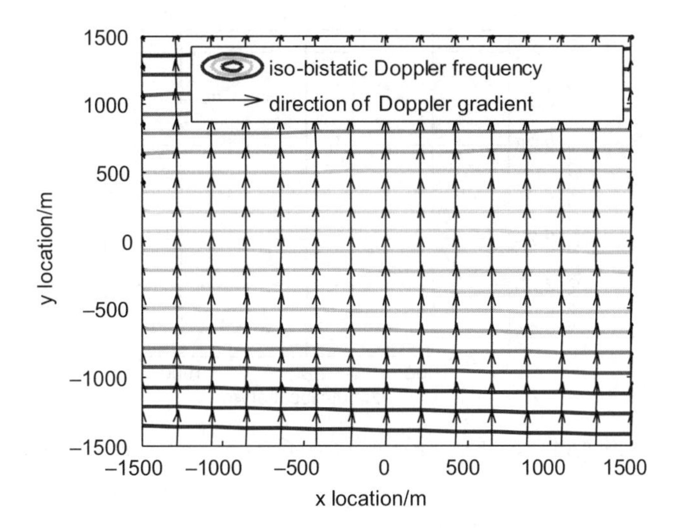

Figure 3.13 Doppler contours and directions of the Doppler gradient when the azimuth time is 0.5 s

which the beam pattern modulation gain (normalized) corresponding to the target's azimuth location in the space–time diagram is larger than −6 dB. Thereby, the median of this period of time (indicated by the white thick line in the figure) is the combined beam center illumination time. As illustrated by the figure, owing to the fact that the moving velocity of the satellite beam is far larger than that of the airplane beam, the illumination time of the airplane beam to the target is far larger than that of the satellite. Thus, in this spaceborne–airborne configuration, the length of the synthetic aperture time of the target is mainly determined by the satellite.

Figure 3.13 shows the Doppler frequency contours and directions of the Doppler gradient within the co-illumination area when the azimuth time is 0.5 s. As can be seen from the figure, in the scenario, the Doppler contours are approximately perpendicular to the y axis, while the Doppler gradient is approximately consistent with the y axis. This is because the airplane and satellite are both in a nonsquint mode in this configuration. Moreover, as a result of the narrowness of the satellite beam itself, the squint angles of the targets within the scenario to the satellite and to the airplane are both very small at their synthetic aperture time centers. Hence, the directions of the Doppler resolution gradient are approximately consistent in the co-illumination area.

From Figure 3.12, the length of the synthetic aperture time corresponding to each target in the scenario can be calculated (which are shown in Figure 3.14), and thus the azimuth resolution can be calculated based on Equation (3.73) (as shown in Figure 3.15). In the spaceborne–airborne mode, the Doppler bandwidth contributed by the airplane during the synthetic aperture time is comparatively small and the total Doppler bandwidth nearly reduces to a half compared with that in the spaceborne monostatic SAR case, as can be seen from Figure 3.15, but the azimuth resolution

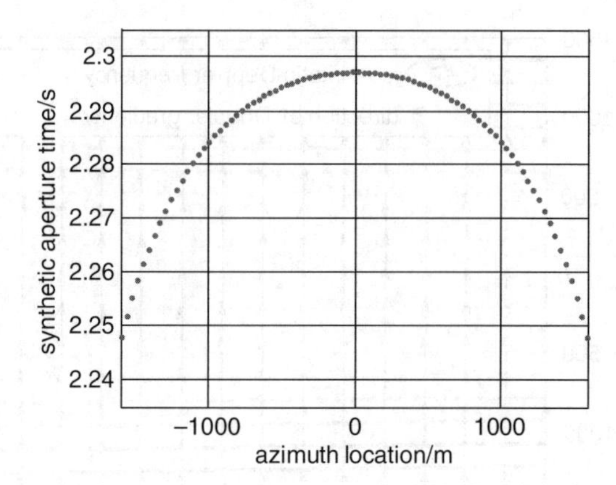

Figure 3.14 Synthetic aperture time of targets at different locations

only decreases a little (here, the theoretic resolution of spaceborne monostatic SAR is 4.58 m). This is because the azimuth resolution is actually determined by the rotation angles of the target to the airplane and to the satellite. The rotation angle of the target to the airplane has a comparable magnitude compared with that of the target to the satellite, thereby resulting in little reduction of resolution. Moreover, the resolution of the targets at the azimuth center is better than that on the azimuth border, which is a result of the variation of the synthetic aperture time with azimuth locations.

Figures 3.16 and 3.17 illustrate the azimuth resolution within the co-illumination area under different airplane velocities and two types of airplane beams, namely narrow beam and wide beam. It can be seen that, when the airplane beam is wide and the beam illumination time of the airplane is larger than that of the satellite, that is, the synthetic aperture time is mainly determined by the satellite, the larger the airplane velocity, the wider is the co-illumination area and thus the higher the azimuth resolution. On the other hand, when the airplane beam is narrow and the synthetic aperture time is mainly determined by the airplane, the larger the airplane velocity,

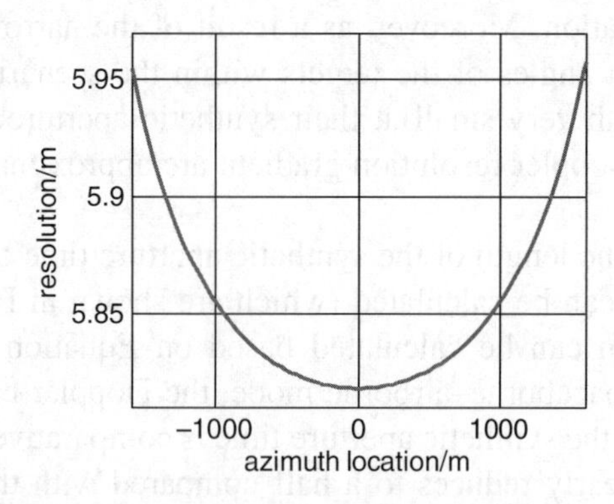

Figure 3.15 Azimuth resolutions of targets at different locations

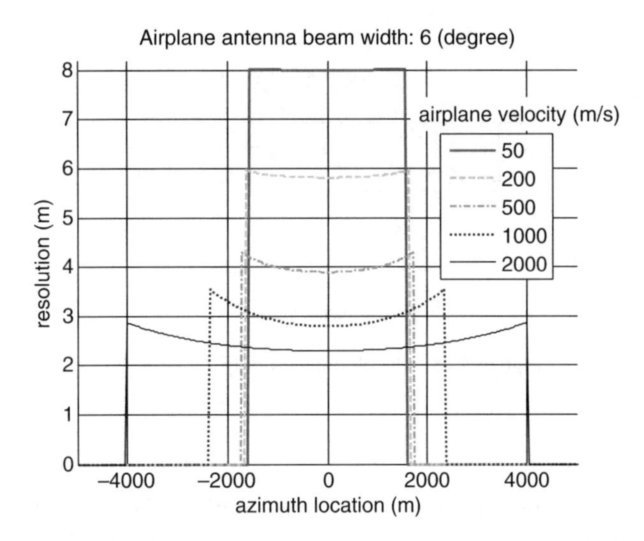

Figure 3.16 Azimuth resolutions under different airplane velocities when the airplane beam is wide

the shorter is the synthetic aperture time and thus the lower the azimuth resolution ability. As a result, the airplane velocity is not always the larger the better. By this token, in spaceborne–airborne mode, for the purpose of enlarging the co-illumination area and improving azimuth resolution, the airplane velocity should be chosen reasonably on the premise that the width of the airplane beam is made at a reasonably wide value.

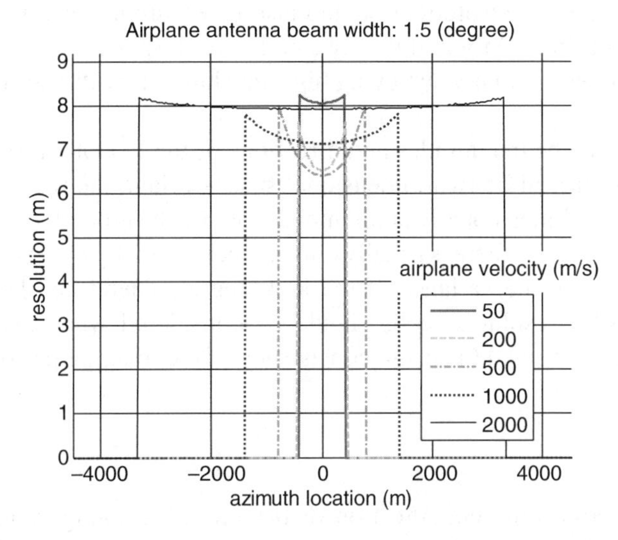

Figure 3.17 Azimuth resolutions under different airplane velocities when the airplane beam is narrow

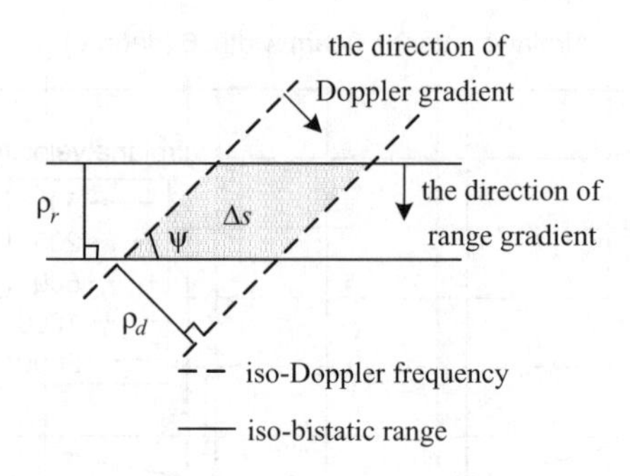

Figure 3.18 Resolution cell of bistatic SAR

3.3.3 Resolution Cell of Bistatic SAR

Sections 3.3.1 and 3.3.2 have analyzed the two resolutions determined by the transmitted wideband signal and the Doppler wideband signal, respectively. In the following, the ground resolution cell of bistatic SAR will be analyzed. Assuming that the magnitude of the Doppler resolution of bistatic SAR somewhere on the ground is ρ_d, the magnitude of the ground range resolution is ρ_r and the angle between the direction of the Doppler gradient and that of the range gradient is ψ (as shown in Figure 3.18), then the area of the resolution cell of bistatic SAR is

$$\Delta s = \frac{\rho_d \rho_r}{|\sin \psi|} \tag{3.87}$$

It can be seen that only to obtain a high range resolution and a high Doppler resolution is not enough to achieve high resolution images of the ground. A certain angle between the directions of the two resolutions should also be ensured, which is better to be as close to 90° as possible. Only in this way can the resolution cell be guaranteed to be small enough, thereby actually achieving high-resolution two-dimensional images of the ground.

Besides, because the magnitudes and directions of the Doppler resolution and the range resolution may all be two-dimensional space-variant, the resolution ability of bistatic SAR is two-dimensionally nonhomogeneous on the ground (on the contrary, in the monostatic SAR case, there is usually only nonhomogeneity in the range resolution that exists because of the variance of the look angle). Therefore, the homogeneity of the bistatic SAR resolution cells should be considered when the bistatic SAR configurations are designed to ensure consistency of the imaging performance.

3.4 Summary

This chapter discusses the classification of bistatic SAR configurations as well as the bistatic SAR radar equation and analyzes the spatial resolution ability of bistatic

SAR with emphasis. Through the content of this chapter, the following conclusions can be drawn: the SNR and spatial resolution of bistatic SAR are affected by both the receive antenna and the transmit antenna, which makes the analysis of its performance more complex. However, on the other hand, it offers bistatic SAR more freedom to achieve better performance, which is also one of the reasons for developing bistatic SAR.

References

[1] Ender, J. (2004) A step to bistatic SAR processing. EUSAR Conference, pp. 359–364.

[2] Cai, F., He, Y. and Tang, X. (2007) Research on bi-static SAR range resolution characteristics. 1st Asian and Pacific Conference on Synthetic Aperture Radar (APSAR), pp. 103–106.

[3] Tang, Z. and Zhang, S. (2003) *Bistatic Synthetic Aperture Radar System Principle*, Science Press, Beijing.

[4] Homer, J., Mojarrabi, B., Palmer, J., *et al.* (2003) Non-cooperative bistatic SAR imaging system: spatial resolution analysis. In: *Proceedings of the IEEE International Geoscience and Remote Sensing Symposium (IGARSS'03)*, pp. 1446–1448.

[5] Cardillo, G.P. (1990) On the use of the gradient to determine bistatic SAR resolution. In: *Proceedings of Antennas and Propagation Society International Symposium*, pp. 1032–1035.

[6] Yan, H. (2005) High Resolution Imaging Based on Distributed Satellites SAR System. PhD thesis, Institute of Electronics, Chinese Academy of Sciences, Beijing.

[7] Zeng, T., Cherniakov, M. and Long, T. (2005) Generalized approach to resolution analysis in BSAR. *IEEE Transactions on Aerospace and Electronic Systems*, **41** (2), 461–474.

[8] Kuang, L., Wan, Q. and Yang, W.L. (2007) The influence of random motion errors on bistatic SAR resolution. International Conference on Communications, Circuits and Systems (ICCCAS), pp. 863–866.

[9] Tang, Z. and Zhang, S. (2004) The azimuth resolution PRF and mapping width of the bistatic-SAR. *Journal of Electronics and Information Technology*, **26** (4), 607–612.

[10] Cai, F., He, Y., Wang, J. and Tang, X. (2008) A new approach to bi-SAR range resolution in the ground plane. *Journal of Electronics and Information Technology*, **30** (9), 2065–2068.

[11] Qiu, X., Hu, D. and Ding, C. (2009) The resolution ability and velocity determination in translational variant bistatic SAR. *Journal of Astronautics*, **30** (4), 1609–1614.

[12] Nico, G. and Tesauro, M. (2007) On the existence of coverage and integration time regimes in bistatic SAR configurations. *IEEE Geoscience and Remote Sensing Letters*, **4** (3), 426–430.

[13] Belcher, D.P. and Baker, C.J. (1996) High resolution processing of hybrid strip-map/spotlight mode SAR. *IEE Proceedings: Radar, Sonar and Navigation*, **1143** (6), 366–374.

[14] Lanari, R., Zoffoli, S., Sansosti, E., *et al.* (2001) New approach for hybrid strip-map/spotlight SAR data focusing. *IEE Proceedings: Radar, Sonar and Navigation*, **148** (6), 363–372.

[15] Dubois-Fernandez, P., Cantalloube, H., Vaizan, B., *et al.* (2006) ONERA-DLR bistatic SAR campaign: planning, data acquisition, and first analysis of bistatic scattering behaviour of natural and urban targets. *IEE Proceedings: Radar, Sonar and Navigation*, **153** (3), 214–223.

4

Echo Simulation of Bistatic SAR

4.1 Introduction

An SAR raw data simulator is an important tool for designing the system parameters and testing the processing algorithms; especially when the real system has not been established, simulated data are the only data basis for parameter analyzing and algorithm validation. Therefore it is quite important to obtain the accurate simulated raw data.

SAR raw data simulation usually contains two parts: one is the simulation of the radar scattering coefficients of the scenario and the other is to establish an accurate SAR transfer function and to simulate the SAR raw data efficiently. Simulation of the radar scattering coefficients is the basis of SAR raw data simulation and it of great significance for interpreting the SAR images. Research has been done in the scattering mechanism both of monostatic SAR and bistatic SAR. As this part of research involves some complicated electromagnetic mechanism, which is beyond the content of this book, here we focus on the second part of SAR raw data simulation, which means we start with a reflectivity map as an input.

Usually, simulating raw data of discrete scattering targets is suitable for testing imaging algorithms and for testing some system parameters such as two-dimensional resolution, peak sidelobe ratio (PSLR) and integral sidelobe ratio (ISLR). However, when noises, jamming, analog-to-digital quantization errors, image ambiguity and interferometric processing methods are concerned, scene raw data have to be simulated. For discrete scattering targets, the scattering coefficients can be set one by one according to the requirement and, as the number of targets is small, raw data can be calculated pulse by pulse and target by target (TBT) according to the SAR imaging geometry. This method works in the time domain and can generate SAR raw data precisely, which means there is no difficulty in simulating raw data of discrete scattering targets; hence it is not discussed in this book.

As to the scene raw data simulation, because that there are a large number of scatters contained in the scene, it is not practical to simulate the raw data by the TBT method

Bistatic SAR Data Processing Algorithms, First Edition. Xiaolan Qiu, Chibiao Ding and Donghui Hu.

due to the large amount of computation, especially in the high-resolution case. For an ordinary PC under the current condition, simulating the raw data of a 1 K×1 K pixel scenario usually takes several hours, so efficient simulating methods are needed. One solution used to achieve high efficiency is to enhance the performance or increase the number of processing equipments [1, 2], such as using parallel or distributed computing to shorten the simulation time. However, this needs a hardware update, which is quite expensive. Another solution for efficient simulation is to find fast batch simulation methods.

For monostatic SAR scene raw data simulation, there are some classic algorithms [3, 4], but with the development of monostatic SAR in all aspects such as imaging modes, resolution and so on, the corresponding scene raw data simulation is still being studied [5–12]. As to the scene raw data simulation for bistatic SAR, which has a more complex geometry than monostatic SAR, the number of literature up to now is relatively small, and the existing literature is mainly about the translational invariant case [13, 14].

In this chapter, a monostatic SAR raw data simulation algorithm is introduced at first, to help the reader to get the basic idea of SAR raw data simulation. On this basis, several simulation algorithms with different accuracies, which are respectively suitable to be applied in different cases of translational invariant configurations, are described.

4.2 Traditional Monostatic SAR Raw Data Simulation

4.2.1 Echo Signal Model and Simulation Theory

The monostatic SAR geometry with a straight trajectory can be shown as in Figure 4.1. Set the scene center to be the coordinates origin and suppose that the SAR flies along the y axis with velocity V and squint angle θ. If the distance between a target-to-radar antenna at the beam center crossing time is denoted as r_c, then this target can be noted

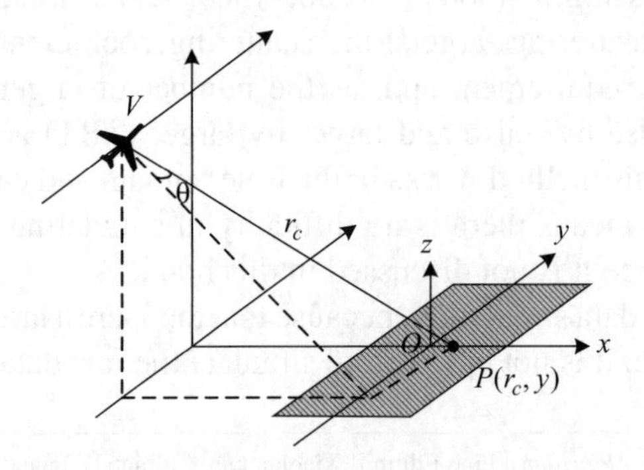

Figure 4.1 Monostatic SAR imaging geometry

as $P(r_c, y)$, where y is the target location in the y axis. Thus the monostatic SAR raw data can be written as follows:

$$s(t, \eta) = \iint \sigma(r_c, y) G_r(r_c) G_a(\eta - y/V; r_c) p[t + t_{dly} - 2R(\eta)/c] \exp\left\{-j\frac{4\pi}{\lambda} R(\eta)\right\} dr_c dy \tag{4.1}$$

where the origin of η is set to be the beam center crossing time of the scene center, which is denoted as $(r_c, 0)$, G_r is the range antenna pattern and G_a is the azimuth antenna pattern; $\sigma(r_c, y)$ indicates the RCS of scatter at (r_c, y) and $p(\cdot)$ is the SAR transmitted signal. To be compatible with the custom of digital signal processing, here the origin of the fast time t is set to be the instant of the first range sample and t_{dly} indicates the delay time of the first sample from the pulse transmission start time. Suppose the transmitted signal is a chirp with bandwidth B and chirp rate k_r, which is

$$p(\tau) = \text{rect}\left[\frac{\tau}{T}\right] \exp\{j\pi k_r \tau^2\} \tag{4.2}$$

Based on POSP, the echo of Equation (4.1) after range Fourier transform can be written as

$$\begin{aligned} S_t(f, \eta) = {} & \frac{f_s}{\sqrt{|k_r|}} \text{rect}\left[\frac{f}{B}\right] \exp\{j\pi/4\} \exp\{-j\pi f^2/k_r\} \exp\{j2\pi f t_{dly}\} \\ & \times \iint \sigma(r_c, y) G_r(r_c) G_a(\eta - y/V; r_c) \exp\left\{-j\frac{4\pi(f_0 + f)}{c} R(\eta)\right\} dr_c dy \end{aligned} \tag{4.3}$$

where

$$R(\eta) = \sqrt{r_c^2 + V^2(\eta - y/V)^2 - 2r_c V(\eta - y/V)\sin\theta} \tag{4.4}$$

Again use POSP and apply the Fourier transform to the azimuth time; then we have

$$\begin{aligned} S(f, f_\eta) = {} & \frac{f_s}{\sqrt{|k_r|}} \text{rect}\left[\frac{f}{B}\right] \exp\{j\pi/4\} \exp\{-j\pi f^2/k_r\} \exp\{j2\pi f t_{dly}\} \\ & \times \iint dr_c dy \left\{ \sigma(r_c, y) G_r(r_c) G_a(f_\eta; r_c) \right. \\ & \times \frac{f_{prf}}{\sqrt{|f_r(r_c)|}} \exp\{j\pi/4\} \exp\{-j2\pi f_\eta y/V\} \\ & \left. \times \exp\left\{-jr_c\left[\cos\theta\sqrt{\left(\frac{4\pi(f_0 + f)}{c}\right)^2 - \left(\frac{2\pi f_\eta}{V}\right)^2} + \frac{2\pi f_\eta}{V}\sin\theta\right]\right\}\right\} \end{aligned} \tag{4.5}$$

where $f_r(r_c)$ is the Doppler FM rate at r_c. Ignoring the change with range of the azimuth beam pattern and the process gain $1/\sqrt{|f_r(r_c)|}$, the above equation can be simplified as follows:

$$S(f, f_\eta) = \frac{f_s f_{prf}}{\sqrt{|k_r f_r|}} G_a(f_\eta) \mathrm{rect}\left[\frac{f}{B}\right] \exp\{j\pi/2\} \exp\{-j\pi f^2/k_r\} \exp\{j2\pi f t_{dly}\}$$
$$\times \iint dr_c dy \left\{ \sigma(r_c, y) G_r(r_c) \exp\{-j2\pi f_\eta y/V\} \right.$$
$$\left. \times \exp\left\{ -jr_c \left[\cos\theta \sqrt{\left(\frac{4\pi(f_0 + f)}{c}\right)^2 - \left(\frac{2\pi f_\eta}{V}\right)^2} + \frac{2\pi f_\eta}{V} \sin\theta \right] \right\} \right\} \tag{4.6}$$

What should be clarified is that, according to continuous time Fourier transform, the processing gain should be $1/\sqrt{|k_r f_r|}$; however it is the discrete time Fourier series that is used in practice, where the integration intervals of t and η, that is, $dt = 1/f_s$ and $d\eta = 1/f_{prf}$, respectively, are replaced by 1. So the processing gain in Equation (4.6) is $f_s f_{prf}/\sqrt{|k_r f_r|}$. It can be seen from Equation (4.6) that there is range–azimuth coupling in the SAR signal in the two-dimensional frequency domain and the coupling is a function of r_c. When the accuracy requirement of the simulated raw data is not strict and the variance of the coupling with r_c can be ignored, simulated raw data can be obtained directly through the convolution of the scene reflectivity map and the echo from the reference target. However, when the accuracy is required to be high and the variance of the coupling cannot be ignored, some algorithms with high precision should be applied.

We know that Equation (4.6) is an exact analytical expression of the echo spectrum in the frequency domain for monostatic SAR of the straight trajectory case. Therefore, if this spectrum is obtained, the SAR raw data can be generated easily by two-dimensional IFFT. Most of the fast SAR raw data simulation algorithms are based on this idea.

4.2.2 Implementation of the Simulation

In order to obtain the spectrum of Equation (4.6), the first step is to project the scenario reflectivity map to the slant range plane and multiply it with the range antenna pattern and the propagation phase corresponding to the Doppler centroid. This step can be expressed by the following equation:

$$s_1(r_c - ct_{dly}/2, y) = \sigma(r_c, y) G_r(r_c) \exp\{-j4\pi r_c/\lambda\} \tag{4.7}$$

Apply two-dimensional FFT to the above signal and then it becomes

$$\begin{aligned}\tilde{S}(\tilde{f}, \tilde{f}_\eta) = \exp\{j2\pi \tilde{f} t_{dly}\} \iint dr_c dy[\sigma(r_c, y) G_r(r_c) \\ \times \exp\{-j4\pi(\tilde{f} + f_0)r_c/c\} \exp\{-j2\pi \tilde{f}_\eta y/V\}]\end{aligned} \tag{4.8}$$

To distinguish from the frequency variables of the SAR echo (i.e., f and f_η), here we use $\tilde{f}$ and $\tilde{f}_\eta$ to indicate the frequency corresponding to the simulated scenario.

4.2.2.1 Inverse Stolt Conversion

Theoretically, the most precise way to obtain Equation (4.6) from Equation (4.8) is to perform the inverse Stolt conversion, which can be expressed as follows:

$$\begin{cases} \tilde{f}_\eta = f_\eta \\ \tilde{f} = \dfrac{c}{4\pi}\left[\cos\theta\sqrt{\left(\dfrac{4\pi(f_0 + f)}{c}\right)^2 - \left(\dfrac{2\pi f_\eta}{V}\right)^2} + \dfrac{2\pi f_\eta}{V}\sin\theta\right] - f_0 \end{cases} \tag{4.9}$$

The desired spectrum $S(f, f_\eta)$ can be obtained through interpolation according to Equation (4.9), which can be illustrated as shown in Figure 4.2. This process is exactly the inverse process of the Stolt conversion in the ω-k algorithm [15], which is described in Chapter 2; therefore, it is called the "inverse Stolt conversion".

The detailed steps used to obtain $S(f, f_\eta)$ from $\tilde{S}(\tilde{f}, \tilde{f}_\eta)$ is as follows:

- For the settled frequency grids $(f_i, f_{\eta j})$, $i = 0, \ldots, N_r - 1$, $j = 0, \ldots, N_a - 1$, calculate the corresponding grid $(\tilde{f}_i, \tilde{f}_{\eta j})$ according to Equation (4.9).

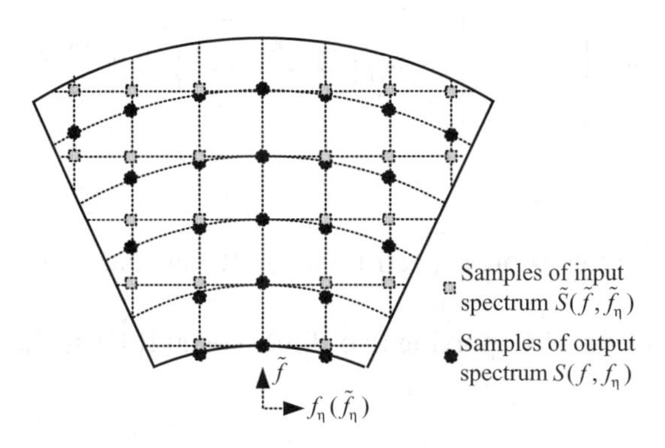

Figure 4.2 Illustration of the inverse Stolt conversion

- Get the value of the spectrum $\tilde{S}$ at $(\tilde{f}_i, \tilde{f}_{\eta j})$ (i.e., $\tilde{S}(\tilde{f}_i, \tilde{f}_{\eta j})$) through interpolation on the known spectrum $\tilde{S}(\tilde{f}, \tilde{f}_\eta)$, which is obtained from Equation (4.8), and consider $\tilde{S}(\tilde{f}_i, \tilde{f}_{\eta j})$ to be the spectrum value of $\widehat{S}(f_i, f_{\eta j})$.
- Multiply $\widehat{S}(f_i, f_{\eta j})$ with the linear frequency modulation term in the range direction, that is, $\mathrm{rect}[f/B]\exp(-j\pi f^2/k_r)$, the azimuth antenna pattern modulation coefficient $G_a(f_\eta)$, the processing gain $f_s f_{prf}/\sqrt{|k_r f_r|}$ and the constant phase term $\exp(j\pi/2)$, and then the desired two-dimensional spectrum $S(f, f_\eta)$ is obtained.

Based on the analysis above, the work flow of the SAR raw data simulation based on the inverse Stolt conversion can be described as follows:

1. Multiply the scene reflectivity map by the range antenna pattern and the propagation delay phase, and then project the multiplication result to the slant range plane (r_c, y), that is, to obtain Equation (4.7).
2. Apply the two-dimensional Fourier transform to Equation (4.7) to obtain Equation (4.8).
3. Compensate the sample delay phase and the reference phase, which is to multiply Equation (4.8) using the following function:

$$H_{1m}(\tilde{f}, \tilde{f}_\eta) = \exp\{-j2\pi \tilde{f} t_{dly}\}\exp\{j4\pi r_{c\_ref}(f_0 + \tilde{f})/c\} \tag{4.10}$$

   where r_{c_ref} is the reference slant range, which is usually chosen to be the slant range of the scene center.
4. Perform the inverse Stolt conversion according to Equation (4.9) to obtain $\widehat{S}(f, f_\eta)$.
5. Multiply $\widehat{S}(f, f_\eta)$ by the following function:

$$H_{2m}(f, f_\eta) = \frac{f_s f_{prf}}{\sqrt{|k_r f_r|}} G_a(f_\eta)\mathrm{rect}\left[\frac{f}{B}\right]\exp\{j\pi/2\}\exp\{-j\pi f^2/k_r\}\exp\{j2\pi f t_{dly}\}$$
$$\times \exp\left\{-jr_{c\_ref}\left[\cos\theta\sqrt{\left(\frac{4\pi(f_0+f)}{c}\right)^2 - \left(\frac{2\pi f_\eta}{V}\right)^2} + \frac{2\pi f_\eta}{V}\sin\theta\right]\right\} \tag{4.11}$$

   and then the two-dimensional spectrum of SAR raw data (i.e., Equation (4.6)) is obtained.
6. Find the SAR raw data by applying two-dimensional IFFT to the two-dimensional spectrum.

The work flow described above is shown in Figure 4.3. Some implementation details should be clarified for this work flow.

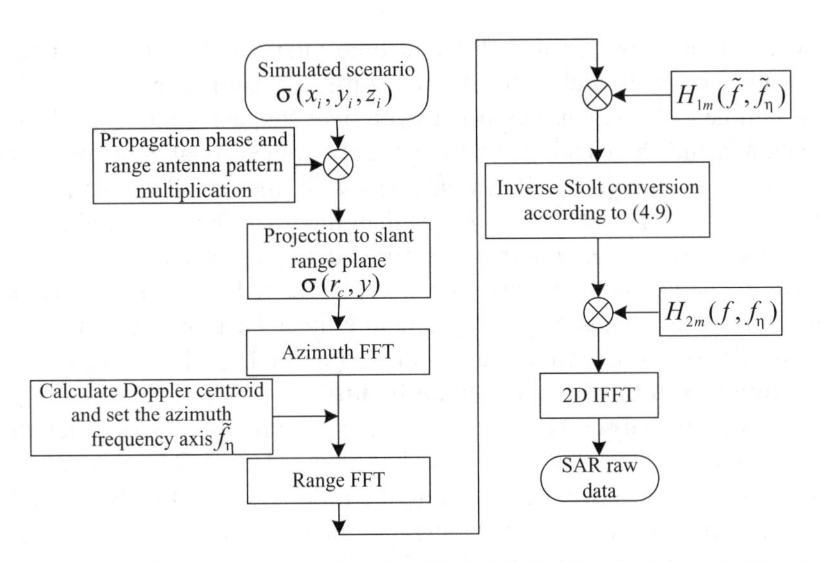

Figure 4.3 Block diagram of the SAR raw data simulation based on the inverse Stolt conversion

- **Azimuth frequency axis and antenna pattern**

  During the simulation, the azimuth frequency axis should be set according to the Doppler centroid, which is

$$f_{\eta dc} = \frac{2V}{\lambda} \sin\theta \tag{4.12}$$

This means that the scope of the frequency axis is $f_\eta \in [f_{\eta dc} - f_{prf}/2, f_{\eta dc} + f_{prf}/2)$ and the position of the Doppler centroid on the axis should agree with the base band Doppler centroid.

For the azimuth antenna pattern modulation, the variance of $G_a(f_\eta)$ along with f should be taken into account in order to simulate the right support field in the frequency domain. The relationship between $G_a(f_\eta)$ and f can be calculated through the following equation:

$$\Delta\theta = \arcsin\left[\frac{cf_\eta}{2V(f_0 + f)}\right] - \theta \tag{4.13}$$

where $\Delta\theta$ indicates the azimuth beam angle deviating from the squint angle θ. If the variance of $G_a(f_\eta)$ along f is ignored, the skewness of the spectrum cannot be simulated. This will finally cause the supporting field of the raw data in the time domain to deviate from the ideal supporting field.

- **Raw data simulation in the spotlight mode**

  The simulation method shown in Figure 4.3 is based on the strip mode. In this mode, the echo signals from different azimuth locations overlap each other in the azimuth frequency domain, but occupy different time intervals. For the spotlight mode, the situation reverses. Echo signals from different azimuth positions are within the same time interval, but they occupy different frequency intervals. Though the simulation method here mainly works on the two-dimensional frequency domain, it can simulate the spotlight SAR raw data only with some small alternations. Firstly, a very large f_{prf} (f_{prf} should be larger than the total Doppler bandwidth of the scenario) is used for the simulation and raw data are obtained by the strip SAR raw data simulation method. Therefore, every target in the scenario has a redundant bandwidth and a redundant synthetic aperture time. Then only a truncation in the slow time domain is needed for generating the spotlight SAR raw data.

- **Simulating raw data with azimuth ambiguity**
  If one intends to analyze the effects of azimuth ambiguity [16–18], then the SAR raw data with azimuth ambiguity should be simulated. Such SAR raw data can also be simulated by the above method using the following tips. Firstly, a large f_{prf} is chosen, which ensures that some ambiguous sidelobes of $G_a(f_\eta)$ can be included in the azimuth frequency axis without alias. Then the raw data with ambiguity can be obtained by desampling the SAR raw data in the slow time domain.

4.2.2.2 Inverse CS Method

The above simulation method based on the inverse Stolt conversion can be considered as the inverse process of the ω-k imaging algorithm. Under the uniform linear motion SAR geometry, this method can simulate the SAR raw data precisely. However, if the precision requirement is not very strong and the high-order coupling term can be ignored, then it is reasonable to use the simulation method proposed by Francechetti [4, 6–12]. This method can be regarded as the inverse process of the CS imaging algorithm, which has been introduced in Chapter 2. Unlike Equation (4.9), the relationship between frequency variables in this method is as follows:

$$\begin{cases} \tilde{f}_\eta^F = f_\eta \\ \tilde{f}^F \approx \alpha + \beta f \end{cases} \tag{4.14}$$

where

$$\alpha = f_0[\cos\theta\sqrt{1-(\lambda f_\eta/2V)^2} + (\lambda f_\eta/2V)\sin\theta] - f_0 \tag{4.15}$$

$$\beta = \frac{\cos\theta}{\sqrt{1-(\lambda f_\eta/2V)^2}} \tag{4.16}$$

Here, $\tilde{f}^F$ is a linear function of f, which means that only the linear coupling term (which corresponds to RCM) is fully simulated in this method, so it cannot simulate variance of the high-order coupling terms. Therefore the application condition of this method is limited to the small squint angle and relatively narrow bandwidth case. Usually, when the transmitted signal bandwidth is larger than 10% of the carrier frequency, this method cannot get a good simulation result. Meanwhile, due to the linear relationship between $\tilde{f}^F$ and f, this method can be realized by using the efficient frequency scaling method other than interpolation. Readers who are interested can refer to the literature [19, 20].

In a word, when the range–azimuth coupling is not serious, such as in those cases where the transmitted bandwidth is relatively narrow and the squint angle is small, the inverse CS method is easy and efficient to use, but when the range–azimuth coupling is serious, the method based on the inverse Stolt conversion is more precise and suitable.

4.2.3 Simulation Results

To gain a better understanding and a visual impression about the simulation algorithm, some simulation results are exhibited in this section. Firstly, raw data of discrete scattering targets is simulated both by the inverse Stolt conversion method (ISC method) and the target-by-target method (TBT method), and results are compared both in amplitude and in phase. Secondly, raw data of the wide transmitted bandwidth and squint mode is simulated by the Francechetti method (FRA method) and the ISC method, to show the difference between these two methods.

- **Raw data simulation of discrete targets in the narrow bandwidth case**
  The simulation parameters are as follows. The wavelength is 0.03125 m, the transmitted bandwidth is 60 MHz and the azimuth length of the antenna is 8 m. Choosing the scene center as the coordinate origin, the SAR platform is at (−928.29 km, 0,650 km) when $\eta = 0$. The squint angle is zero and the SAR velocity is 7000 m/s. The locations and the scattering coefficients of the targets are listed in Table 4.1. Here target 1 is at scene center, targets 2 and 3 are far away from the scene center, which are used to test the wide swath simulation ability, and targets 4 and 5 are close to the scene center, which are used to interfere with the center target to test the precision of the method.

  Figure 4.4 shows the raw data of target 2 simulated by the TBT method (noted as S_{TBT}) and the ISC method (noted as S_{ISC}). The two subfigures are quite similar.

Table 4.1 Information of simulated discrete scattering targets

Target number	Location (x, y, z) km	Back-scattering coefficient
1	(0, 0, 0)	1
2	(−11.273, 0, 0)	0.71–0.71i
3	(11.228, 0, 0)	i
4	(0.007, 0.166, 0.01)	−1
5	(0.021, −0.166, 0.03)	i

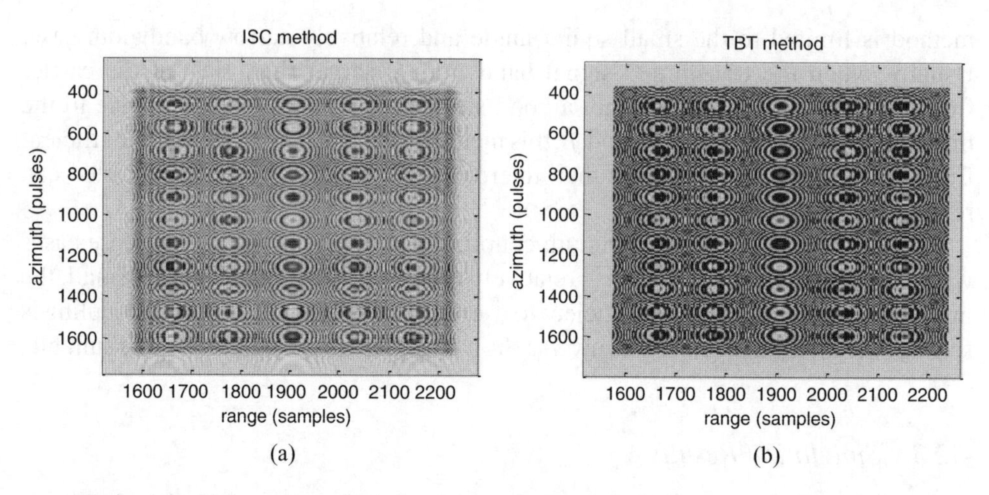

Figure 4.4 SAR raw data simulation results of target 2: (a) ISC method, (b) TBT method

Figure 4.5 shows the amplitude of the range signal and the azimuth signal. It can be seen that, in the range direction, except for the Fresnel undulation caused by generating chirp pulse through the frequency domain method, the amplitude of the ISC method matches the ideal amplitude of the TBT method quite well. In the azimuth direction, the magnitude slightly deviates from the ideal rectangular because of the interpolation (here, 8 points sinc interpolation is used). The phase map of $S_{ISC} \times S_{TBT}^{*}$(where $*$ stands for conjunction) is shown in Figure 4.6, which represents the phase difference between these two methods. It can be seen from Figure 4.6(a) that the differential phase of these two methods is close to zero. In

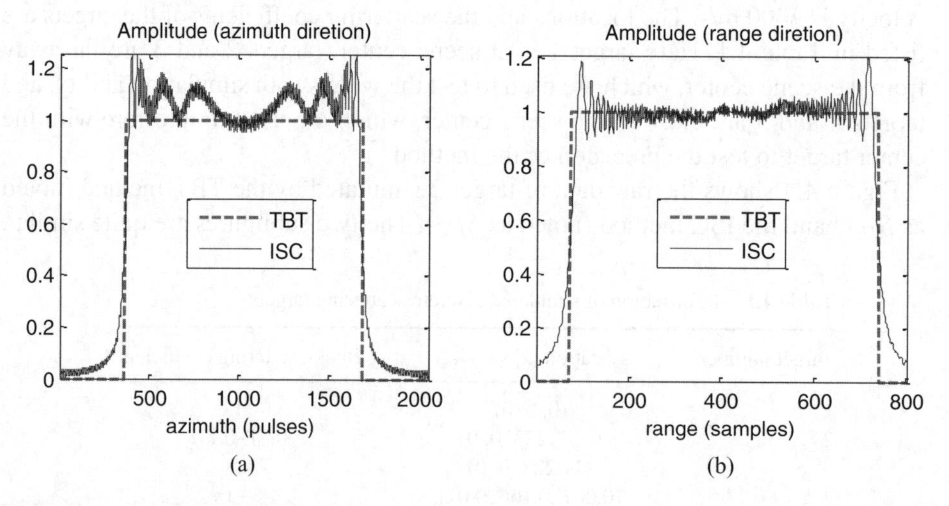

Figure 4.5 Amplitude of the simulated raw data of target 2: (a) amplitude in the azimuth direction, (b) amplitude in the range direction

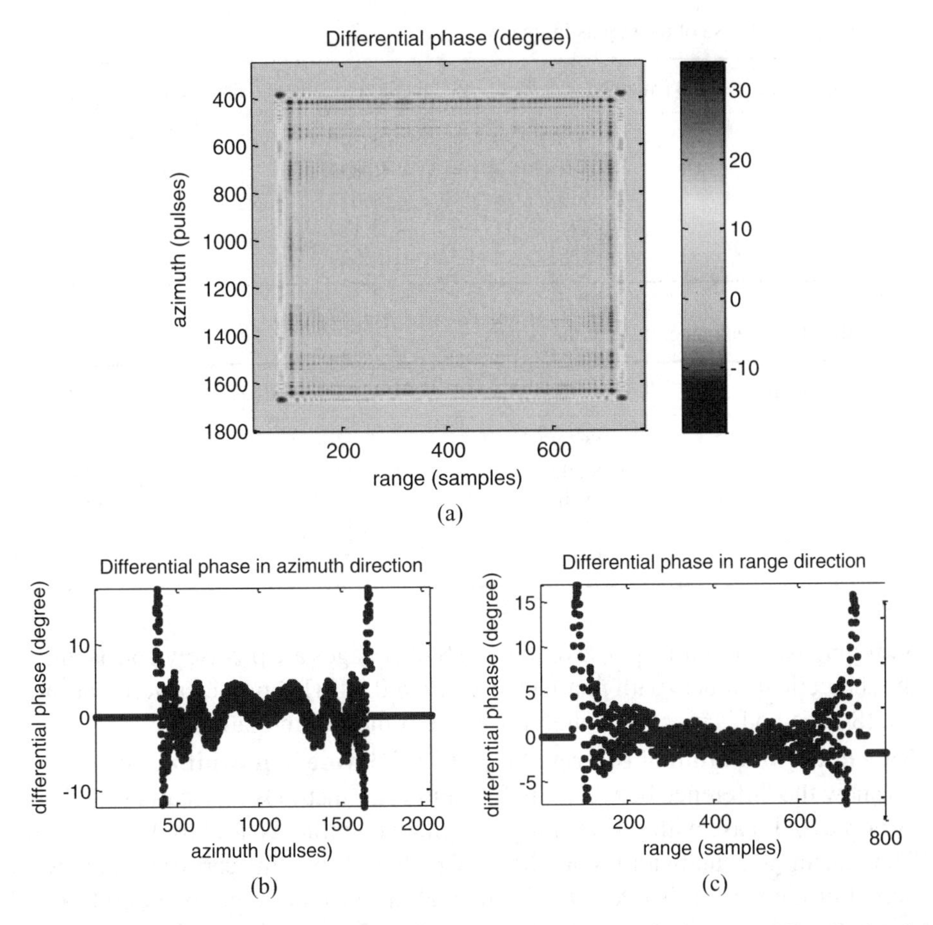

Figure 4.6 Differential phase between SAR raw data of target 2 simulated by the ISC method and the TBT method: (a) differential phase, (b) differential phase in the azimuth direction, (c) differential phase in the range direction

addition, from Figure 4.6(b) and (c), it can be seen more clearly that the differential phase is less than 5° both in the range and azimuth except for the fringes, which validates the precision of the ISC method.

To inspect whether this simulation method is precise enough for interferometric application, raw data of another SAR close to the SAR simulated above is generated by the ISC and TBT methods. The baseline between the two SARs is 60 m and is within the x–z plane. The angle between the baseline and the x direction is 20°. The CS imaging algorithm is used to focus the simulated data. The peak magnitude, phase and differential phase are examined in the imaging results and are listed in Tables 4.2, 4.3 and 4.4, respectively. It can be seen from the tables that, though slight errors exist because of the interpolation process, the magnitudes and phases of all the targets match the values obtained using the TBT method very well. In

Table 4.2 Phase of the targets in image

Target number	TBT method (degree)	ISC method (degree)	Phase error (degree)
1	−109.343	108.349	−0.994
2	−47.445	−48.857	−1.412
3	−58.450	−59.058	−0.609
4	−70.371	−71.112	−0.741
5	−160.328	−161.329	−0.678

Table 4.3 Peak magnitude of the targets

Target number	TBT method (dB)	ISC method (dB)	Error (dB)
1	58.082	57.884	−0.197
2	58.020	58.679	0.659
3	58.107	58.691	0.584
4	58.090	57.883	−0.207
5	58.126	57.824	−0.302

addition, the differential phase of the two SAR images are precisely consistent with the theoretical values, with phase errors below 0.1°. This provides verification that the ISC method is precise enough for interferometric simulation.

- **Raw data simulation of discrete targets in the wide bandwidth case**
  To show the difference between the ISC and FRA methods in accuracy, an L-band airborne SAR case with serious range–azimuth coupling is used for the simulation. The transmitted bandwidth is 500 MHz, which is 40% of the carrier frequency. The height of the aircraft is 4 km, the horizontal distance between the aircraft and the scene center is 6.928 km and the squint angle is 3°. The distance between the near and far targets to the scene center is −1311.3 m and 1250.5 m, respectively.

Figure 4.7 shows the differential phase between the two SAR raw datasets of the near target, which are simulated by the ISC and FRA methods. It can be seen that a significant differential phase (see Figure 4.7(a)) exists between the echo of the FRA method and the echo of the TBT method. This shows the phase error of the

Table 4.4 Differential phase of the targets in two SARs

Target number	Theoretical value (degree)	ISC method (degree)	Error (degree)
1	167.508	167.570	0.062
2	119.698	119.696	−0.002
3	58.983	58.985	0.002
4	179.707	179.611	−0.096
5	−155.895	−155.959	−0.064

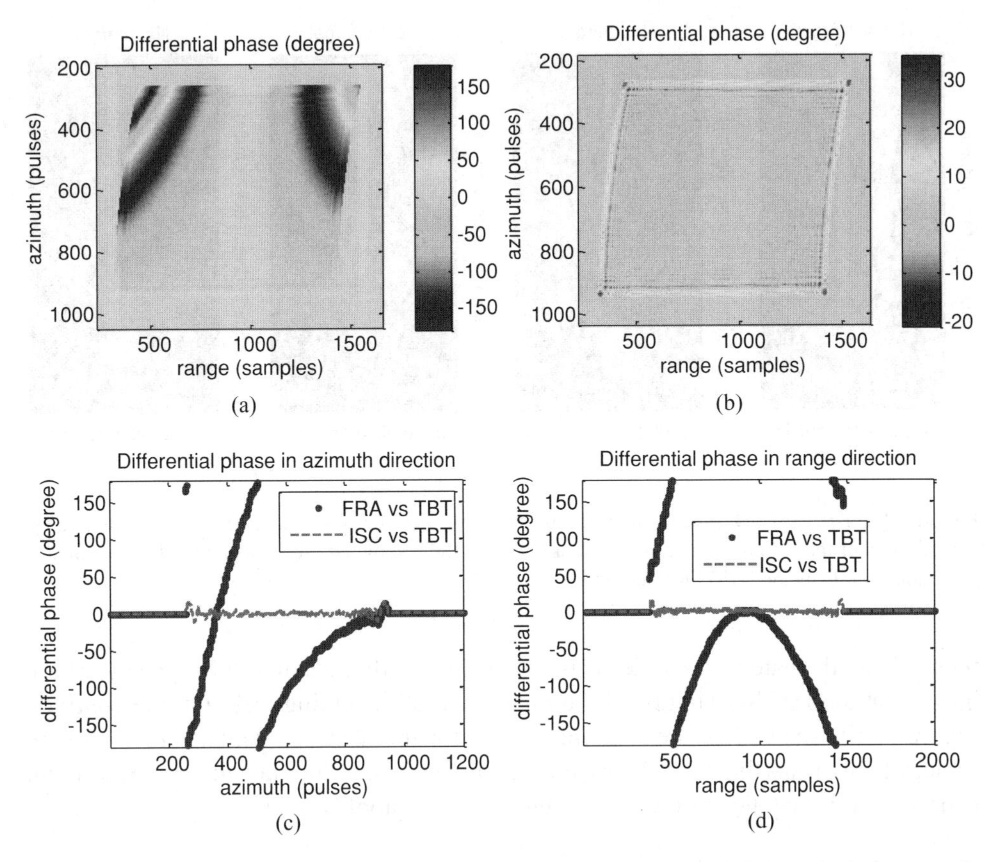

Figure 4.7 Comparing the raw data phase simulated by the ISC, FRA and TBT methods for target 2: (a) differential phase between FRA and TBT, (b) differential phase between ISC and TBT, (c) differential phase in the azimuth direction, (d) differential phase in the range direction

FRA method caused by lack of consideration of the high-order coupling variance. Meanwhile, the differential phase between the simulation results of the ISC method and the TBT method remains quite small (see Figure 4.7(b)), while the differential phase both in the azimuth and the range directions is still within 5° (see Figure 4.7(c) and (d)). The difference between the ISC method and the FRA method can also be seen from Figure 4.8. Using the traditional CS imaging algorithm to process the simulated raw data, after the chirp scaling process, RCMC and range compression, the results of the near target and the center target are exactly the same in the FRA case (see Figure 4.8(b)), while in the ISC case the results are obviously different (see Figure 4.8(a)). Because the traditional CS imaging algorithm only copes with the variance of the linear coupling term, if the variance of the high-order coupling terms are appropriately simulated, the CS imaging results of the near and center targets should be different. As we can see from Figure 4.8(a), the center target is well focused in the range direction at all azimuth frequencies, while the near target is not well

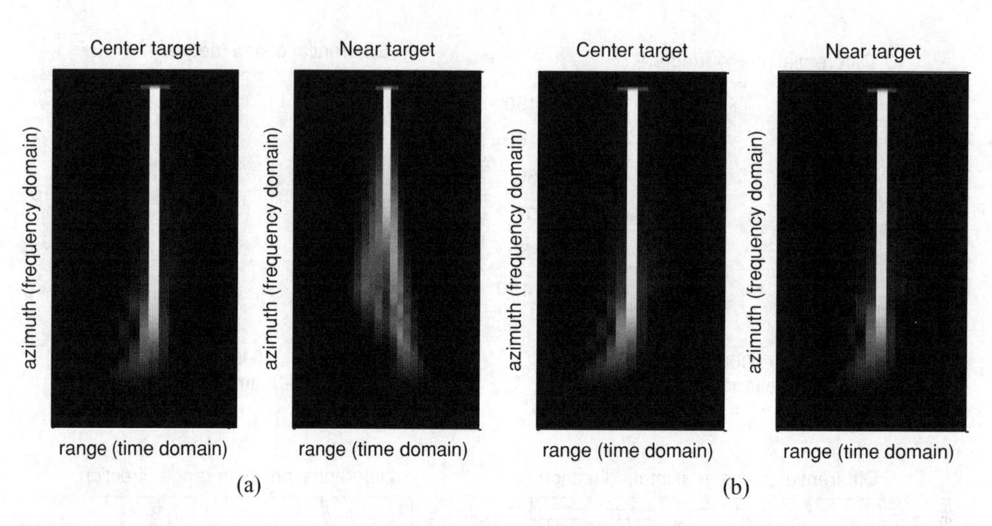

Figure 4.8 The range Doppler domain focusing results for the near and center targets after range compression of the CS imaging algorithm, based on the ISC simulated raw data and the FRA simulated raw data: (a) ISC method; (b) FRA method

focused and the extent of the defocusing changes with the azimuth frequency. These imply that the ISC has simulated the range-azimuth coupling well. On the contrary, Figure 4.8(b) shows that the imaging results for the near and center targets are the same, which means that the FRA method cannot precisely simulate the range–azimuth coupling, so it may not be precise enough for some applications.

4.3 Raw Data Simulation for Translational Invariant Bistatic SAR

The previous section describes the raw data simulation methods for monostatic SAR. These methods take the analytical expression of the two-dimensional spectrum of the scene echo as a basis and use FFT to achieve high efficiency. For bistatic SAR, the expression of the echo spectrum in the two-dimensional frequency domain is hard to obtain, even for the relatively simple configuration, the translational invariant configuration, because the bistatic SAR geometry is more complex than monostatic SAR geometry. However, as the Doppler parameters do not change along the azimuth direction in the translational invariant case, theoretically the raw data of this configuration can be generated by batch processing. When the bistatic baseline is relatively short, some approximations can be applied and the bistatic SAR raw data can be simulated by algorithms that are similar to the algorithms of monostatic SAR raw data simulation. For example, in dual-antenna interferometric SAR raw data simulation, such ideas are usually applied [21]. However, when the bistatic baseline is relatively long and the bistatic angle is quite large, the efficient bistatic SAR scene raw data simulation becomes more difficult. The authors extended the idea of the inverse Stolt conversion to the bistatic SAR raw data simulation to get a more universal method for

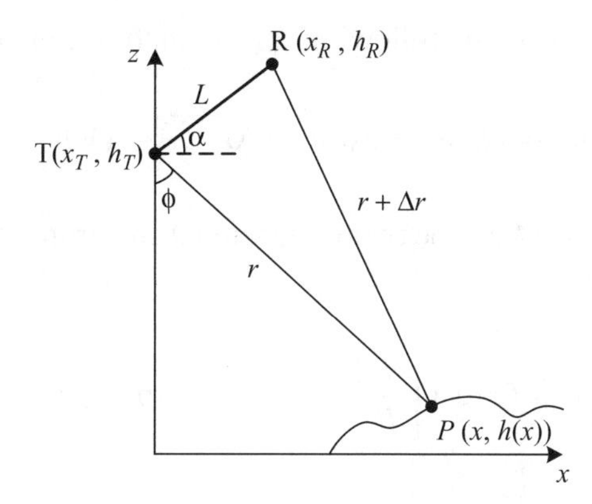

Figure 4.9 Dual-antenna interferometric SAR geometry

bistatic SAR of the translational invariant configuration [22]. The simulation method based on the monostatic SAR approximation and the more universal method based on the idea of the inverse Stolt conversion will be introduced in the following subsections.

4.3.1 *Short Bistatic Baseline Case*

Take the dual-antenna interferometric SAR system as an example. In dual-antenna interferometric SAR, the transmit antenna and the receive antenna are physically separate, so it is a kind of bistatic SAR. However, the two antennas are mounted on the same platform, so the baseline is limited and cannot be long. In this case, some approximations can be applied and then some monostatic-like SAR raw data simulation methods can be adopted. The geometry of dual-antenna interferometric SAR in the nonsquint case is shown in Figure 4.9. Suppose that the SAR moves along the y axis, where the length of the baseline is L, the angle between the baseline and the x axis is α and the side-look angle of the transmit antenna is ϕ. Then, similar to Equation (4.1), the received SAR echo can be written as follows:

$$s(t,\eta)=\iint \sigma(r,y)G_r(r)G_a(\eta-y/V;r)p[t+t_{dly}-R_{bi}(\eta)/c]\exp\left\{-j\frac{2\pi}{\lambda}R_{bi}(\eta)\right\}drdy \tag{4.17}$$

where

$$R_{bi}(\eta)=R_T(\eta)+R_R(\eta) \tag{4.18}$$

$$R_T(\eta)=\sqrt{r^2+(V\eta-y)^2} \tag{4.19}$$

$$R_R(\eta)=\sqrt{(r+\Delta r)^2+(V\eta-y)^2} \tag{4.20}$$

Due to the fact that $\Delta r \ll r$, the following approximation can be applied:

$$R_{bi}(\eta) \approx 2R(\eta) = 2\sqrt{(r + \Delta r/2)^2 + (V\eta - y)^2} \tag{4.21}$$

Thus, similar to Equation (4.5), the two-dimensional spectrum of Equation (4.17) can be expressed as follows:

$$\begin{aligned} S(f, f_\eta) = & \frac{f_s f_{prf}}{\sqrt{|k_r f_r|}} G_a(f_\eta) \mathrm{rect}\left[\frac{f}{B}\right] \exp\{j\pi/2\} \exp\{-j\pi f^2/k_r\} \exp\{j2\pi f t_{dly}\} \\ & \times \iint drdy \left\{ \sigma(r, y) G_r(r) \exp\{-j2\pi f_\eta y/V\} \right. \\ & \left. \times \exp\left\{ -j(r + \Delta r/2)\sqrt{\left(\frac{4\pi(f_0 + f)}{c}\right)^2 - \left(\frac{2\pi f_\eta}{V}\right)^2} \right\} \right\} \end{aligned} \tag{4.22}$$

where

$$\Delta r = \sqrt{(x - x_R)^2 + (h_R - h)^2} - \sqrt{(x - x_T)^2 + (h_T - h)^2} \tag{4.23}$$

When $|x_R - x_T| \ll r$ and $|h_R - h_T| \ll r$, it can be approximated as

$$\begin{aligned} \Delta r & \approx -\frac{x - x_T}{r}(x_R - x_T) + \frac{h_T - h}{r}(h_R - h_T) \\ & = -L\cos\alpha\sin\phi + L\sin\alpha\cos\phi \\ & = L\sin(\alpha - \phi) \end{aligned} \tag{4.24}$$

If the transmit range and the differential range between the receive range and the transmit range for the reference target $P(x_{ref}, h_{ref})$ are denoted by r_{ref} and Δr_{ref}, respectively, then the differential range between the receive and transmit ranges of other targets can be approximately expressed as follows:

$$\Delta r \approx \Delta r_{ref} + k_s(r - r_{ref}) \tag{4.25}$$

where k_s is a constant and can be calculated by the derivative of Δr with r at r_{ref}, which is

$$k_s = \frac{L}{r_{ref}}\sin(\alpha - \theta_{ref}) - \frac{L}{r_{ref}}\frac{\cos\alpha}{\sin\theta_{ref}} \tag{4.26}$$

Then Equation (4.22) can be written as

$$S(f, f_\eta) = \frac{f_s f_{prf}}{\sqrt{|k_r f_r|}} G_a(f_\eta) \mathrm{rect}\left[\frac{f}{B}\right] \exp\{j\pi/2\} \exp\{-j\pi f^2/k_r\} \exp\{j2\pi f t_{dly}\}$$
$$\times \iint drdy \left\{ \sigma(r, y) G_r(r) \exp\{-j2\pi f_\eta y/V\} \exp\{-j\Phi_{ref}(f, f_\eta)\} \right.$$
$$\left. \times \exp\left\{ -j(1 + k_s/2)(r - r_{ref}) \sqrt{\left(\frac{4\pi(f_0 + f)}{c}\right)^2 - \left(\frac{2\pi f_\eta}{V}\right)^2} \right\} \right\} \tag{4.27}$$

where

$$\Phi_{ref}(f, f_\eta) = (r_{ref} + \Delta r_{ref}/2) \sqrt{\left(\frac{4\pi(f_0 + f)}{c}\right)^2 - \left(\frac{2\pi f_\eta}{V}\right)^2} \tag{4.28}$$

Therefore, the raw data simulation of bistatic SAR with a short baseline can be carried out as follows:

1. Multiply the scene reflectivity map with the range antenna pattern and part of the propagation delay phase, which means

$$s_1(x, y) = \sigma[x, y; h(x, y)] G_r(x) \exp\left\{-j\frac{4\pi}{\lambda}(r - r_{ref})\right\} \tag{4.29}$$

2. Project the above signal to the (r', y) plane, where $r' = r - r_{ref}$, and get

$$s_2[r' - r'_{near}, y] = \sigma(r', y) G_r(r') \exp\left\{-j\frac{4\pi}{\lambda} r'\right\} \tag{4.30}$$

   where $r'_{near} = r_{near} - r_{ref}$ and r_{near} indicates the transmit range corresponding to the sample delay.
3. Apply the range Fourier transform to the above signal; then we get

$$S_3(\tilde{f}, \tilde{f}_\eta) =$$
$$\iint_{r', y} \sigma(r', y) G_r(r') \exp\left\{j\frac{4\pi \tilde{f}}{c} r'_{near}\right\} \exp\left\{-j\frac{4\pi(f_0 + \tilde{f})}{c} r'\right\} \tag{4.31}$$
$$\exp\left\{-j\frac{2\pi f_\eta}{V} y\right\} dr'dy$$

4. Multiply the above signal by the following function:

$$H_1(\tilde{f}, \tilde{f}_\eta) = \exp\left\{-j\frac{4\pi\tilde{f}}{c}r'_{near}\right\} \tag{4.32}$$

Then we get

$$S_4(\tilde{f}, \tilde{f}_\eta) = \iint_{r',y} \sigma(r', y)G_r(r')\exp\left\{-j\frac{4\pi(f_0+\tilde{f})}{c}r'\right\}\exp\left\{-j\frac{2\pi f_\eta)}{V}y\right\}dr'dy \tag{4.33}$$

5. According to the following relationship:

$$\begin{cases} \tilde{f}_\eta = f_\eta \\ \tilde{f} = \dfrac{c}{4\pi}(1+k_s/2)\sqrt{\left(\dfrac{4\pi(f_0+f)}{c}\right)^2 - \left(\dfrac{2\pi f_\eta}{V}\right)^2} - f_0 \end{cases} \tag{4.34}$$

$S_4(f, f_\eta)$ can be obtained by interpolation on $S_4(\tilde{f}, \tilde{f}_\eta)$.

6. Multiply $S_4(f, f_\eta)$ with the function composed by the reference phase, transmitted LFM signal, antenna pattern modulation and so on, which is

$$\begin{aligned} H_2(f, f_\eta) = {} & \frac{f_s f_{prf}}{\sqrt{|k_r f_r|}}G_a(f_\eta)\mathrm{rect}\left[\frac{f}{B}\right]\exp\{j\pi/2\}\exp\{-j\pi f^2/k_r\} \\ & \times \exp\{-j\Phi_{ref}(f, f_\eta)\}\exp\{j2\pi f t_{dly}\} \end{aligned} \tag{4.35}$$

Then the spectrum of Equation (4.27) is obtained.

7. Finally, only the two-dimensional inverse FFT is needed to obtain the scene raw data of this kind of bistatic SAR.

It can be seen that the above processing steps are quite similar to the monostatic SAR raw data simulation. The main difference is that the relationship between the frequency variables used for conversion here has a factor of $1 + k_s/2$. However, what should be kept in mind is that the derivation above is based on some approximations, which are reasonable only when $|x_R - x_T| \ll r$, $|h_R - h_T| \ll r$ and $\Delta r \ll r$. Therefore, when the bistatic baseline is long and the bistatic angle is large, some simulation methods with high precision should be explored.

4.3.2 Long Bistatic Baseline Case

In this section, a bistatic SAR raw data simulation algorithm that can be used in the long-baseline and big bistatic angle case of the translational invariant configuration is introduced. Though the derivation of this algorithm is based on some conclusion about the bistatic SAR two-dimensional frequency domain spectrum, which will be described in Chapter 5, this simulation algorithm is arranged here in order to make the introduction of bistatic SAR processing systematical. Therefore, readers do not have to understand the derivation of the bistatic SAR echo spectrum expression here, but focus on the idea of generating raw data based on this spectrum. Alternately, readers could firstly read the content about the imaging algorithm for translational invariant bistatic SAR in Chapter 5 and then come back to this section.

According to Equations (5.276) and (5.291) in Chapter 5, the bistatic SAR echo spectrum in the two-dimensional frequency domain without considering the antenna pattern modulation can be written as follows:

$$\begin{aligned} S(k_u, k) &= \exp\{j2\pi f t_{dly}\} \iint\limits_{(x,y)} dxdy\sigma(x, y)\exp\{-jk_u y\} \\ &\quad \times \exp\{-jkh(C)q(x) - jkg(C) - jkE(C, x)\} \\ &= \exp\{j2\pi f t_{dly}\} \iint\limits_{(x,y)} dxdy\sigma(x, y)\exp\{-jk_u y\} \\ &\quad \times \exp\{-jk_q q(x) - jG(k, C) - j\Phi_{res}(k, C, x)\} \end{aligned} \tag{4.36}$$

where k_q is a function of k_u and k, which represents the range–azimuth coupling. According to the SAR raw data simulation theory introduced in Section 4.2.1, the bistatic SAR echo spectrum $S(k_u, k)$ can be obtained by firstly applying the two-dimensional Fourier transform to $\sigma[q(x), y]\exp\{-jk_{q0}q\}$, which is

$$\tilde{S}(k_u, k_q) = \iint\limits_{(q,y)} dqdy\sigma[q(x), y]\exp\{-jk_u y\}\exp\{-j(\tilde{k}_q + k_{q0})q(x)\} \tag{4.37}$$

and then by wavenumber coordinates conversion, that is, $(k_u, k_q) \to (k_u, k)$, through interpolation in the (k_u, k) domain, and finally by multiplying with phase $G(k, C)$. In Equation (4.37), $k_q = k_{q0} + \tilde{k}_q$ and k_{q0} is the median of k_q. However, as will be mentioned in Section 5.5.3, the phase of the bistatic SAR echo spectrum is linearly inseparable, which means that the residual phase $\Phi_{res}(k, C, x)$ is unavoidable; especially in large bistatic angle or long-baseline configurations, the residual phase will become more noticeable. From the analysis in Section 5.5.3 it can be found that the variance of Φ_{res} with k_q is trivial while its variance with k_u and q is remarkable. Therefore, it is reasonable to add in the residual phase in the (k_u, q) domain to get more precise simulated raw data for bistatic SAR of the big bistatic angle and long-baseline configuration.

Based on the above analysis, the raw data simulation of translational invariant bistatic SAR can be carried out as follows:

1. The reflectivity map is multiplied with a phase term that represents a part of the propagation time delay and with a modulation term of the range antenna pattern.

This step can be written as

$$s_1(x, y) = \sigma[x, y; h(x, y)]G_r(x)\exp\{-jk_{q0}q(x)\} \tag{4.38}$$

2. The above signal is then projected on to the (q, y) plane, which becomes

$$s_2[q - q_{near}, y] = \sigma(q, y)G_r(q)\exp\{-jk_{q0}q\} \tag{4.39}$$

where $q_{near} = q(x_{near})$ and x_{near} indicates the location in the x direction that corresponds to the sample delay time, that is, the x direction location of the near range target.

3. Azimuth Fourier transform is performed to the above signal to obtain the signal in the $(\tilde{k}_u, q)$ domain, which is

$$S_3[q - q_{near}, \tilde{k}_u] = \int_y \sigma(q, y)G_r(q)\exp\{-jk_{q0}q\}\exp\{-j\tilde{k}_u y\}dy \tag{4.40}$$

Then the bistatic SAR Doppler centroid $f_{\eta dc}$ is calculated according to the SAR geometry and the scope of $\tilde{k}_u$ is decided as

$$\tilde{k}_u \in \left(\frac{2\pi(f_{\eta dc} - f_{prf}/2)}{V}, \frac{2\pi(f_{\eta dc} + f_{prf}/2)}{V}\right)$$

4. The residual phase is added into the signal by multiplying the signal with the following function:

$$H_{1c}(\tilde{k}_u, q; k_{q0}) = \exp\{-j\Phi_{res}(\tilde{k}_u, q; k_{q0})\} \tag{4.41}$$

which then gives

$$S_4[q - q_{near}, \tilde{k}_u] = \int_y \sigma(q, y)G_r(q)\exp\{-jk_{q0}q - j\tilde{k}_u y - j\Phi_{res}(\tilde{k}_u, q; k_{q0})\}dy \tag{4.42}$$

5. Range Fourier transform is then performed to the above signal, giving

$$S_5[\tilde{k}_q, \tilde{k}_u] = \iint_{q\ y} \sigma(q, y)G_r(q)\exp\{-j(\tilde{k}_q + k_{q0})q + j\tilde{k}_q q_{near} - j\tilde{k}_u y - j\Phi_{res}(\tilde{k}_u, q; k_{q0})\}dydq \tag{4.43}$$

6. The above signal is then multiplied with the following function:

$$H_{2c}(\tilde{k}_q) = \exp\{-j\tilde{k}_q q_{near}\} \tag{4.44}$$

7. The wavenumber coordinates conversion, that is, $(\tilde{k}_q, \tilde{k}_u)$ to (k, k_u) conversion, is performed according to Equation (5.280), (5.286) or (5.289) through interpolation.
8. The bistatic SAR echo spectrum can be obtained by multiplying the converted signal with the following function, which is composed of the reference phase, LFM signal in the range direction, antenna pattern modulation and so on:

$$\begin{aligned} H_{3c}(k, k_u) = &\frac{f_s f_{prf}}{\sqrt{|k_r f_r|}} G_a(f_\eta)\text{rect}\left[\frac{f}{B}\right] \exp\{j\pi/2\} \exp\{-j\pi f^2/k_r\} \\ &\times \exp\{-jG(k, k_u; q_{ref})\} \exp\{j2\pi f t_{dly}\} \end{aligned} \tag{4.45}$$

9. Finally, a two-dimensional IFFT is needed to obtain the simulated SAR scene raw data.

Figure 4.10 is the block diagram of the raw data simulation algorithm for translational invariant bistatic SAR. It is very similar to the block diagram of Figure 4.3. However, the wavenumber conversion and the image plane projection are more complex in bistatic SAR than in monostatic SAR, and an extra step of residual phase multiplication is needed here. Therefore the implementation of the bistatic SAR raw data simulation

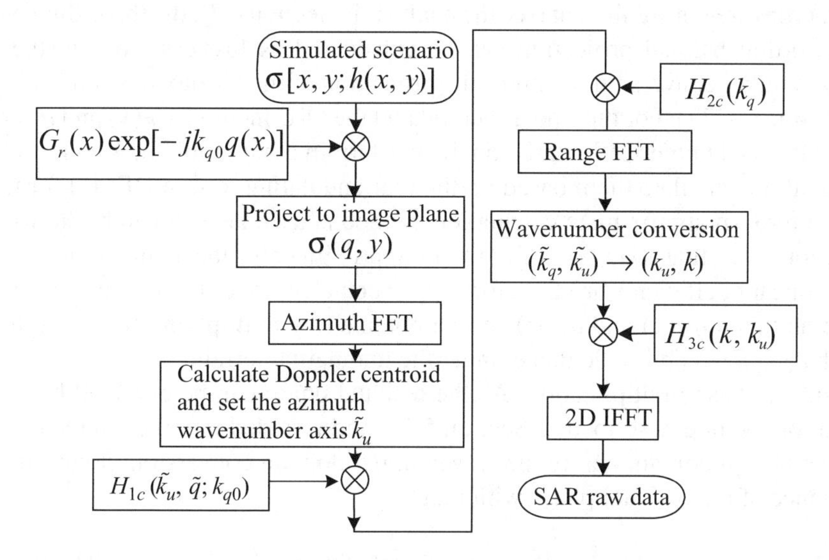

Figure 4.10 Block diagram of the raw data simulation algorithm for translational invariant bistatic SAR

is relatively more complex. Besides, some key steps should be further clarified, which are:

1. Wavenumber domain conversion. The wavenumber domain conversion from $(\tilde{k}_q, \tilde{k}_u)$ to (k, k_u) can be performed according to the three methods described in Chapter 5, which are the Ender method, QIU method and Giroux method. Among these methods, neither the wavenumber variable $k_{q,E}(k, k_u)$ nor the range variable $q_E(x)$ (here the subscript "$_E$" represents the Ender method) in the Ender method have analytical expressions and need to be calculated using the numerical fitting method. The analytical expression of the wavenumber variable $k_{q,Q}(k, k_u)$ in the QIU method is shown in Equation (5.289), while $q_Q(x)$ has to be calculated by the numerical method. The Giroux method cannot achieve the high precision produced by the Ender and QIU methods, but both $k_{q,G}(k, k_u)$ and $q_G(x)$ have analytical expressions, which is that $k_{q,G}(k, k_u)$ can be expressed by Equation (5.280) and $q_G(x) = r(x) - r(x_{ref})$(where $r(x)$ is in Equation (5.277) and the subscript "ref" means reference). Therefore, implementation of the Giroux method is relatively simple and thus it is reasonable to use the Giroux wavenumber conversion method when the requirement for precision is not so strict.
2. Reflectivity map projection. As will be mentioned in the introduction to Chapter 5, the imaging geometry of bistatic SAR has a three-dimensional property, so it cannot be projected on to a two-dimensional plane without any change of the signal properties. The SAR raw data simulation algorithm described here is based on a two-dimensional plane assumption, so the projection of the reflectivity map $\sigma[x, y; h(x, y)]$ on to the $[q(x), y]$ plane will cause a slight deviation in the simulated raw data. However, in many configurations, when the three-dimensional properties are not significant (see the analysis in Section 6.3), the three-dimensional to two-dimensional projection can be considered as lossless. Among the three wavenumber conversion methods, the projection of the Giroux method is given by Equation (5.277). For the Ender method and the QIU method, $q_E(x)$ and $q_Q(x)$ can be calculated numerically based on the flat-terrain assumption; however, it is quite difficult to calculate them based on the real undulating terrain DEM information. Therefore, an approximate calculation method is used here. It can be shown from Section 5.5.5 that $q_E(x)$ and $q_Q(x)$ are approximately linear functions of $q_G(x)$. The linear coefficients are therefore first calculated under the flat terrain assumption and then $q_E(x)$ and $q_Q(x)$ can be obtained by multiplying these coefficients with $q_G(x)$, which is calculated under the undulating terrain.
3. Residual phase multiplication. As the residual phase can be obtained by the calculation method described in Section 5.5.3 and is added into the simulated signal in the (k_u, q) domain before the wavenumber domain conversion, it considers the variance of the residual phase with range.

What should be noticed is that, due to the fact that there are errors caused by the three-dimensional to two-dimensional projection and by ignoring the variance residual

Table 4.5 Simulation system parameters. (© 2010 IEEE. Reprinted with permission from X. Qiu, D. Hu, L. Zhou, C. Ding, "A Bistatic SAR raw data simulator based on inverse omega algorithm," *IEEE Transactions on Geoscience and Remote Sensing*, **48**, 3, pp. 1540–1547, 2010.)

Wavelength λ	0.031 m
Frequency bandwidth B	500 MHz
Sample frequency f_s	550 MHz
Pulse repeat frequency f_{prf}	922.78 Hz
Platform velocity V	200 m/s
Transmitter location (X_1, Y_1, Z_1)	(−1616.0, 250, 2759.1) m
Receiver location (X_2, Y_2, Z_2)	(−2578.6, 0, 1549.4) m
Baseline length L	1597.1 m
Azimuth antenna size	0.5 m

phase of k_q, the simulated bistatic SAR raw data by the algorithm introduced here cannot be exactly accurate. However, as can be seen from Chapter 5, the imaging results of the Ender or QIU algorithm with residual phase compensation are of good quality in the long-baseline and quite wide imaging swath configuration under the flat terrain assumption. Therefore, the raw data simulation method, which is the inverse process of these wavenumber domain imaging algorithms, could be of high precision and could be a good choice for the translational invariant bistatic SAR raw data simulation.

To get a better understanding of the simulation algorithm, two simulation examples are exhibited in the following:

- **Discrete scattering target simulation**
  The simulation system parameters for the long-baseline airborne bistatic SAR configuration are listed in Table 4.5 and the locations and scattering coefficients of the simulated targets are shown in Table 4.6.
  Figure 4.11 shows the phase map of the result of $S_{TBT} \times S^*_{IOK}$ with and without residual phase multiplication, where S_{TBT} indicates the simulated data by the

Table 4.6 Information of simulated discrete targets. (© 2010 IEEE. Reprinted with permission from X. Qiu, D. Hu, L. Zhou, C. Ding, "A Bistatic SAR raw data simulator based on inverse omega algorithm," *IEEE Transactions on Geoscience and Remote Sensing*, **48**, 3, pp. 1540–1547, 2010.)

Target number	Location (x, y, z) m	Scattering coefficient
1	(0, 0, 0)	1
2	(–662.41, 0, 0)	$0.71 - 0.71i$
3	(569.14, 0, 0)	i
4	(0, 10.84, 0)	-1
5	(0, –10.84, 0)	$-i$

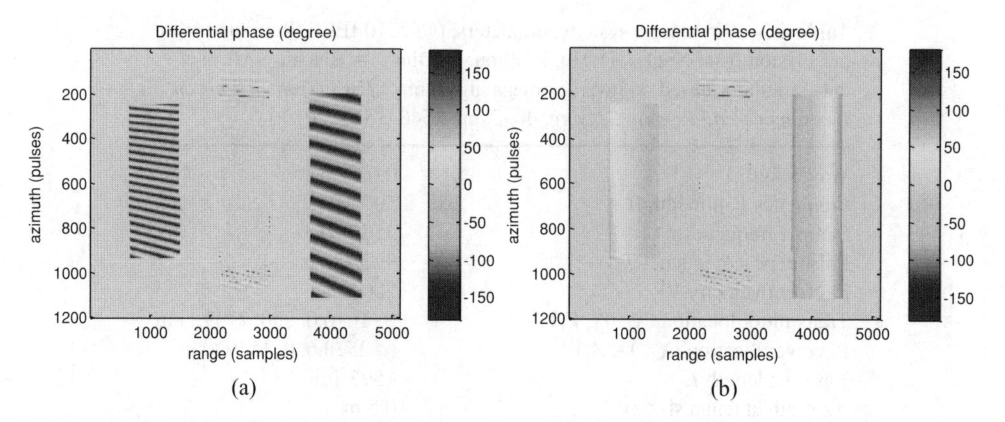

Figure 4.11 Differential phase map of the simulated SAR raw data by the TBT method and the IOK method: (a) without residual phase multiplication, (b) with residual phase multiplication (© 2010 IEEE. Reprinted with permission from X. Qiu, D. Hu, L. Zhou, C. Ding, "A Bistatic SAR raw data simulator based on inverse omega algorithm," *IEEE Transactions on Geoscience and Remote Sensing*, **48**, 3, pp. 1540–1547, 2010.)

target-by-target (TBT) method, S_{IOK} indicates the simulated data by the method introduced here (named the IOK method because it is the inverse process of the ω-k imaging algorithm) and "$*$" stands for conjunction. It can be seen that without the residual phase multiplication, the differential phase of the center targets (which are at the reference range) is close to zero, while the phase of the near and far targets are quite different between S_{TBT} and S_{IOK}. However, with the residual phase multiplication, the differential phases of the near and far targets are minimized and only a small linear phase remains in the range direction. The remaining phase is caused by ignoring the variance of residual phase Φ_{res} with k_q. The detailed differential

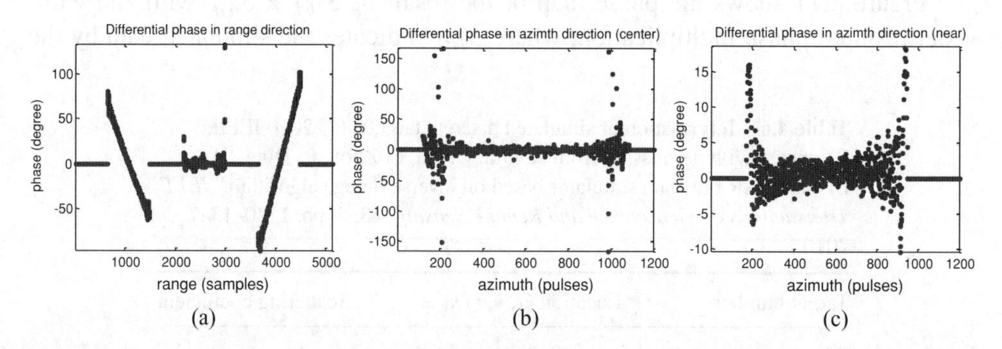

Figure 4.12 Differential phase between the TBT method and the IOK method in the azimuth and range directions: (a) differential phase in the range direction, (b) differential phase for the central targets in the azimuth direction, (c) differential phase for the near target in the azimuth direction (© 2010 IEEE. Reprinted with permission from X. Qiu, D. Hu, L. Zhou, C. Ding, "A Bistatic SAR raw data simulator based on inverse omega algorithm," *IEEE Transactions on Geoscience and Remote Sensing*, **48**, 3, pp. 1540–1547, 2010.)

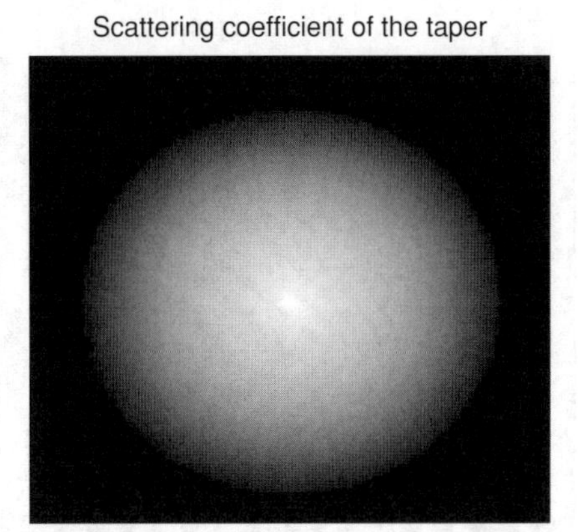

Figure 4.13 Scattering coefficient of the simulated cone (© 2010 IEEE. Reprinted with permission from X. Qiu, D. Hu, L. Zhou, C. Ding, "A Bistatic SAR raw data simulator based on inverse omega algorithm," *IEEE Transactions on Geoscience and Remote Sensing*, **48**, 3, pp. 1540–1547, 2010.)

phases for azimuth and range signals are shown in Figure 4.12. It can be seen that the differential phase of the TBT method and the IOK method is less then 10° except for the fringes for the central targets, both in the range and azimuth directions. For the near and far targets, there is a linear phase term within $\pm 100^\circ$ in the range direction, while in the azimuth direction the differential phase is quite small.

- **Simple scene raw data simulation**
In order to show the feasibility of the simulation algorithm for bistatic SAR interferometric usage, both monostatic and bistatic SAR raw signals for a cone are simulated by the IOK method. The radius and the height of the cone are 250 m and 80 m, respectively, and the scattering coefficient of the cone is shown in Figure 4.13. The simulation system parameters are shown in Table 4.7. The simulated raw signals are focused by the CS algorithm based on the hyperbolic approximation,

Table 4.7 System parameters for scene raw data simulation. (© 2010 IEEE. Reprinted with permission from X. Qiu, D. Hu, L. Zhou, C. Ding, "A Bistatic SAR raw data simulator based on inverse omega algorithm," *IEEE Transactions on Geoscience and Remote Sensing*, **48**, 3, pp. 1540–1547, 2010.)

Wavelength λ	0.031 m
Frequency bandwidth B	150 MHz
Sample frequency f_s	165 MHz
Pulse repeat frequency f_{prf}	3514.6 Hz
Platform velocity V	7000 m/s
Transmitter location (X_1, Y_1, Z_1)	(−613.51, 0, 514.8) km
Receiver location (X_2, Y_2, Z_2)	(−612.11, 0, 515.31) km
Baseline length L	1500 m
Azimuth antenna size	4.78 m

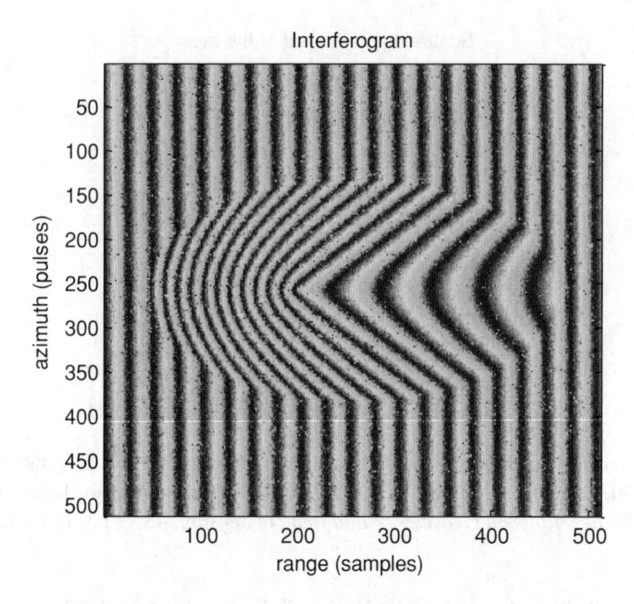

Figure 4.14 Interferogram of bistatic and monostatic SAR images (© 2010 IEEE. Reprinted with permission from X. Qiu, D. Hu, L. Zhou, C. Ding, "A Bistatic SAR raw data simulator based on inverse omega algorithm," *IEEE Transactions on Geoscience and Remote Sensing*, **48**, 3, pp. 1540–1547, 2010.)

which will be introduced in Chapter 5. The interferogram between the monostatic and bistatic images is shown in Figure 4.14 and the DEM information obtained from the interferogram for this cone is shown in Figure 4.15. It can be seen that the cone is well rebuilt. This makes the IOK method valid for use in the bistatic interferometric SAR simulation.

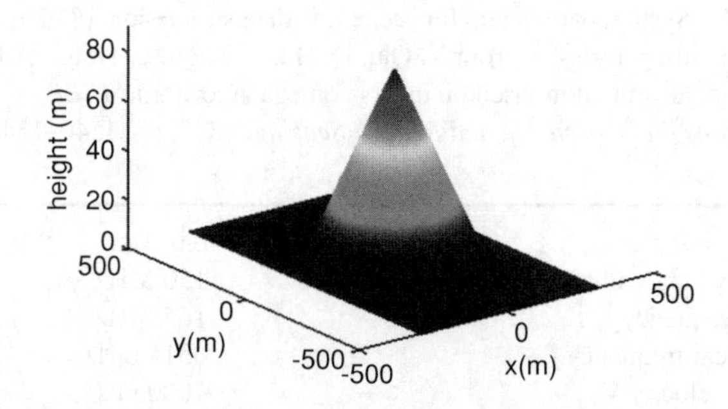

Figure 4.15 DEM calculated by interferometry (© 2010 IEEE. Reprinted with permission from X. Qiu, D. Hu, L. Zhou, C. Ding, "A Bistatic SAR raw data simulator based on inverse omega algorithm," *IEEE Transactions on Geoscience and Remote Sensing*, **48**, 3, pp. 1540–1547, 2010.)

4.4 Summary

This chapter has firstly introduced the traditional monostatic SAR raw data simulation methods and then describes the translational invariant bistatic SAR raw data simulation algorithms based on the monostatic SAR raw data simulation theory. The simulation algorithms introduced here can provide the simulated data for the study of bistatic SAR imaging algorithms. However, the algorithms here can only simulate the range variance of the SAR echo and thus can only be applied in the translational invariant configuration. However, for the translational variant case, the efficient scene raw data simulation methods need further study.

References

[1] Ma, J. and Song, H. (2005) MPI based simulation algorithm for distributed targets in SAR. *Review of Electronic Science and Technology*, **4**, 20–24 (In Chinese).

[2] Zhang, F., Lin, Y. and Hong, W. (2008) SAR echo distributed simulation based on grid computing. *Journal of System Simulation*, **20** (12), 3165–3170 (In Chinese).

[3] Franceschetti, G., Migliaccio, M., Riccio, D., *et al.* (1990) A three dimensional SAR simulator. 20th European Microwave Conference, vol. 1, pp. 887–892.

[4] Franceschetti, G., Migliaccio, M., Riccio, D., *et al.* (1992) SARAS: a synthetic aperture radar (SAR) raw signal simulator. *IEEE Transactions on Geoscience and Remote Sensing*, **30** (1), 110–123.

[5] Horrell, J.M. and Inggs, M.R. (1993) Satellite and airborne SAR simulator. In: *Proceedings of the 1993 IEEE South African Symposium on Communications and Signal Processing*, pp. 193–198.

[6] Franceschetti, G., Iodice, A., Migliaccio, M., *et al.* (1997) Efficient simulation of SAR interferometric raw signal pairs. IEEE International Geoscience and Remote Sensing Symposium (IGARSS'97), vol. 4, pp. 1701–1703.

[7] Franceschetti, G., Iodice, A., Riccio, D., *et al.* (2002) A 2-D Fourier domain approach for spotlight SAR raw signal simulation of extended scenes. IEEE International Geoscience and Remote Sensing Symposium (IGARSS'02), vol. 2, pp. 853–55.

[8] Cimmino, S., Franceschetti, G., Iodice, A., *et al.* (2003) Efficient spotlight SAR raw signal simulation of extended scenes. *IEEE Transactions on Geoscience and Remote Sensing*, **41** (10): 2329–2337.

[9] Franceschetti, G., Iodice, A., Riccio, D., *et al.* (2003) SAR raw signal simulation for urban structures. *IEEE Transactions on Geoscience and Remote Sensing*, **41** (9), 1986–1995.

[10] Franceschetti, G., Iodice, A., Perna, S., *et al.* (2006) Efficient simulation of airborne SAR raw data of extended scenes. *IEEE Transactions on Geoscience and Remote Sensing*, **44** (10), 2851–2860.

[11] Franceschetti, G., Guida, R., Iodice, A., *et al.* (2007) Simulation tools for interpretation of high resolution SAR images of urban areas. Urban Remote Sensing Joint Event, pp. 1–5.

[12] Franceschetti, G., Iodice, A., Perna, S., *et al.* (2006) SAR sensor trajectory deviations: Fourier domain formulation and extended scene simulation of raw signal. *IEEE Transactions on Geoscience and Remote Sensing*, **44** (9), 2323–2334.

[13] Wang, X., Yu, Y., Chen, Y., *et al.* (2007) Bistatic SAR raw data simulation for ocean. IEEE International Geoscience and Remote Sensing Symposium (IGARSS'07), pp. 871–874.

[14] Wang, M., Lu, X. and Liang, D. (2007) Raw data simulation of ground scene in spaceborne bistatic SAR. *Modern Radar*, **29** (11), 22–24 (In Chinese).

[15] Bamler, R. (1992) A comparision of range-Doppler and wavenumber domain SAR focusing algorithms. *IEEE Transaction on Geoscience and Remote Sensing*, **30** (4), 706–713.

[16] Wei, Z. (2001) *Synthetic Aperture Radar*, Science Press, Beijing (In Chinese).

[17] Raney, R.K. and Princz, G.J. (1987) Reconsideration of azimuth ambiguities in SAR. *IEEE Transactions on Geoscience and Remote Sensing*, **25** (6), 783–787.
[18] Moreira, A. (1993) Suppressing the azimuth ambiguities in synthetic aperture radar images. *IEEE Transactions on Geoscience and Remote Sensing*, **31** (4), 885–895.
[19] Wang, Y., Zhang, Z. and Deng, Y. (2008) Squint spotlight SAR raw signal simulation in the frequency domain using optical principles. *IEEE Transactions on Geoscience and Remote Sensing*, **46** (8), 2208–2215.
[20] Natroshvili, K., Loffeld, O., Nies, H., *et al.* (2006) Focusing of general bistatic SAR configuration data with 2-D inverse scaled FFT. *IEEE Transactions on Geoscience and Remote Sensing*, **44** (10), 2718–2727.
[21] Wei, L. (2005) Studies on Designs, Implementations and Applications of Computer Simulation System for Airborne Dual-Antenna Interferometric SAR. Doctoral thesis, Institute of Electronics, Chinese Academy of Sciences, Beijing.
[22] Qiu, X., Hu, D., Zhou, L., *et al.* (2010) A bistatic SAR raw data simulator based on inverse omega-K algorithm. *IEEE Transactions on Geoscience and Remote Sensing*, **48** (3), 1540–1547.

5

Imaging Algorithms for Translational Invariant Bistatic SAR

5.1 Introduction

From this chapter on, algorithms for bistatic SAR of different configurations will be introduced step by step. In this chapter, imaging algorithms for translational invariant bistatic SAR are presented. We have tried to be comprehensive and the algorithms are classified according to the different ideas they based on.

Translational invariant configuration means that the aperture phase centers of the transmitter and the receiver have the same constant velocity vector. This means that the transmitter and receiver are relatively stationary with data acquisition. The translational invariant configuration can be further classified into the tandem case and the case with parallel paths according to the trajectories of the transmitter and receiver.

The main difficulties of the translational invariant bistatic SAR come from the following two aspects:

1. The range history has the form of a double square root (DSR). The range history of the traditional monostatic SAR with a straight trajectory has the form of a single square root, which is a hyperbola. However, in bistatic SAR, the range history is the summarization of the contribution of the transmitter movement and that of the receiver movement, so the bistatic range history has the form of a DSR, which is a flat-top hyperbola [1] (see Figure 5.1). If the transmitter closest range experiences a different time from the receiver closest range, the top of the hyperbolic transmitter range history will deviate from the top of the hyperbolic receiver range history in the time dimension. Thus, summarizing these two hyperbolas usually results in a flat-top hyperbola-type curve, which is actually not a strict hyperbola. The bistatic range history in the DSR form results in the difficulty of finding the analytical solution

Bistatic SAR Data Processing Algorithms, First Edition. Xiaolan Qiu, Chibiao Ding and Donghui Hu.

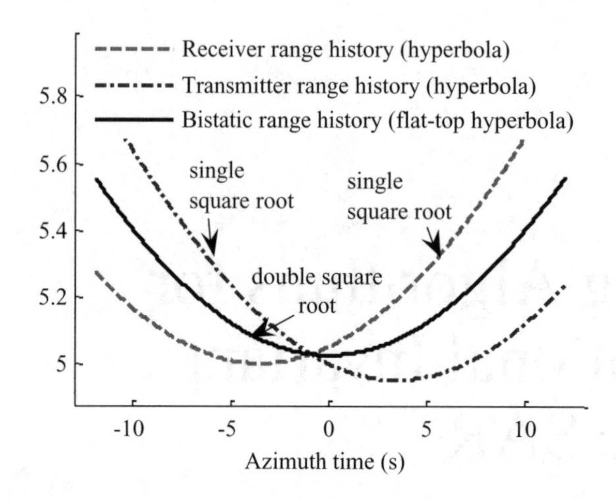

Figure 5.1 Illustration of the topography sensitivity of the bistatic SAR imaging geometry

of the stationary phase. Consequently, it is difficult to get an analytical expression of the two-dimensional bistatic point target reference spectrum (BPTRS) in the frequency domain. As can be seen in Chapter 2, classical SAR imaging algorithms (like the RD/CS/ω-k algorithm) are all based on an analytical point target reference spectrum in the frequency domain. Therefore, bistatic SAR imaging encounters a new challenge of a lack of the frequency domain signal spectrum basis.

2. The imaging geometry is topography sensitive. In those bistatic SAR configurations, which have a separate transmitter trajectory and receiver trajectory, the imaging geometries have elliptical cylindrical symmetry, while in monostatic and tandem cases, where only one track exists, the imaging geometries have circular cylindrical symmetry. The elliptical cylindrical symmetry means that the imaging geometry is topography sensitive. Those targets having the same closest bistatic range at the same azimuth time but different heights will experience different bistatic range histories. Therefore the bistatic range history and the transmitter and receiver trajectories can uniquely identify a point in the three-dimensional space (except for the mirror point) (illustrated in Figure 5.2). The topography sensitivity of the bistatic SAR signal forces us to build the signal model according to the target position in three-dimensional space, while in monostatic SAR the signal model can be built according to the target position in a so-called two-dimensional slant range plane, that is, the (r, y) plane, which is mentioned in Chapter 2. Therefore, the imaging algorithms of bistatic SAR are more complex than those of monostatic SAR, as well as the bistatic SAR image geocoding.

To handle the above difficulties, many studies have been undertaken and a considerable number of research results have been published. In this chapter, we make a classification of the algorithms according to the different ideas they are based on. Then, one or two algorithms of each category are described in detail. Finally, different

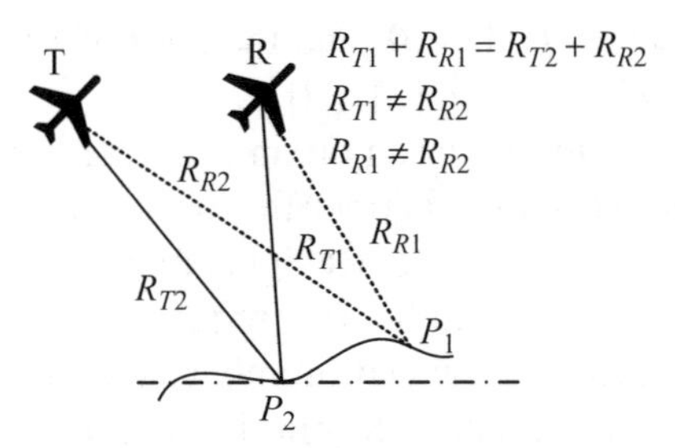

Figure 5.2 Bistatic range history

algorithms are compared and summarized in order to show the different performances of different algorithms.

Here the imaging algorithms for bistatic SAR of the translational invariant configuration are classified by their basic ideas, which are sorted as follows:

1. Raw data monostatic transform. This means that the bistatic SAR raw data are firstly preprocessed so that they can be transformed into monostatic or monostatic-like SAR raw data, and then traditional monostatic SAR imaging algorithms can be directly used for processing. A representative method based on this idea is Dip-Move-Out (DMO), which was first introduced by Rocca [1] from seismic signal processing into bistatic SAR imaging, and was further explained and implemented from different points of view in [2] to [4].
2. Bistatic range history simplification. Here the aim was to find some simple form (usually a monostatic-like form) to approximate the DSR bistatic range history, so that monostatic-like imaging algorithms can be obtained based on this approximated range history through similar derivations as those of monostatic algorithms. There have been quite a lot of approaches published based on this idea [5–15], for example, the baseline-middle-point monostatic SAR approximation method [5], the quadratic polynomial approximation method [6, 9, 11, 12], the hyperbola approximation (or single square root approximation) [8, 13] and the advanced hyperbola approximation [15].
3. Bistatic SAR spectrum approximation by an analytical explicit expression. Some approximations in the two-dimensional frequency domain were made in order to obtain an approximated BPTRS, so that the bistatic SAR imaging algorithms can be derived based on this two-dimensional spectrum basis. A representative analytical explicit expression of BPTRS is Loffeld's bistatic formula (LBF) [3, 16, 17], and a number of algorithms for different configurations have been proposed based on LBF. Other methods to obtain an approximated analytical explicit expression in the frequency domain have also been proposed [18–27], such as the method of series

reversion (MSR) by Neo *et al.* [18–20] and the method of instantaneous Doppler wavenumber (IDW) by Zhang *et al.* [21, 22].

4. Wavenumber domain decoupling based on an accurate implicit expression of the bistatic SAR spectrum. This idea adopts the basic idea of the traditional ω-k algorithm, which is to decompose the SAR signal two-dimensional phase spectrum into a linear series of some wavenumber (or frequency) domain variables and then transform the spectrum into this new wavenumber space and obtain the final image by Fourier transform. Here an accurate implicit expression bistatic SAR signal in two-dimensional frequency is used as a basis, and approximations are made later while linear decomposition of the phase spectrum takes place. This idea was firstly proposed by Ender [28] and a numerical phase decomposition method was addressed in [29] to [31]. Analytical phase decomposition methods were further proposed by Giroux *et al.* [32, 33] and Qiu *et al.* [34], while in [34] the residual phase error after phase decomposition is analyzed and compensated to enhance the precision of the algorithm.
5. Numerical SAR (NuSAR). Here all the transfer functions of imaging processing pass through numerical calculation. This idea was proposed by Bamler and Boerner [13] and a numerical SAR processor was built according to this idea [35].
6. The universal back-projection (BP) method and fast BP [36–40].

Among the above approaches, 5 and 6 can be widely used in more complex configurations (such as curved SAR, translational variant bistatic SAR and so on) at the cost of a huge computation burden. As these two approaches do not take advantage of the translational invariant property of this particular configuration, they are usually not the best choice for this case. Therefore, in this chapter the first four approaches are introduced.

5.2 Imaging Algorithms Based on Monostatic Transform

It is a very attractive concept to transform the bistatic SAR raw data into monostatic SAR raw data, because after this transform the mature monostatic SAR algorithms can then be used for bistatic SAR imaging. However, this concept is hard to realize because there is a great difference in imaging geometry between the bistatic SAR and the monostatic SAR. Among the bistatic configurations, the tandem configuration is the one most similar to the monostatic case: in case the "stop–go" assumption is invalid, the monostatic SAR turns into the tandem one; thus research of monostatic transform starts with the tandem configuration. Rocca introduced the DMO method, which is well known in the seismic signal processing field, into bistatic SAR imaging for the first time, and studied how to transform the tandem bistatic SAR raw data into the monostatic one [1]. The transfer function is a short operator and has the shape of a "smile"; hence, in the field of SAR signal processing, the DMO method is usually called "Rocca's smile". In [41] the "smile" concept has been further extended to the general bistatic case. The idea of "Rocca's smile" holds the attention of many

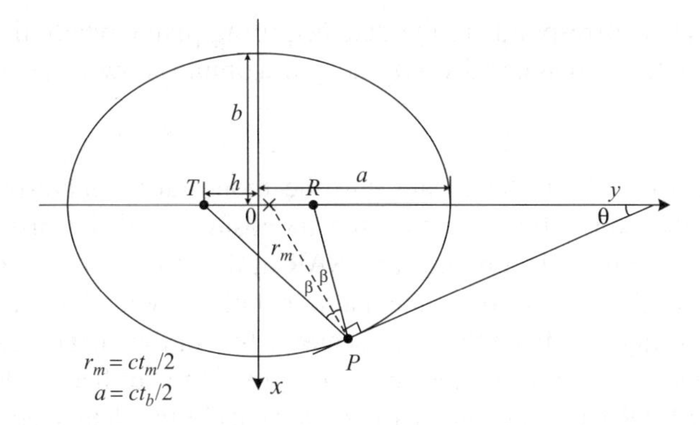

Figure 5.3 Imaging geometry of the tandem bistatic SAR in the slant range plane

researchers since it was proposed [2, 4]. Spectral aspects of explanation and frequency domain methods of approximate implementation about "Rocca's smile" in the tandem case have been further put forward based on point target spectrums, which are derived through different methods (the LBF method and the IDW method). Spectral explanations about "Rocca's smile" will be addressed when the LBF method and the IDW method are introduced later in this chapter. Therefore, this section focuses on the idea of the DMO method and shows an intuitive explanation of the DMO method from the aspect of synthesizing narrow beams.

5.2.1 DMO Method

To get the basic idea, we start with the tandem configuration. As mentioned in Section 5.1, the tandem bistatic SAR has a circular cylindrical symmetry, that is, it is rotationally invariant along the flying path, and thus its imaging geometry can be projected on to a certain slant range plane without any distortion. Its imaging geometry in the two-dimensional plane is illustrated in Figure 5.3, where T denotes the transmitting antenna, R denotes the receiving antenna and P denotes the point target.

Let 2β be the magnitude of the bistatic angle and $2h$ be the phase center distance between the transmitting and the receiving antennas. According to the bistatic SAR imaging geometry, at the range gate where the time delay is t_b, all the point targets at a certain part of the ellipse in the slant range plane contribute to the echo. The foci of this ellipse are respectively located in the phase center of the transmitting and the receiving antennas. The horizontal semiaxis, a, is equal to $ct_b/2$. The effective part of this ellipse is codetermined by the transmitting and receiving antenna beams.

For the convenience of the following description, we define two spaces at the very start:

1. Model space. This corresponds to the imaging scene in the slant range plane, where the location of a point target is denoted by the slant range and the azimuth position.

2. Data space. This corresponds to the data acquiring plane, where the two axes are the range fast time (denoted by t_b) and the azimuth slow time (denoted by η) respectively.

In order to transform the bistatic data into the monostatic counterpart, we should firstly find out the connection between the monostatic and the bistatic SAR data acquisition. Suppose there is a monostatic SAR at the middle point of the line from the phase center of the transmitting antenna to that of the receiving antenna (see O in Figure 5.3), acquiring data from the same scene as that of the bistatic SAR. To simply the following analyses, the data after range compression is utilized below.

In the bistatic SAR case, the signal value at (t_b, η) in the data space is the sum of the echo from targets at a given ellipse, the foci of which are located in the phase center of the transmitting and the receiving antennas at η, and the horizontal semiaxis is equal to $ct_b/2$. Thus, the signal value at (t_b, η) can be written as

$$s_b(t_b, \eta) = \int_l \sigma(x, y) \exp\{-j2\pi f_0 t_b\} dl \tag{5.1}$$

where f_0 is the carrier frequency, (x, y) is the coordinate in the model space and the curve l is

$$l : \sqrt{x^2 + (V\eta - h - y)^2} + \sqrt{x^2 + (V\eta + h - y)^2} = ct_b \tag{5.2}$$

or

$$l : \frac{x^2}{b^2} + \frac{(y - V\eta)^2}{a^2} = 1, \quad \text{where} \quad a = ct_b/2, \qquad b = \sqrt{a^2 - h^2} \tag{5.3}$$

Then we consider what happens in the supposed monostatic SAR data acquisition. The targets $\sigma(x, y)$, $(x, y) \in l$ in the model space contribute a set of hyperbolic loci in the monostatic data (as shown by the thin solid line in Figure 5.4). For each hyperbola, if the piece where the corresponding monostatic line-of-sight direction is coincident with the bisector of the bistatic angle at η is taken into account (as shown by the black "+" in Figure 5.4), an elliptic curve is obtained, which turns out to be the envelope of these hyperbolas. This curve is the so-called "Rocca's smile" (as shown by the thick solid line in Figure 5.4). Thus, the connections between the bistatic and the monostatic data can be described as follows: one strike in the bistatic data space corresponds to an elliptic curve in the monostatic data space.

The above conclusion shown by the simulation results in Figure 5.4 can be verified more strictly using the following geometrical analysis. Considering the case when the azimuth time is zero, the signal at t_b in the data space is created by the targets along the corresponding elliptic curve in the model space, as illustrated in Figure 5.5. If T and R are considered as two illuminators, the equivalent propagation direction at a certain target $P_1(x_1, y_1)$ in the model space is the same as the bistatic angle bisector (referring

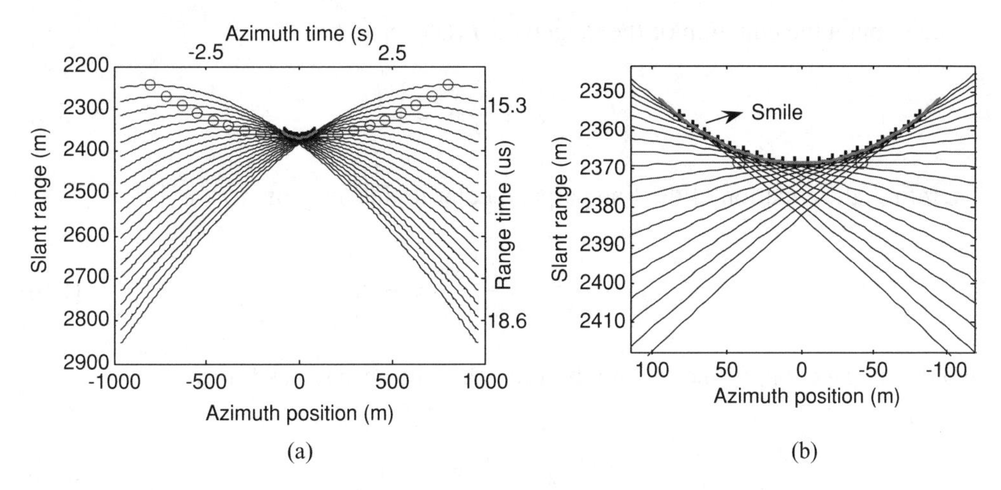

Figure 5.4 Relation between monostatic and bistatic SAR data: (a) whole illustration, (b) expanded illustration of Rocca's smile. The circles denote the scatters in the model space, whose echo locates at $\eta = 0$ and $t_b = 4.166$ μs in the bistatic data space. The thin solid lines denote the hyperbolic range histories of these scatters in the monostatic SAR data space. The thick solid lines denote the envelope of these range histories, where the monostatic line of sight direction is coincident with the bistatic angular bisector, that is, the so-called Rocca's smile

to [22]). Hence, an equivalent monostatic transmitter for target P_1 can be supposed to be located at $(0, y_m(\theta_1))$. According to the geometric property of an ellipse, the line perpendicular to the bistatic angle bisector is the very tangent of this ellipse at P_1. Moreover, by the equation of the ellipse

$$\frac{y^2}{a^2} + \frac{x^2}{b^2} = 1 \tag{5.4}$$

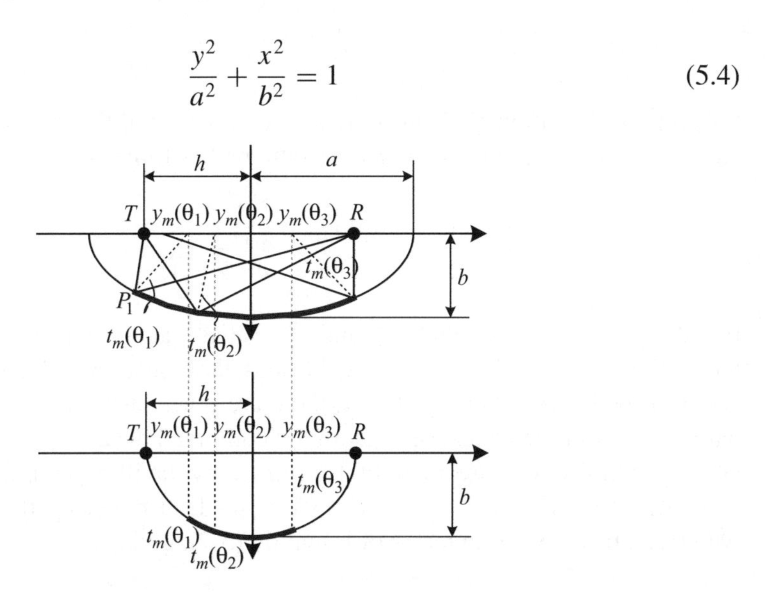

Figure 5.5 Geometry of Rocca's smile. (© 2004 IEEE. Reprinted with permission from D'Aria, D.; Guarnieri, A.M.; Rocca, F. "Focusing bistatic synthetic aperture radar using dip move out". *IEEE Transactions on Geoscience and Remote Sensing*, **42**(7), pp. 1362–1376, 2004.)

we can obtain the equation of the tangent at $P_1(x_1, y_1)$

$$l_1 : \frac{y_1 y}{a^2} + \frac{x_1 x}{b^2} = 1 \tag{5.5}$$

Therefore, the equation of the line perpendicular to this tangent at P_1 is

$$l_{1q} : y - y_1 = \frac{b^2}{a^2} \frac{y_1}{x_1} (x - x_1) \tag{5.6}$$

and the intersection of line l_{1q} and the y axis has the abscissa as follows:

$$y_m = \frac{h^2}{a^2} y_1 \tag{5.7}$$

Therefore, the equivalent monostatic range is equal to the range from the point target $P_1(x_1, y_1)$ to this intersection $(0, h^2 y_1/a^2)$, which is given by

$$r_m^2 = x_1^2 + \left(y_1 - \frac{h^2}{a^2} y_1 \right)^2 \tag{5.8}$$

According to Equation (5.4), the above expression can be simplified to

$$\frac{r_m^2}{b^2} + \frac{h^2 y_1^2}{a^4} = 1 \tag{5.9}$$

Substituting Equation (5.7) into the above expression, the relationship between the equivalent monostatic range and y_m is obtained as follows:

$$\frac{r_m^2}{b^2} + \frac{y_m^2}{h^2} = 1 \tag{5.10}$$

Equation (5.10) shows that r_m and y_m satisfy an elliptic equation and y_m has the limitation of $|y_m| \leq h$. In fact, besides the constraint of the elliptic equation, $|y_1| \leq a$, which results in $|y_m| \leq h$, there are other restrictions. Because the illuminating area is limited because of a limited combined bistatic antenna beam, the upper and lower limits of y_1 are actually determined by the illuminating area, $|y_1| \leq y_{1\,\max}$. Consequently, the value extent of y_m is also determined by the illuminating area, which can be obtained in terms of Equation (5.7), giving

$$|y_m| \leq \frac{h^2}{a^2} y_{1\,\max} \tag{5.11}$$

Therefore, we have proved through the tandem bistatic SAR geometry that the connection between monostatic and bistatic data is that a strike in the bistatic SAR data space corresponds to a section of an elliptic curve in the monostatic data space.

In terms of this connection, the monostatic transform can be implemented by convolving the bistatic data with a range-varying short operator in the time domain or the range-Doppler domain [1]. Moreover, with some approximations, the monostatic transform process can also be achieved approximately by a phase multiplied in the two-dimensional frequency domain [2, 4].

For other configurations, the relationship between the bistatic SAR survey and the monostatic SAR survey can also be found through a similar approach, but without the property of circular cylindrical symmetry, the model space should be three dimensional and the smiles will be more complex.

5.2.2 An Explanation of the DMO Method by Synthesizing Narrow Beams

The interpretation of the DMO method given above is based on geometrical derivation. A more intuitional interpretation is given below from a narrow beams synthesizing point of view, allowing one to intuitively understand how tandem bistatic SAR data can be transformed into monostatic SAR data.

Figure 5.6 illustrates the concept of narrow beams synthesizing for both a monostatic and a tandem bistatic SAR survey. For the monostatic case, the whole azimuth antenna beam can be decomposed into many narrow beams with various pointing directions originating at the SAR position of each instant in time; thus, at any instant, the received signal of the whole beam can be regarded as the sum of the signal obtained

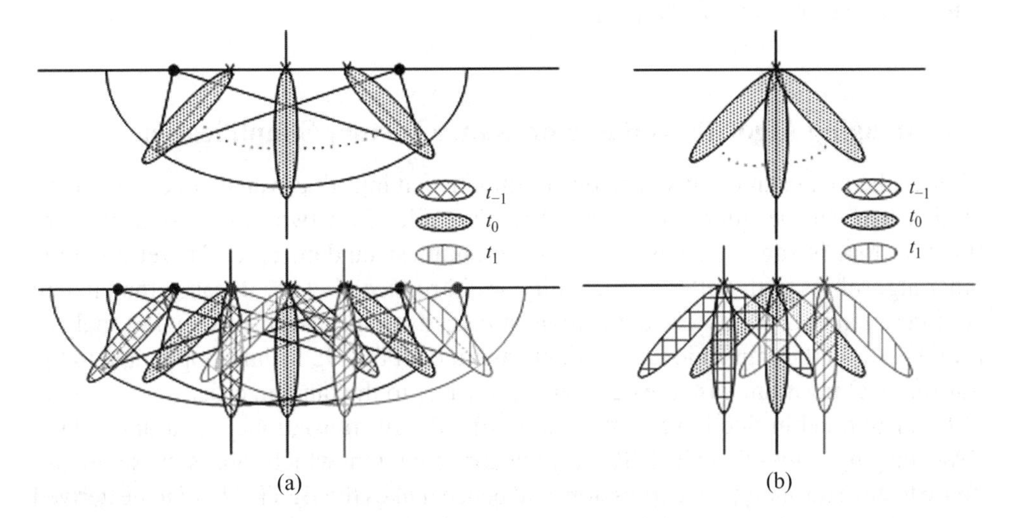

Figure 5.6 The concept of narrow beams synthesizing data acquisition in monostatic and tandem bistatic SARs: (a) tandem bistatic case, (b) monostatic case

by each narrow beam. As this bunch of narrow beams moves along the SAR path, the entire imaging area will be illuminated and the backscattered signal will be received and monostatic SAR raw data are formed (as shown in Figure 5.6(b)). For the tandem bistatic SAR case, the combined bistatic SAR antenna beam can be decomposed into many narrow beams with various pointing directions and different originating points. Each narrow beam points to the direction of the bistatic angle bisector corresponding to the relevant target position. The origin is the intersection of the bistatic angle bisector and a chosen equivalent monostatic flying path, which in the tandem bistatic SAR is the same flying path for the transmitter and the receiver (as shown in Figure 5.6(a)). Over a period of time, this bunch of narrow beams moves along the flying track. Each position in the flying track will be passed by the originating points of all the narrow beams. Therefore, considering a certain position in the flying track, when the data acquisition is finished, there are equivalent narrow beams in the bistatic case and in the monostatic case at this position. While all the narrow beams appear at this position at the same time in the monostatic case, the narrow beams appear at this position at different times in the bistatic case. Conversely, while narrow beams appear at different positions at a single instant in the bistatic case, these narrow beams can be obtained by monostatic SAR at different instants. Therefore, from a narrow beam synthesizing point of view, we can infer that the bistatic SAR data of the tandem configuration can be transformed into monostatic SAR data if the difference in the RCS between the monostatic SAR and the bistatic SAR is not taken into account (this can be valid while only imaging algorithms are considered).

For other bistatic configurations without the superposition of the transmitter flying track and the receiver flying track, the whole combined bistatic beam should be decomposed into narrow beams in a three-dimensional space, and a proper equivalent monostatic track should be chosen when looking for a relatively simple "smile". Interested readers can refer to [41].

5.3 Imaging Algorithms Based on Range History Simplification

Though the monostatic transform idea is attractive, it introduces an extra computation burden that can be quite large, for "Rocca's smile" is a two-dimensional transfer function that is range variant, even in the simplest tandem case. Therefore some other algorithms with better applicability are introduced below. Because the tandem configuration is a special case of the translational invariant case, the following analysis is all based on the translational invariant case and all the algorithms applicable to the translational invariant case are certainly applicable to the tandem case.

As mentioned in Section 5.1, the main difficulty of translational invariant bistatic SAR imaging comes from its DSR range history equation, which causes the stationary phase to have no analytical expression, and hence makes the BPTRS hard to be derived and the imaging algorithm difficult to be developed. However, as long as the bistatic range equation can be approximated by a simpler one (e.g., a monostatic-like one),

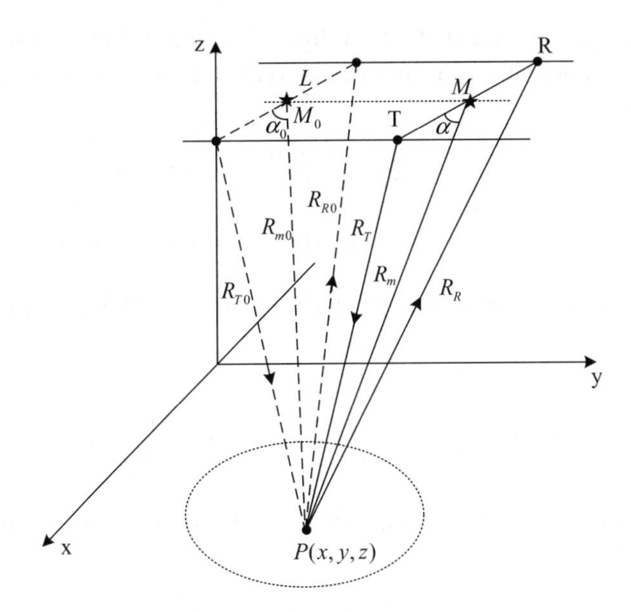

Figure 5.7 Geometry of the baseline middle-point approximation

the quasi-monostatic algorithm can be developed following the mature monostatic algorithms. Many researchers have adopted this idea and developed some appropriate imaging algorithms [9, 11, 42–46]. Among them, three approaches will be introduced below.

5.3.1 Baseline Middle-Point Monostatic SAR Approximation

The idea of a baseline middle-point monostatic SAR approximation is to make some approximation directly on the bistatic geometry. An equivalent monostatic SAR on the middle point of the baseline (as shown in Figure 5.7) is supposed to be acquiring the SAR echo, and the approximating precision is further enhanced by range compensation.

At the zero azimuth time, we suppose that the phase center of the transmitting antenna is at $(X_{\mathrm{T}}, Y_{\mathrm{T}}, Z_{\mathrm{T}})$, the phase center of the receiving antenna is at $(X_{\mathrm{R}}, Y_{\mathrm{R}}, Z_{\mathrm{R}})$, the baseline length is L, the velocity vectors of transmitting and receiving antennas are parallel to the y axis and the magnitude is V. At η, the sum of transmitting and receiving ranges to point target $P(x, y, z)$ is

$$\begin{aligned} R_{bi}(\eta) &= |TP| + |RP| \\ &= \sqrt{(x - X_T)^2 + (Y_T + V\eta - y)^2 + (z - Z_T)^2} \\ &\quad + \sqrt{(x - X_R)^2 + (Y_R + V\eta - y)^2 + (z - Z_R)^2} \end{aligned} \tag{5.12}$$

This is a double square root equation. In Figure 5.7, the included angle between line MP and line RT is denoted by α and can be related to other geometric variables by the law of cosine

$$|TP| = \sqrt{(L/2)^2 + |MP|^2 + L|MP|\cos\alpha} \tag{5.13}$$

$$|RP| = \sqrt{(L/2)^2 + |MP|^2 - L|MP|\cos\alpha} \tag{5.14}$$

Let $|TP| + |RP| = 2a$; the length of MP can be calculated in terms of Equations (5.13) and (5.14) by

$$|MP| = a\sqrt{\frac{4a^2 - L^2}{4a^2 - L^2\cos^2\alpha}} \approx a - \frac{\sin^2\alpha}{8a}L^2 \tag{5.15}$$

The approximation in the above equation is tenable when $L \ll a$. Furthermore, we arrive at

$$|TP| + |RP| = 2a \approx 2|MP| + \frac{\sin^2\alpha}{4a}L^2 \tag{5.16}$$

This means that the bistatic range equation approximates to the sum of a monostatic range equation at the baseline middle point and a small range component proportional to $L^2\sin^2\alpha/a$, which is not constant. First of all, it is a function of a and α changes with range, so $L^2\sin^2\alpha/a$ is range variant. Secondly, the relative position between the antennas and target varies with the movement of the antennas; hence α and a are variant within one synthetic aperture time, so it is Doppler frequency variant. If the synthetic aperture length, L_{syn}, satisfies $L_{syn} \ll a$, and the range swath is quite narrow, $\alpha \approx \alpha_0$ can be valid, where α_0 represents α of the scene center at the central time of its synthetic aperture. Moreover, if $L_{syn} \ll a$, $L/a \approx L/a_0$ can be obtained, where a_0 is usually chosen to be the average of transmitting and receiving ranges to target at the scene center. Using these approximations, the bistatic SAR is equivalent to a monostatic SAR at the middle point of the baseline, only with an extra constant time delay $L^2\sin^2\alpha_0/(4a_0c)$ when receiving data (this can be imagined as the effect of a delay line in the receiver system), which is

$$|TP| + |RP| = 2a \approx 2|MP| + \frac{\sin^2\alpha_0}{4a_0}L^2 \tag{5.17}$$

Then the bistatic range, $R_{bi}(\eta)$, becomes a single square root, similar to that of the monostatic SAR:

$$R_{bi}(\eta) \approx 2\sqrt{(x - X_m)^2 + (Y_m + V\eta - y)^2 + (z - Z_m)^2} + \frac{\sin^2\alpha_0}{4a_0}L^2 \tag{5.18}$$

where $X_m = (X_{\mathrm{T}} + X_{\mathrm{R}})/2$, $Y_m = (Y_{\mathrm{T}} + Y_{\mathrm{R}})/2$ and $Z_m = (Z_{\mathrm{T}} + Z_{\mathrm{R}})/2$.

Based on above approximations and after compensating for the constant time delay, firstly (if the bistatic time delay is t_{dly}, the monostatic time delay by the baseline middle-point approximation should be altered to $t_{dly} - L^2 \sin^2 \alpha_0/(4a_0c)$), we can follow the monostatic SAR algorithms to process the bistatic SAR data. However, due to the simplifications used in Equation (5.15) and the range and azimuth invariant assumptions of α and a, the approximating precision of Equation (5.18) is not very good:

1. Firstly, the baseline middle-point approximation results in constant error, especially when the target is far from the scene center (i.e., $a \neq a_0$). It makes the value of $|MP|$ calculated according to the altered propagation time severely deviate from the actual value of $|MP|$ and thus leads to a great higher-order error of the range history. However, since the monostatic algorithms have a phase preservation property, that is, the peak constant phase is not destroyed, the corresponding bistatic algorithms based on the monostatic approximation also have a phase preservation property.
2. Secondly, since the Doppler frequency variance of α and a is ignored, the baseline middle-point approximation induces linear and higher-order azimuth phase errors, which further impair the image quality. The capability of this approach is investigated by a simplified analysis of the target at the scene center. In order to avoid significant deterioration of the image quality, the phase error induced by a higher approximating error (mainly the quadratic error) should be kept within $\pi/4$, which is

$$\frac{\sin^2 \alpha_0}{4} L^2 \left| \frac{1}{a_{0c}} - \frac{1}{a_{0syn}} \right| < \frac{\lambda}{16} \tag{5.19}$$

where a_{0c} is the average of transmitting and receiving ranges of the scene center at the central time of its synthetic aperture and $a_{0\mathrm{syn}}$ is the average of transmitting and receiving ranges of the scene center at the edges of its synthetic aperture time. Let D be the antenna size in the azimuth direction and L_{syn} be the synthetic aperture length; then we have

$$|a_{0syn} - a_{0c}| \approx \frac{(L_{syn}/2)^2}{2a_{0c}} \approx \frac{(a_{0c}\lambda/2D)^2}{2a_{0c}} = \frac{a_{0c}\lambda^2}{8D^2} \tag{5.20}$$

Accordingly, Equation (5.19) approximately turns into

$$\frac{L^2 \sin^2 \alpha_0}{a_{0c}^2} < \frac{2D^2}{\lambda a_{0c}} \tag{5.21}$$

Taking airborne SAR, for instance, in which the antenna length is 0.3 m, a_{0c} is 10 km approximately and the wavelength is 0.03 m; then $L \sin \alpha_0 / a_{0c} < 0.024$ should be satisfied, that is, a_{0c} should be larger than L by a few orders of magnitude. Hence, for the long baseline case, if the above condition is not satisfied, a more precise imaging algorithm must be adopted.

5.3.2 *Hyperbolic Approximation*

As discussed in Section 5.3.1, the bistatic algorithm based on the baseline middle-point monostatic SAR approximation is simple, because the monostatic SAR algorithms can be completely followed by altering the time delay. Though this method is suitable in short baseline cases, it results in severely unfocused images in long baseline cases, for example, when the baseline is comparable to the average bistatic range. Thus, we wonder if there is another approximation method that enables the quasi-monostatic algorithms and also achieves higher precision. The hyperbolic approximation [13, 45, 46] has these advantages. By introducing three equivalent variables, that is, the equivalent slant range, equivalent velocity and equivalent squint angle, the flat-top hyperbolic bistatic range history is approximated by a hyperbolic one and has higher approximation precision than the baseline middle-point equivalent. In this section, the hyperbolic approximation is introduced first and then proper quasi-monostatic algorithms based on the hyperbolic approximation are described.

5.3.2.1 Method of Hyperbolic Approximation

As shown in Figure 5.8, the velocities of transmitted and received antennas are V_T and V_R in the direction of the *y* axis and are related by $V_T = V_R = V$ in the translational invariant case. The transmitted and received beam footprints determine the imaging area together. For a given target $P(x_p, y_p, z_p)$ in the illuminating area, the central time of its synthetic aperture, η_{pc}, can be obtained according to the composite bistatic

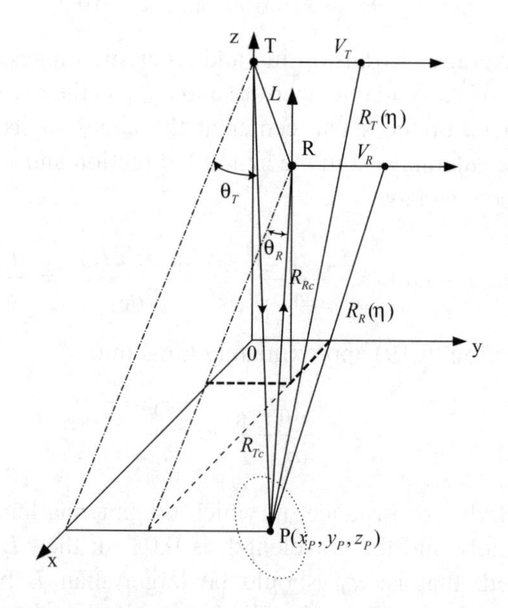

Figure 5.8 Imaging geometry of bistatic SAR in the translational invariant mode

beam. At this time, the transmitted and received ranges to the target are denoted by R_{Tc} and R_{Rc} and the squint angles are denoted by θ_T and θ_R. Then at the moment of η, the transmitted and received ranges to the target P are given by

$$R_T(\eta) = \sqrt{R_{Tc}^2 + V_T^2(\eta - \eta_{pc})^2 - 2R_{Tc}V_T(\eta - \eta_{pc})\sin(\theta_T)} \tag{5.22}$$

$$R_R(\eta) = \sqrt{R_{Rc}^2 + V_R^2(\eta - \eta_{pc})^2 - 2R_{Rc}V_R(\eta - \eta_{pc})\sin(\theta_R)} \tag{5.23}$$

The bistatic range is

$$R_{bi}(\eta) = R_T(\eta) + R_R(\eta) \tag{5.24}$$

In order to adopt the derivation of monostatic SAR algorithms, $R_{bi}(\eta)$ is approximated by a hyperbolic form as follows:

$$R_{bi}(\eta) \approx 2R_{mono}(\eta) = 2\sqrt{R_{mc}^2 + V_m^2(\eta - \eta_{pc})^2 - 2R_{mc}V_m(\eta - \eta_{pc})\sin(\theta_m)} \tag{5.25}$$

where R_{mc}, V_m and θ_m are variables waiting to be resolved. Taking the Taylor series expansion of $R_T(\eta)$, $R_R(\eta)$ and $R_{mono}(\eta)$ at $\eta = \eta_{pc}$, we get

$$\begin{aligned} R_T(\xi) = R_{Tc} - V_T\xi\sin\theta_T + \frac{V_T^2\xi^2\cos^2\theta_T}{2R_{Tc}} + \frac{V_T^3\xi^3\sin\theta_T\cos^2\theta_T}{2R_{Tc}^2} \\ + \frac{V_T^4\xi^4\cos^2\theta_T(5\sin^2\theta_T - 1)}{8R_{Tc}^3} + o\left(\frac{V_T^4\xi^4}{R_{Tc}^3}\right) \end{aligned} \tag{5.26}$$

$$\begin{aligned} R_R(\xi) = R_{Rc} - V_R\xi\sin\theta_R + \frac{V_R^2\xi^2\cos^2\theta_R}{2R_{Rc}} + \frac{V_R^3\xi^3\sin\theta_R\cos^2\theta_R}{2R_{Rc}^2} \\ + \frac{V_R^4\xi^4\cos^2\theta_R(5\sin^2\theta_R - 1)}{8R_{Rc}^3} + o\left(\frac{V_R^4\xi^4}{R_{Rc}^3}\right) \end{aligned} \tag{5.27}$$

$$\begin{aligned} R_{mono}(\xi) = R_{mc} - V_m\xi\sin\theta_m + \frac{V_m^2\xi^2\cos^2\theta_m}{2R_{mc}} + \frac{V_m^3\xi^3\sin\theta_m\cos^2\theta_m}{2R_{mc}^2} \\ + \frac{V_m^4\xi^4\cos^2\theta_m(5\sin^2\theta_m - 1)}{8R_{mc}^3} + o\left(\frac{V_m^4\xi^4}{R_{mc}^3}\right) \end{aligned} \tag{5.28}$$

where $\xi = \eta - \eta_{pc}$. Equating the first three order terms of Equation (5.25), we have

$$R_{Tc} + R_{Rc} = 2R_{mc} \tag{5.29}$$

$$V_T \sin\theta_T + V_R \sin\theta_R = 2V_m \sin\theta_m \tag{5.30}$$

$$\frac{V_T^2 \cos^2\theta_T}{2R_{Tc}} + \frac{V_R^2 \cos^2\theta_R}{2R_{Rc}} = 2\frac{V_m^2 \cos^2\theta_m}{2R_{mc}} \tag{5.31}$$

Then the three variables can be resolved:

$$R_{mc} = (R_{\mathrm{Tc}} + R_{\mathrm{Rc}})/2 \tag{5.32}$$

$$V_m = \sqrt{A^2 + B} \tag{5.33}$$

$$\theta_m = \arcsin(A/V_m) \tag{5.34}$$

where

$$A = (V_T \sin\theta_T + V_R \sin\theta_R)/2 \tag{5.35}$$

$$B = \left(\frac{V_T^2 \cos^2\theta_T}{R_{Tc}} + \frac{V_R^2 \cos^2\theta_R}{R_{Rc}}\right) R_{mc}/2 \tag{5.36}$$

It can be seen that the equivalent velocity V_m can only be range invariant when the transmitter and the receiver coincide (i.e., bistatic SAR degrades to monostatic SAR). In another case, even in the tandem configuration when $\theta_T = -\theta_R$, the equivalent velocity is dependent on the range via $\cos\theta_R$ (i.e., $\cos\theta_T$).

As can be seen from the derivation of these equivalent variables, the hyperbolic approximation can equalize the constant, linear and quadratic terms of the flat-top hyperbola. The approximation error is analyzed below. Note in Equation (5.26) that the phase induced by the higher-order terms of the transmitting range history within the synthetic aperture time, T_{syn}, can be written by

$$\phi_{Terro3} = -\frac{2\pi}{\lambda} O\left(\frac{V^3 T_{syn}^3}{R_{Tc}^2}\right) = -\frac{2\pi}{\lambda} O\left(\frac{V^3}{R_{Tc}^2}\left(\frac{\lambda R_{Tc}}{D_a V}\right)^3\right) = O\left(\frac{\lambda^2 R_{Tc}}{D_a^3}\right) \tag{5.37}$$

This shows that the higher-order phase is directly proportional to λ^2 and R_{Tc} but inversely proportional to the cube of the antenna size in the azimuth direction, D_a. Hence, if the wavelength becomes shorter, the flying height becomes lower and if the azimuth resolution becomes coarser, the higher-order phase becomes smaller. In general, the smaller the higher order phase, the smaller is the approximation error. Consequently, the hyperbolic approximation usually works well for airborne SAR with high frequency and moderate resolution.

In squint cases, the cubic term is usually predominant in the approximation error. Hence, the cubic error term is considered here, which is expressed as follows:

$$R_{err3}(\xi) = \frac{V_T^3\xi^3 \sin\theta_T \cos^2\theta_T}{2R_{Tc}^2} + \frac{V_R^3\xi^3 \sin\theta_R \cos^2\theta_R}{2R_{Rc}^2} - 2\frac{V_m^3\xi^3 \sin\theta_m \cos^2\theta_m}{2R_{mc}^2} \tag{5.38}$$

By means of Equations (5.29) to (5.31), Equation (5.38) can be simplified as follows:

$$R_{err3}(\xi) = \frac{R_{Tc}V_R \sin\theta_R - R_{Rc}V_T \sin\theta_T}{R_{Tc} + R_{Rc}} \left(\frac{V_R^2 \cos^2\theta_R}{2R_{Rc}^2} - \frac{V_T^2 \cos^2\theta_T}{2R_{Tc}^2} \right) \xi^3 \tag{5.39}$$

This shows that, if $R_{Tc}V_R \sin\theta_R = R_{Rc}V_T \sin\theta_T$ or $R_{Tc}^2 V_R^2 \cos^2\theta_R = R_{Rc}^2 V_T^2 \cos^2\theta_T$ is satisfied, the cubic approximation error reduces to zero.

Some simulations have been done to testify the hyperbolic approximation. The simulation results indicate that the hyperbolic approximation not only completely equalizes the constant, linear and quadratic terms of the DSR bistatic range history but also partly compensates higher-order terms (as illustrated by Figure 5.9(a)). Thus, the hyperbolic approximation is more precise than the baseline middle-point approximation. Under the simulation parameters listed in Table 5.1, the approximated range error of the scene center within a synthetic aperture time is illustrated in Figure 5.9(b), where the maximum range error is of the order of magnitude of 10^{-3} m. The azimuth profile compressed by the match filter based on the hyperbolic approximation is shown in Figure 5.9(c), where only slight dissymmetry of the sidelobes can be seen. By contrast, if the baseline middle-point approximation is used, the approximated range error is shown in Figure 5.10(a) and the compressed azimuth profile is shown in Figure 5.10(b), where the signal severely disperses in the azimuth direction. The simulation results prove that the hyperbolic approximation has a significant advantage in the long baseline case compared with the baseline middle-point approximation.

Furthermore, we will explain the three equivalent variables, which are derived by the algebraic operation above, from a relatively intuitional point of view. As illustrated in Figure 5.11, the velocity vectors $\boldsymbol{V}$ of transmitted and received antennas can be decomposed into two components respectively in the transmitting and receiving slant range planes, which are the velocity vector parallel to the slant range vector $\boldsymbol{V}_{//}$ (named by the translational component) and the velocity vector perpendicular to the slant range vector $\boldsymbol{V}_{\perp}$ (named by the rotational component)

$$\begin{cases} V_{//T} = V\sin\theta_T \\ V_{\perp T} = V\cos\theta_T \end{cases}, \quad \begin{cases} V_{//R} = V\sin\theta_R \\ V_{\perp R} = V\cos\theta_R \end{cases} \tag{5.40}$$

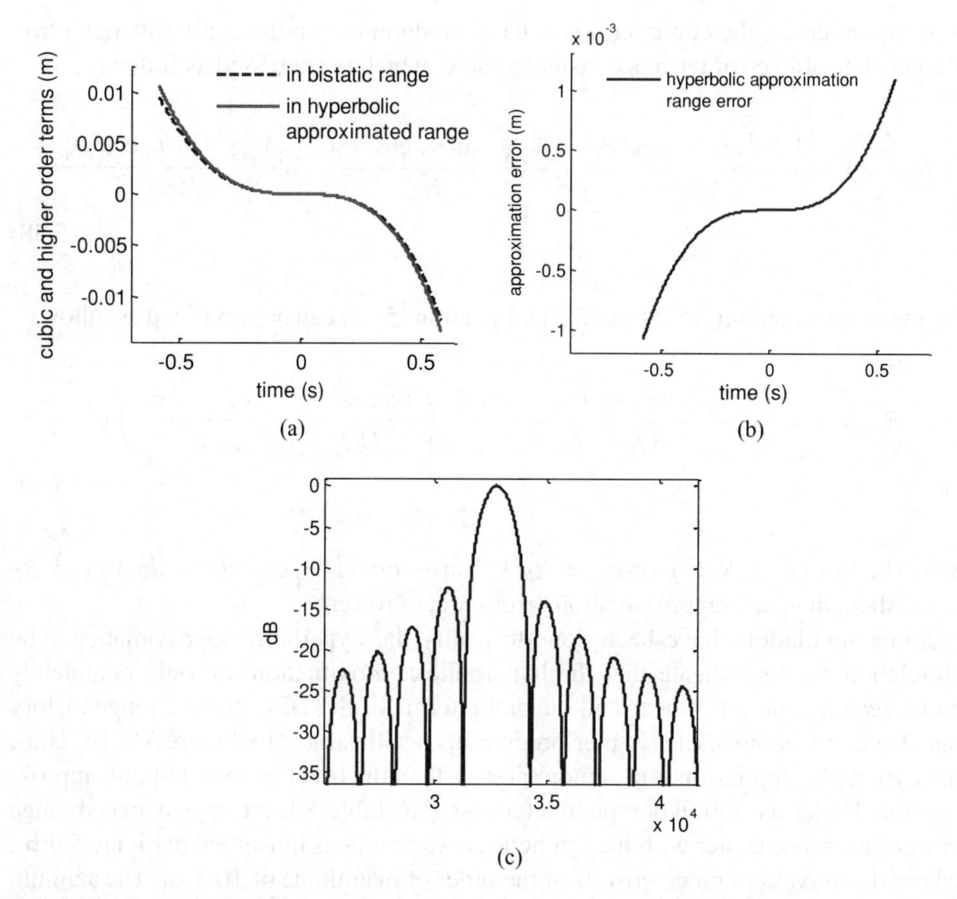

Figure 5.9 Properties of the hyperbolic approximation: (a) cubic and higher-order terms, (b) approximation error, (c) compressed azimuth profile

Table 5.1 Simulation parameters of the translational invariant bistatic SAR

Parameter name	Value	Units
Carrier frequency	9.6	GHz
Transmitted pulse bandwidth	500	MHz
Transmitted pulse duration	1	μs
Range sampling rate	550	MHz
Original position of transmitting antenna	(−1616.0, 0, 2799.1)	m
Original position of receiving antenna	(−2578.6, 0, 1549.4)	m
Antenna size in azimuth direction	0.35	m
Flying velocity (T&R)	200	m/s
Pulse repetition frequency	1371.43	Hz
Azimuth processing bandwidth	937.14	Hz

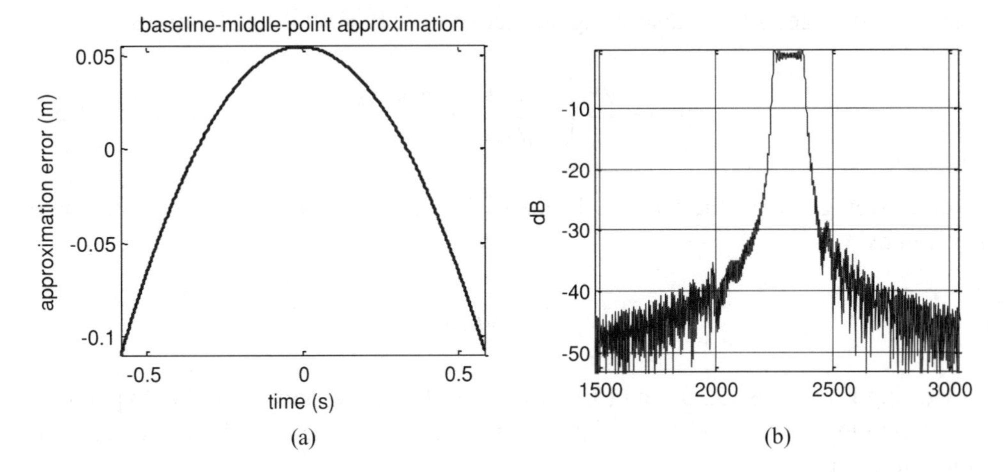

Figure 5.10 Properties of the hyperbolic approximation: (a) quadratic and higher-order approximated error, (b) compressed azimuth profile

Supposing that the rotational velocity is caused by centripetal force, we have

$$F_T = ma_T = mV_{\perp T}^2 / R_{Tc} \tag{5.41}$$

where m is the gravitational mass of an imaginary particle. Thus the centripetal acceleration is

$$a_T = V_{\perp T}^2 / R_{Tc} \tag{5.42}$$

Similarly,

$$a_R = V_{\perp R}^2 / R_{Rc} \tag{5.43}$$

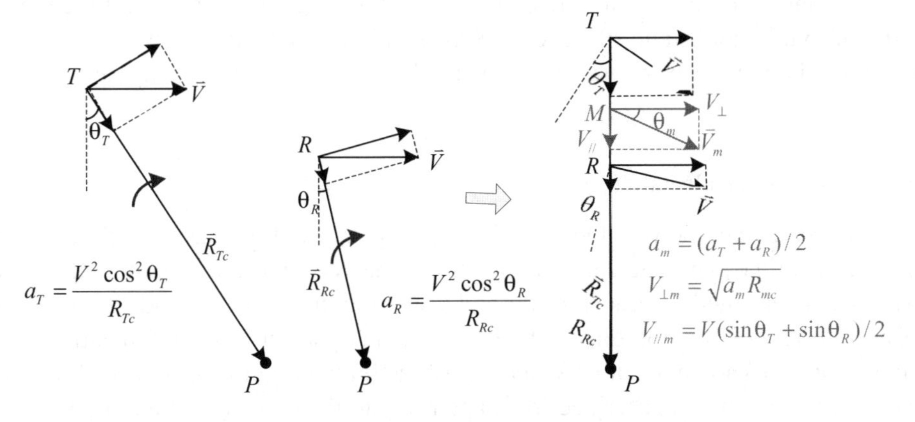

Figure 5.11 Intuitional explanation of the hyperbolic approximation

Then, the average of the transmitting and receiving centripetal accelerations is

$$a_m = \frac{V^2}{2}\left(\frac{\cos^2\theta_T}{R_{Tc}} + \frac{\cos^2\theta_R}{R_{Rc}}\right) \tag{5.44}$$

Over an average range, the rotational velocity induced by the average centripetal force is given by

$$V_{\perp m} = \sqrt{a_m R_{mc}} \tag{5.45}$$

Reviewing Equations (5.35) and (5.36), we find that the variable B in the hyperbolic approximation is $V_{\perp m}^2$ and the variable A is equal to the average of the translational velocities, that is,

$$V_{//m} = (V_{//T} + V_{//R})/2 \tag{5.46}$$

Therefore, the hyperbolic approximation can be explained as follows: it makes the bistatic SAR geometry approximate to some monostatic geometry for a given target. At the center of the target's synthetic aperture time, the equivalent slant range from monostatic SAR to the target is equal to the average of transmitting and receiving ranges in bistatic geometry; the equivalent velocity is the composition of a translational velocity and a rotational velocity, where the translational component is equal to the average of the transmitting and receiving translational velocities, and the rotational component is equal to the rotational velocity at the equivalent monostatic slant range caused by the average of transmitting and receiving centripetal forces; the equivalent squint angle is the complement of the included angle between the equivalent velocity vector and the equivalent slant range vector.

It should be noticed that, in the hyperbolic approximation of the bistatic range, both the equivalent velocity V_m and the equivalent squint angle θ_m vary with range. This is similar to the monostatic spaceborne SAR result, though the reasons are different. Therefore, the range variant variables need to be considered in the imaging process.

5.3.2.2 Variables Calculation

As discussed above, the variables θ_T, R_{Tc}, θ_R, R_{Rc}, θ_m, and R_{mc} are all defined at the central time of the synthetic aperture, η_{pc}. Since the synthetic aperture time in bistatic SAR is codetermined by transmitting and receiving antennas, θ_T and θ_R of targets at different ranges may be different. If the antenna patterns, antenna positions and attitudes are all known accurately, the η_{pc} of each target can be obtained through a space–time diagram, as described in Chapter 3, and then the corresponding θ_T, R_{Tc}, θ_R and R_{Rc} can be obtained.

However, the attitude measurements are not accurate enough in some cases; therefore, some other estimation methods are adopted to acquire more precise values of the Doppler centroid frequency f_{dc} (namely, the instantaneous Doppler frequency at η_{pc}) and the azimuth FM rate f_r. As a result, how to calculate the equivalent variables from f_{dc} and f_r for each range gate, and then implement the hyperbolic approximation introduced in the last section, is a practical problem of the hyperbolic approximation method. Besides, the variables (θ_T, R_{Tc}, θ_R and R_{Rc}) need to be calculated for geocoding after focusing. Thus, in the following text, the calculation methods of these variables based on known f_{dc} and f_r will be shown.

If the space interval between the transmitter and receiver is given by $(\Delta x_b, \Delta y_b, \Delta z_b)^T$, then, when the transmitting antenna is located at T, the received antenna is located at $T + (\Delta x_b, \Delta y_b, \Delta z_b)^T$. In the translational invariant mode, the space interval is constant. For a given range gate, the bistatic range R_{bi} can be precisely figured out according to the time delay t_{bi}, which is

$$R_{bi} = R_{Tc} + R_{Rc} = ct_{bi} \tag{5.47}$$

The Doppler parameters of bistatic SAR have the formulas as follows:

$$f_{dc} = \frac{V}{\lambda}(\sin\theta_T + \sin\theta_R) \tag{5.48}$$

$$f_r = -\frac{V^2}{\lambda}\left(\frac{\cos^2\theta_T}{R_{Tc}} + \frac{\cos^2\theta_R}{R_{Rc}}\right) \tag{5.49}$$

According to the imaging geometry, we have

$$\sin\theta_R = \frac{R_{Tc}\sin\theta_T - \Delta y_b}{R_{bi} - R_{Tc}} \tag{5.50}$$

Substituting Equation (5.50) into Equation (5.48) and setting $D = \lambda f_{dc}/V$, we have

$$\sin\theta_T = \frac{\Delta y_b + D(R_{bi} - R_{Tc})}{R_{bi}} \tag{5.51}$$

$$\sin\theta_R = \frac{DR_{Tc} - \Delta y_b}{R_{bi}} \tag{5.52}$$

Using Equation (5.51) and Equation (5.52), we can transform Equation (5.49) into

$$(R_{bi} - R_{Tc})\left\{1 - \frac{[\Delta y_b + D(R_{bi} - R_{Tc})]^2}{R_{bi}^2}\right\} + R_{Tc}\left\{1 - \frac{(DR_{Tc} - \Delta y_b)^2}{R_{bi}^2}\right\}$$
$$+\frac{\lambda f_r}{V^2}R_{Tc}(R_{bi} - R_{Tc}) = 0 \tag{5.53}$$

This expression can be simplified into a quadratic equation of the unknown variable R_{Tc}, which is

$$P_2 R_{Tc}^2 + P_1 R_{Tc} + P_0 = 0 \tag{5.54}$$

where

$$P_2 = -\frac{\lambda f_r}{V^2} - \frac{3D^2}{R_{bi}} \tag{5.55}$$

$$P_1 = 4Dy_R + 3D^2 + \frac{\lambda f_r}{V^2} R_{bi} \tag{5.56}$$

$$P_0 = (1 - D^2)R_{bi} - y_R^2 - 2DR_{bi}y_R \tag{5.57}$$

By finding out the root of this quadratic equation, R_{Tc} can be obtained. Then, the other variables θ_T, θ_R and R_{Rc} can also be calculated by Equations (5.51), (5.52) and (5.49).

On the other hand, after hyperbolic approximation, the relations between the Doppler parameters and the equivalent parameters V_m and θ_m are as follows:

$$f_{dc} = \frac{2V_m}{\lambda} \sin\theta_m \tag{5.58}$$

$$f_r = -\frac{2V_m^2}{\lambda R_{m,c}} \cos^2\theta_m \tag{5.59}$$

where V_m and θ_m in the expressions are as follows:

$$V_m = \sqrt{\left(\frac{\lambda f_{dc}}{2}\right)^2 - \frac{\lambda R_{mc} f_r}{2}} \tag{5.60}$$

$$\theta_m = \arcsin\left(\frac{\lambda f_{dc}}{2V_m}\right) \tag{5.61}$$

5.3.2.3 Quasi-Monostatic Algorithms Based on the Hyperbolic Approximation

The hyperbolic approximation changes the DSR bistatic range equation into a single square root form, which is the same as the monostatic range equation form, and, accordingly, the bistatic algorithms can be derived following the derivations of the monostatic SAR algorithms. However, since the equivalent parameters vary with range, the bistatic algorithms need careful consideration of this range variance (usually, this range variance is more severe than that of monostatic spaceborne SAR induced by the curved orbit and the earth's rotation). In the following text, the RD, CS, nonlinear CS (NLCS) and ω-k algorithms are derived one by one, and the implementation details are clarified.

RD Algorithm Based on the Hyperbolic Approximation

Supposing the linear FM signal to be the transmitted signal of SAR, the received signal of the target at (R_{mc}, η_{pc}) after the hyperbolic approximation can be written as

$$s(t, \eta; R_{mc}, \eta_{pc}) \approx \sigma_p \text{rect}\left[\frac{t - 2R_m(\eta)/c + t_{dly}}{T}\right] \times \exp\{j\pi b[t - 2R_m(\eta)/c + t_{dly}]^2\} \exp\left\{-j\frac{4\pi R_m(\eta)}{\lambda}\right\} \quad (5.62)$$

where b is the FM rate of the linear FM signal, t_{dly} is the sample time delay and $t \in [0, R_{\text{size}}/f_s)$, where R_{size} is the number of range samples. In order to make the definition of range time accord with the convention of digital signal processing, the origin of t is not set at the time of signal transmission but at the time of the first sample of the received signal. By applying the POSP to evaluate the Fourier transform with respect to t and η, the two-dimensional spectrum can be obtained as follows [47]:

$$S(f, f_\eta; R_{mc}, \eta_{pc}) = \sigma_p \exp\{j2\pi f t_{dly}\} \exp\{-j2\pi f_\eta (R_{mc} \sin\theta_m / V_m + \eta_{pc})\} \times \exp\{-j\pi f^2/b\} \exp\left\{-jR_{mc}\cos\theta_m \sqrt{\left[\frac{2\pi(f + f_0)}{c/2}\right]^2 - \left(\frac{2\pi f_\eta}{V_m}\right)^2}\right\} \quad (5.63)$$

By expanding the square root term in the above phase expression around $f = 0$ and keeping terms up to f^3, Equation (5.63) becomes

$$\begin{aligned} S(f, f_\eta; R_{mc}, \eta_{pc}) = {} & \sigma_p \exp\{j2\pi f t_{dly}\} \exp\{-j\pi f^2/b\} \\ & \times \exp\{-j2\pi f_\eta (R_{mc} \sin\theta_m / V_m + \eta_{pc})\} \\ & \times \exp\left\{-j\frac{4\pi}{\lambda} R_{mc} \cos\theta_m \sqrt{1 - (\lambda f_\eta / 2V_m)^2}\right\} \text{ (zeroth-order term)} \\ & \times \exp\left\{-j\frac{4\pi f}{c} \frac{R_{mc}\cos\theta_m}{\sqrt{1 - (\lambda f_\eta/2V_m)^2}}\right\} \text{ (linear term)} \\ & \times \exp\left\{j\frac{2\pi f^2}{c f_0} R_{mc}\cos\theta_m \frac{(\lambda f_\eta/2V_m)^2}{[\sqrt{1 - (\lambda f_\eta/2V_m)^2}]^3}\right\} \text{ (quadratic term)} \\ & \times \exp\left\{-j\frac{2\pi f^3}{c f_0^2} R_{mc}\cos\theta_m \frac{(\lambda f_\eta/2V_m)^2}{[\sqrt{1 - (\lambda f_\eta/2V_m)^2}]^5}\right\} \text{ (cubic term)} \end{aligned} \quad (5.64)$$

1. The zeroth-order phase term is the azimuth phase, which only depends on the azimuth frequency. Thus the azimuth matched filter is designed as

$$H_a(f_\eta; R_{mc}) = \exp\left\{ j\frac{4\pi}{\lambda} R_{mc} \left[\cos\theta_m \sqrt{1-(\lambda f_\eta/2V_m)^2} - 1 \right] + j2\pi f_\eta \frac{R_{mc}\sin\theta_m}{V_m} \right\} \tag{5.65}$$

2. The linear phase term of f determines the signal position after inverse Fourier transform with respect to f (i.e., determines the RCM). Thus the amount of RCM needing to be corrected is as follows:

$$\Delta R_{RCM}(f_\eta; R_{mc}) = \frac{R_{mc}\cos\theta_m}{\sqrt{1-(\lambda f_\eta/2V_m)^2}} - R_{m,c} \tag{5.66}$$

3. The quadratic phase term of f will cause defocusing in the range dimension and thus needs to be compensated for. The compensation of this quadratic phase term is usually named as the second range compression (SRC). If the range variance of R_{mc}, θ_m and V_m is ignored, the SRC can be implemented in the two-dimensional frequency domain, for which the corresponding matched filter is

$$H_{src}(f, f_\eta) = \exp\left\{ -j\frac{2\pi f^2}{c f_0} R_{mc\_ref}\cos\theta_{m\_ref} \frac{(\lambda f_\eta/2V_{m\_ref})^2}{[\sqrt{1-(\lambda f_\eta/2V_{m\_ref})^2}]^3} \right\} \tag{5.67}$$

where R_{mc_ref}, θ_{m_ref} and V_{m_ref} correspond to the reference range (generally at the scene center).

4. The cubic phase term of f induces sidelobe raising and dissymmetry of the compressed range profile. When the squint angle is large, this phase term usually becomes large and should be compensated in order to refine the image quality. In the same way, if the range variance of R_{mc}, θ_m and V_m is ignored, the following matched filter can be used in the two-dimensional frequency domain to accomplish the phase compensation:

$$H_{3err}(f, f_\eta) = \exp\left\{ j\frac{2\pi f^3}{c f_0^2} R_{mc\_ref}\cos\theta_{m\_ref} \frac{(\lambda f_\eta/2V_{m\_ref})^2}{[\sqrt{1-(\lambda f_\eta/2V_{m\_ref})^2}]^5} \right\} \tag{5.68}$$

In the above RD algorithm, the azimuth compression and RCMC operations are implemented in the range-Doppler domain, where they can accommodate the range variance of the three equivalent parameters. The SRC and cubic phase compensation operations are implemented in the two-dimensional frequency domain, where they cannot accommodate the range variance and thus present an error to a certain extent. To deal with this problem, a more precise and complicated scaling function has to be adopted in the time domain, which will be introduced later with the NLCS algorithm.

CS Algorithm Based on the Hyperbolic Approximation

The CS algorithm [48–52] is a widely used algorithm in monostatic SAR, because it can avoid the interpolation and hence enhance the efficiency. Based on the hyperbolic approximation, the CS algorithm, which is suitable for bistatic SAR imaging, can also be derived, the key expressions of which are shown below.

Reviewing the two-dimensional spectrum in Equation (5.64), if the range variance of the three equivalent parameters in the cubic phase term is ignored, the cubic phase compensation function is identical with Equation (5.68). After the cubic phase compensation, considering the left linear and quadratic terms, the inverse Fourier transform with respect to f takes the signal into

$$\begin{aligned} S_\eta(t', f_\eta; R_{mc}, \eta_{pc}) = \sigma_p \mathrm{rect}\left[\frac{t' - t_m(f_\eta; R_{mc})}{T}\right] \\ \times \exp\left\{-j2\pi f_\eta (R_{mc}\sin\theta_m / V_m + \eta_{pc})\right\} \\ \times \exp\left\{j\pi b_r(f_\eta; R_{mc})[t' - t_m(f_\eta; R_{mc})]^2\right\} \\ \times \exp\left\{-j\frac{4\pi}{\lambda} R_{mc}\cos\theta_m \sqrt{1 - (\lambda f_\eta / 2V_m)^2}\right\} \end{aligned} \tag{5.69}$$

where $t' = t + t_{dly}$ and

$$b_r(f_\eta; R_{mc}) = \frac{b}{1 - \dfrac{2b}{cf_0} R_{mc}\cos\theta_m \dfrac{(\lambda f_\eta/2V_m)^2}{[\sqrt{1-(\lambda f_\eta/2V_m)^2}]^3}} \tag{5.70}$$

$$t_m(f_\eta; R_{mc}) = 2R_{mc}\left[1 + C_s(f_\eta; R_{mc})\right] \big/ c \tag{5.71}$$

$$C_s(f_\eta; R_{mc}) = \frac{\cos\theta_m}{\sqrt{1 - (\lambda f_\eta/2V_m)^2}} - 1 \tag{5.72}$$

It can be seen that the range migration is $R_{mc}C_s(f_\eta; R_{mc})$, which varies with range. If the range variance of V_m in C_s and the dependence of $b_r(f_\eta; R_{mc})$ on R_{mc} are ignored, we have

$$\frac{1}{\sqrt{1-(\lambda f_\eta/2V_m)^2}} \approx \frac{1}{\sqrt{1-(\lambda f_\eta/2V_{m\_ref})^2}} \tag{5.73}$$

$$b_r(f_\eta; R_{mc}) \approx b_r(f_\eta; R_{mc\_ref}) = b_{r\_ref}(f_\eta) \tag{5.74}$$

Thus, the differential RCM can be corrected by multiplication with the scaling function, $H_{cs}(t', f_\eta)$, which is

$$H_{cs}(t', f_\eta) = \exp\left\{j\pi b_{r\_ref}(f_\eta) C_{s\_ref}(f_\eta)\left[t' - \frac{2}{c}R_{mc\_ref}\left(1 + C_{s\_ref}(f_\eta)\right)\right]^2\right\} \tag{5.75}$$

Then, the signal in Equation (5.69) turns out to be

$$
\begin{aligned}
S_{\eta cs}(t', f_\eta; R_{mc}, \eta_{pc}) = \sigma_p \text{rect}\left[\frac{t' - t_m(f_\eta; R_{mc})}{T}\right] \\
\times \exp\{-j2\pi f_\eta(R_{mc}\sin\theta_m / V_m + \eta_{pc})\} \\
\times \exp\left\{ j\pi b_{r\_ref}(f_\eta)\left[1 + C_{s\_ref}(f_\eta)\right] \right. \\
\left. \times \left[t' - \frac{2R_{mc}}{c}\frac{\cos\theta_m}{\cos\theta_{m\_ref}} - \frac{2R_{mc\_ref}}{c}C_{s\_ref}(f_\eta)\right]^2 \right\} \\
\times \exp\left\{ j\frac{4\pi}{c^2} b_{r\_ref}(f_\eta) C_{s\_ref}(f_\eta)\left[1 + C_{s\_ref}(f_\eta)\right] \right. \\
\left. \times \left[R_{mc}\frac{\cos\theta_m}{\cos\theta_{m\_ref}} - R_{mc\_ref}\right]^2 \right\} \\
\times \exp\left\{ -j\frac{4\pi}{\lambda} R_{mc}\cos\theta_m\sqrt{1 - (\lambda f_\eta / 2V_m)^2} \right\}
\end{aligned}
\tag{5.76}
$$

The two-dimensional spectrum of Equation (5.76) is

$$
\begin{aligned}
S_{cs}(f, f_\eta; R_{mc}, \eta_{pc}) = \sigma_p \exp\{j2\pi f t_{dly}\} \exp\{-j2\pi f_\eta (R_{mc}\sin\theta_m / V_m + \eta_{pc})\} \\
\times \exp\left\{ -j\frac{4\pi f}{c}\left[R_{mc}\frac{\cos\theta_m}{\cos\theta_{m\_ref}} + R_{mc\_ref}C_{s\_ref}(f_\eta)\right] \right\} \\
\times \exp\left\{ -j\frac{\pi f^2}{b_{r\_ref}(f_\eta)[1 + C_{s\_ref}(f_\eta)]} \right\} \\
\times \exp\left\{ j\frac{4\pi}{c^2} b_{r\_ref}(f_\eta) C_{s\_ref}(f_\eta)\left[1 + C_{s\_ref}(f_\eta)\right] \right. \\
\left. \times \left[R_{mc}\frac{\cos\theta_m}{\cos\theta_{m\_ref}} - R_{mc\_ref}\right]^2 \right\} \\
\times \exp\left\{ -j\frac{4\pi}{\lambda} R_{mc}\cos\theta_m\sqrt{1 - (\lambda f_\eta / 2V_m)^2} \right\}
\end{aligned}
\tag{5.77}
$$

Thus, the range compression and RCMC can be accomplished by the following phase multiplication in the two-dimensional frequency domain:

$$
H_r(f, f_\eta) = \exp\left\{ j\frac{\pi f^2}{b_{r\_ref}(f_\eta)[1 + C_{s\_ref}(f_\eta)]} \right\} \exp\left\{ j\frac{4\pi f}{c} R_{mc\_ref} C_{s\_ref}(f_\eta) \right\}
\tag{5.78}
$$

In the end, the matched filter for azimuth compression is

$$\begin{aligned} H_a(f_\eta; R_{mc}) = \exp&\left\{ j\frac{4\pi}{\lambda}R_{mc}\left[\cos\theta_m\sqrt{1-(\lambda f_\eta/2V_m)^2}-1\right] + j2\pi f_\eta \frac{R_{mc}\sin\theta_m}{V_m}\right\} \\ &\times \exp\left\{ -j\frac{4\pi}{c^2}b_{r\_ref}(f_\eta)C_{s\_ref}(f_\eta)\left[1+C_{s\_ref}(f_\eta)\right]\right. \\ &\left.\times\left[R_{mc}\frac{\cos\theta_m}{\cos\theta_{m\_ref}} - R_{mc\_ref}\right]^2\right\} \end{aligned} \tag{5.79}$$

For the above CS algorithm, the range variance of V_m is ignored in differential RCMC (less precise than the RD algorithm in the last subsection from this aspect); meanwhile, the range variance of R_{mc}, θ_m, and V_m are also ignored in SRC and cubic phase compensation, but adapted in azimuth compression, which is comparable to the RD algorithm from these two aspects.

NLCS Algorithm Based on the Hyperbolic Approximation

As discussed above, both the RD and the CS algorithms cannot accommodate the dependence of SRC on R_{mc}, θ_m and V_m. Thus, the image quality of the targets at the edges of the swath worsens as the range swath extends, due to the range FM rate mismatch. Davidson [53] proposed a range NLCS technique for monostatic SAR to mitigate the range variance of the range FM rate. This concept is followed in this section to derive the bistatic NLCS algorithm and the practical details to be noticed are also pointed out.

The range FM rate is related to R_{mc}, θ_m and V_m by Equation (5.70), which is rewritten as follows:

$$b_r(f_\eta; R_{mc}) = \frac{b}{1 - \frac{2b}{cf_0}R_{mc}\cos\theta_m \frac{(\lambda f_\eta/2V_m)^2}{\left[\sqrt{1-(\lambda f_\eta/2V_m)^2}\right]^3}} \tag{5.80}$$

Combining Equation (5.71) and Equation (5.72), what can be seen from Equation (5.69) is that, because of the range migration, the middle position of the range FM signal envelope (which corresponds to the zero frequency) is given by

$$t_m(f_\eta; R_{mc}) = \frac{2}{c}\frac{R_{mc}\cos\theta_m}{\sqrt{1-(\lambda f_\eta/2V_m)^2}} \tag{5.81}$$

Thus, Equation (5.80) can be written as

$$b_r(f_\eta; R_{mc}) = \frac{b}{1 - \dfrac{b}{f_0} \dfrac{(\lambda f_\eta / 2V_m)^2}{1 - (\lambda f_\eta / 2V_m)^2} t_m(f_\eta; R_{mc})} \tag{5.82}$$

This means that the range variance of $b_r(f_\eta; R_{mc})$ with R_{mc} is mainly dominated by that of $t_m(f_\eta; R_{mc})$. By expanding $b_r(f_\eta; R_{mc})$ around $t_{m\_ref}(f_\eta)$ at a certain reference range, and keeping terms up to the linear one, $b_r(f_\eta; R_{mc})$ becomes

$$b_r(f_\eta; R_{mc}) = b_{r\_ref}(f_\eta) + k_s[t_m(f_\eta; R_{mc}) - t_{m\_ref}(f_\eta)] \tag{5.83}$$

where

$$b_{r\_ref}(f_\eta) = \frac{b}{1 - \dfrac{2b}{cf_0} R_{m,c\_ref} \cos\theta_{m\_ref} \dfrac{(\lambda f_\eta / 2V_{m\_ref})^2}{[\sqrt{1 - (\lambda f_\eta / 2V_{m\_ref})^2}]^3}} \tag{5.84}$$

$$t_{m\_ref}(f_\eta) = \frac{2}{c} \frac{R_{mc\_ref} \cos\theta_{m\_ref}}{\sqrt{1 - (\lambda f_\eta / 2V_{m\_ref})^2}} \tag{5.85}$$

$$k_s = \frac{2b_{r\_ref}^2(f_\eta)}{cf_0} \frac{(\lambda f_\eta / 2V_{m\_ref})^2}{1 - (\lambda f_\eta / 2V_{m\_ref})^2} \tag{5.86}$$

At a given azimuth frequency f_η, k_s denotes the variation slope of the range FM rate with respect to the range time t_m.

In order to compress the targets at different ranges with the same range matched filter, $b_r(f_\eta; R_{mc})$ should be identified. The nonlinear scaling operation in the time domain can be adopted for this purpose, but it would inevitably cause the nonlinear shift to target a position in the range direction (the related explanation will be given in Chapter 6), which would destroy the result of RCMC and deteriorate the quality of azimuth compression. Therefore, a cubic phase is multiplied in the frequency domain in the first place and the range shift thus induced can compensate for the counterpart induced by nonlinear scaling in the time domain. Then the range FM rates and RCMs at different ranges are both identified after NLCS operation.

The cubic phase function to be multiplied in the frequency domain is represented by

$$H_{Y3}(f, f_\eta) = H_{3err}(f, f_\eta) \exp\left\{ j \frac{2\pi}{3} Y_3(f_\eta) f^3 \right\} \tag{5.87}$$

where $H_{3err}(f, f_\eta)$ is denoted by Equation (5.68). Multiplied by the above cubic phase function, the two-dimensional point target spectrum in Equation (5.64) turns to

$$\begin{aligned} S(f, f_\eta; R_{mc}, \eta_{pc}) = \sigma_p \exp\{j2\pi f t_{dly}\} \\ \times \exp\{-j2\pi f_\eta(R_{mc}\sin\theta_m / V_m + \eta_{pc})\} \\ \times \exp\left\{-j\frac{4\pi}{\lambda}R_{mc}\cos\theta_m\sqrt{1-(\lambda f_\eta/2V_m)^2}\right\} \text{(zeroth-order term)} \\ \times \exp\left\{-j\frac{4\pi f}{c}\frac{R_{m,c}\cos\theta_m}{\sqrt{1-(\lambda f_\eta/2V_m)^2}}\right\} \text{(linear term)} \\ \times \exp\left\{-j\pi\frac{f^2}{b_r(f_\eta; R_{mc})}\right\} \text{(quadratic term)} \\ \times \exp\left\{j\frac{2\pi}{3}Y_3(f_\eta)f^3\right\} \text{(cubic term)} \end{aligned} \quad (5.88)$$

The above expression is obtained by ignoring the dependence of the cubic phase term on R_{mc}, θ_m and V_m. Moreover, the cubic term is generally small enough that its effect on the stationary point can be ignored. Simulation results show that, for a linear FM signal having a time bandwidth product near 2000, even though the maximum cubic phase is 1.5% of the maximum quadratic phase (i.e., the maximum cubic phase approaches 1336°), the phase error in the transform domain caused by ignoring the effect of the cubic term on the stationary point is still less than 45°. As a result, the signal after the inverse Fourier transform with respect to f becomes

$$\begin{aligned} S_\eta(t', f_\eta; R_{mc}, \eta_{pc}) = \sigma_p \mathrm{rect}\left[\frac{t' - t_m(f_\eta; R_{mc})}{T}\right] \\ \times \exp\{-j2\pi f_\eta(R_{mc}\sin\theta_m / V_m + \eta_{pc})\} \\ \times \exp\left\{-j\frac{4\pi}{\lambda}R_{mc}\cos\theta_m\sqrt{1-(\lambda f_\eta/2V_m)^2}\right\} \\ \times \exp\left\{j\pi b_{r\_ref}(f_\eta)\left[t' - t_m(f_\eta; R_{mc})\right]^2\right\} \\ \times \exp\left\{j\pi k_s\left[t_m(f_\eta; R_{mc}) - t_{m\_ref}(f_\eta)\right]\left[t' - t_m(f_\eta; R_{mc})\right]^2\right\} \\ \times \exp\left\{j\frac{2\pi}{3}Y_3(f_\eta)b^3_{r\_ref}(f_\eta)\left[t' - t_m(f_\eta; R_{mc})\right]^3\right\} \end{aligned} \quad (5.89)$$

The last term in this expression uses the approximation $b_r(f_\eta; R_{m,c}) \approx b_{r\_ref}(f_\eta)$.

Afterwards, the following NLCS function is adopted for time domain scaling:

$$H_{NLCS}(f_\eta, t) = \exp\left\{j\pi p(f_\eta)\left[t - t_{m\_ref}(f_\eta)\right]^2 + j\frac{2\pi}{3}q(f_\eta)\left[t - t_{m\_ref}(f_\eta)\right]^3\right\} \tag{5.90}$$

For the purpose of identifying the range FM rate and correcting the differential RCM, the signal after multiplying $S_\eta(t', f_\eta; R_{mc}, \eta_{pc})$ by $H_{NLCS}(f_\eta, t')$ must satisfy the following two restrictions:

1. The RCM trajectory of a given target is identical with that of the reference target, that is, the zero frequency position (where the first derivative of phase term is zero) of the range FM signal with respect to f_η of both the given target and the reference target are identical. To simplify the following derivations and expressions, let

$$\alpha(f_\eta) = \frac{\cos\theta_d}{\sqrt{1 - (\lambda f_\eta / 2V_{m\_ref})^2}} \tag{5.91}$$

Let the zero frequency position of the range signal locate at

$$t_{m,new}(f_\eta; R_{mc}) = \frac{2}{c}\frac{R_{mc}\cos\theta_m}{\cos\theta_d} + \frac{2}{c}\frac{R_{mc\_ref}\cos\theta_{m\_ref}}{\cos\theta_d}[\alpha(f_\eta) - 1] \tag{5.92}$$

that is, $f_\eta = 2V_{m\_ref}\sin\theta_d/\lambda$ is chosen as the reference frequency for RCMC. The choice of θ_d will be interpreted later.

2. The range FM rate of any target reduces to a constant, that is, the second derivative of the phase term of $S_\eta(t', f_\eta; R_{mc}, \eta_{pc})H_{NLCS}(f_\eta, t')$ at $t_{m,new}$ is a constant.

According to restriction (1), we have

$$\begin{aligned} &b_{r\_ref}(f_\eta)\left[1 - \alpha(f_\eta)\right]\Delta t + p(f_\eta)\Delta t \\ &\quad + k_s\alpha(f_\eta)\left[1 - \alpha(f_\eta)\right]\Delta t^2 + Y_3(f_\eta)b^3_{r\_ref}(f_\eta)\left[1 - \alpha(f_\eta)\right]^2\Delta t^2 + q(f_\eta)\Delta t^2 \equiv 0 \end{aligned} \tag{5.93}$$

where

$$\Delta t = \frac{2}{c}\left[\frac{R_{mc}\cos\theta_m}{\cos\theta_d} - \frac{R_{mc\_ref}\cos\theta_{m\_ref}}{\cos\theta_d}\right] \tag{5.94}$$

To satisfy (5.93), a set of two equations in $p(f_\eta)$, $q(f_\eta)$ and $Y_3(f_\eta)$ are given as follows:

$$\begin{cases} b_{r\_ref}(f_\eta)\left[1-\alpha(f_\eta)\right]+p(f_\eta)=0 & (1) \\ k_s\alpha(f_\eta)\left[1-\alpha(f_\eta)\right]+Y_3(f_\eta)b^3_{r\_ref}(f_\eta)\left[1-\alpha(f_\eta)\right]^2+q(f_\eta)=0 & (2) \end{cases} \tag{5.95}$$

According to restriction (2), we have

$$\begin{aligned} &b_{r\_ref}(f_\eta)+p(f_\eta)+k_s\alpha(f_\eta)\Delta t+2Y_3(f_\eta)b^3_{r\_ref}(f_\eta)\left[1-\alpha(f_\eta)\right]\Delta t \\ &\quad +2q(f_\eta)\Delta t=b_{const}(f_\eta) \end{aligned} \tag{5.96}$$

where $b_{const}(f_\eta)$ is a time-invariant FM rate. To satisfy (5.96), another equation in $p(f_\eta)$, $q(f_\eta)$ and $Y_3(f_\eta)$ is given by

$$k_s\alpha(f_\eta)+2Y_3(f_\eta)b^3_{r\_ref}(f_\eta)\left[1-\alpha(f_\eta)\right]+2q(f_\eta)=0 \qquad (3) \tag{5.97}$$

Combining restrictions (1), (2) and (3), we have

$$Y_3=\frac{k_s(f_\eta)[\alpha(f_\eta)-0.5]}{b^3_{r\_ref}(f_\eta)[\alpha(f_\eta)-1]} \tag{5.98}$$

$$p(f_\eta)=b_{r\_ref}(f_\eta)[\alpha(f_\eta)-1] \tag{5.99}$$

$$q(f_\eta)=\frac{k_s(f_\eta)[\alpha(f_\eta)-1]}{2} \tag{5.100}$$

$$b_{const}(f_\eta)=b_{r\_ref}(f_\eta)\alpha(f_\eta) \tag{5.101}$$

It can be seen from (5.98) that $|\alpha(f_\eta)-1|$ cannot approach zero. Thus, the choice of θ_d should meet the requirements that $|\alpha(f_\eta)-1|\gg 0$ holds for each f_η in the processed bandwidth. Only by this can the effect of the cubic phase (the last term in (5.88)) on the stationary point in the frequency domain be considered as negligible. After cubic phase compensation and multiplication with the NLCS function, the signal becomes

$$\begin{aligned} S_{\eta 2}(t',f_\eta;R_{mc},\eta_{pc}) &= \sigma_p \mathrm{rect}\left[\frac{t'-t_m(f_\eta;R_{mc})}{T}\right] \\ &\quad\times\exp\left\{-j2\pi f_\eta(R_{mc}\sin\theta_m/V_m+\eta_{pc})\right\} \\ &\quad\times\exp\left\{-j\frac{4\pi}{\lambda}R_{mc}\cos\theta_m\sqrt{1-(\lambda f_\eta/2V_m)^2}\right\} \\ &\quad\times\exp\left\{j\pi b_{const}(f_\eta)\left[t'-t_{m,new}(f_\eta;R_{mc})\right]^2\right\} \\ &\quad\times\exp\left\{j\Theta(f_\eta;R_{mc})\right\} \end{aligned} \tag{5.102}$$

where

$$\Theta(f_\eta; R_{mc}) = j\pi b_{r\_ref}(f_\eta)\alpha(f_\eta)\left[\alpha(f_\eta) - 1\right]\Delta t^2 + j\frac{\pi}{3}k_s\alpha(f_\eta)^2\left[\alpha(f_\eta) - 1\right]\Delta t^3 \tag{5.103}$$

The range Fourier transform turns Equation (5.102) into

$$\begin{aligned} S_2(f, f_\eta; R_{mc}, \eta_{pc}) = \sigma_p \exp\left\{-j2\pi f_\eta(R_{mc}\sin\theta_m / V_m + \eta_{pc})\right\} \\ \times \exp\left\{-j\frac{4\pi}{\lambda}R_{mc}\cos\theta_m\sqrt{1 - (\lambda f_\eta/2V_m)^2}\right\} \\ \times \exp\left\{-j2\pi f t_{m,new}(f_\eta; R_{mc})\right\} \\ \times \exp\left\{-j\pi f^2/b_{const}(f_\eta)\right\} \\ \times \exp\left\{j\Theta(f_\eta; R_{mc})\right\} \end{aligned} \tag{5.104}$$

Accordingly, the range processing equation is given by

$$H_r(f, f_\eta) = \exp\left\{j\frac{\pi f^2}{b_{const}(f_\eta)} + j\frac{4\pi f}{c}\frac{R_{mc\_ref}\cos\theta_{m\_ref}}{\cos\theta_d}[\alpha(f_\eta) - 1]\right\} \tag{5.105}$$

The azimuth processing equation is given by

$$\begin{aligned} H_a(f_\eta, R_{mc}) = \exp\left\{j\frac{4\pi}{\lambda}R_{mc}\left[\cos\theta_m\sqrt{1 - (\lambda f_\eta/2V_m)^2} - 1\right] + j2\pi f_\eta R_{mc}\sin\theta_m / V_m\right\} \\ \times \exp\left\{-j\Theta(f_\eta; R_{mc})\right\} \end{aligned} \tag{5.106}$$

Up to now, the block diagram of the bistatic NLCS algorithm based on the hyperbolic approximation can be shown in Figure 5.12.

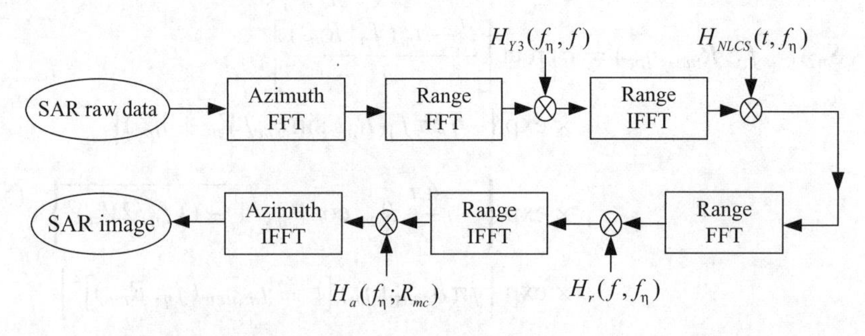

Figure 5.12 Block diagram of the bistatic NLCS algorithm

ω-k Algorithm Based on the Hyperbolic Approximation

As mentioned in Chapter 2, the ω-k algorithm is a kind of two-dimensional frequency domain algorithm, the key processing steps of which are performed in the two-dimensional frequency domain. For the hyperbolic range equation, it corrects the cross-coupling between the range and azimuth in the square root term in Equation (5.63) completely. This gives the ω-k algorithm better precision than the RD and CS algorithms in high squint angle cases and high-resolution cases [54, 55], such as the spotlight mode. However, the conventional ω-k algorithm cannot accommodate the range variance of V_m, and so generally it is more suitable for small scenes. In Equation [56] the ω-k algorithm is modified by using special approximations and a compensation filter in the range time and azimuth wavenumber domain, so as to accommodate the range variance of V_m in azimuth compression. In this book, a stricter derivation is used to improve the precision of the compensation filter and to handle the range variant V_m.

Let $2\pi(f + f_0)/c = k$ and $2\pi f_\eta / V_{m\_ref} = k_u$; then Equation (5.63) can be rewritten as

$$\begin{aligned} S(k, k_u; R_{mc}, \eta_{pc}) = {} & \sigma_p \exp\left\{j2\pi f t_{dly}\right\} \exp\left\{-j2\pi f_\eta \eta_{pc}\right\} \exp\left\{-j\pi f^2/k\right\} \\ & \times \exp\left\{-jk_u R_{mc} \sin\theta_m \frac{V_{m\_ref}}{V_m}\right\} \\ & \times \exp\left\{-jR_{mc}\cos\theta_m \sqrt{(2k)^2 - \left(k_u \frac{V_{m\_ref}}{V_m}\right)^2}\right\} \end{aligned} \tag{5.107}$$

The reference function multiply (RFM) is first implemented and the targets at the reference range, R_{mc_ref}, are well focused. The RFM filter is given by

$$\begin{aligned} H_{\text{ref}}(k, k_u; R_{mc\_ref}) = {} & \exp\left\{-j2\pi f t_{dly}\right\} \exp\left\{j\pi f^2/k\right\} \\ & \times \exp\left\{jR_{mc\_ref}\cos\theta_{m\_ref}\sqrt{(2k)^2 - k_u^2}\right\} \end{aligned} \tag{5.108}$$

Then the signal turns to

$$\begin{aligned} S_H(k, k_u; R_{mc}, \eta_{pc}) = {} & \sigma_p \exp\left\{-j2\pi f_\eta \eta_{pc}\right\} \exp\left\{-jk_u R_{mc}\sin\theta_m \frac{V_{m\_ref}}{V_m}\right\} \\ & \times \exp\left\{-j\left[R_{mc}\cos\theta_m\sqrt{(2k)^2 - \left(k_u\frac{V_{m\_ref}}{V_m}\right)^2}\right.\right. \\ & \left.\left. - R_{mc\_ref}\cos\theta_{m\_ref}\sqrt{(2k)^2 - k_u^2}\right]\right\} \end{aligned} \tag{5.109}$$

By expanding the first square root term around $V_{m\_ref}/V_m = 1$ and keeping the first three order terms, we arrive at

$$\begin{aligned} S_H(k, k_u; R_{mc}, \eta_{pc}) &\approx \sigma_p \exp\left\{-jk_u V_m \eta_{pc}\right\} \exp\left\{-jk_u R_{mc} \sin\theta_m \frac{V_{m\_ref}}{V_m}\right\} \\ &\times \exp\left\{-j\left(R_{mc}\cos\theta_m - R_{mc\_ref}\cos\theta_{m\_ref}\right)\sqrt{(2k)^2 - k_u^2}\right\} \\ &\times \exp\left\{jR_{mc}\cos\theta_m \frac{k_u^2}{\sqrt{(2k)^2 - k_u^2}}\left(\frac{V_{m\_ref}}{V_m} - 1\right)\right\} \\ &\times \exp\left\{j\frac{1}{2}R_{mc}\cos\theta_m \frac{k_u^2(2k)^2}{[(2k)^2 - k_u^2]^{3/2}}\left(\frac{V_{m\_ref}}{V_m} - 1\right)^2\right\} \end{aligned} \tag{5.110}$$

Let $k_x = \sqrt{(2k)^2 - k_u^2}$; then Equation (5.110) becomes

$$\begin{aligned} S_H(k_x, k_u; R_{mc}, \eta_{pc}) &\approx \sigma_p \exp\left\{-jk_u V_m \eta_{pc}\right\} \exp\left\{-jk_u R_{mc} \sin\theta_m \frac{V_{m\_ref}}{V_m}\right\} \\ &\times \exp\left\{-j\left(R_{mc}\cos\theta_m - R_{mc\_ref}\cos\theta_{m\_ref}\right)k_x\right\} \\ &\times \exp\left\{jR_{mc}\cos\theta_m \frac{k_u^2}{k_x}\left(\frac{V_{m\_ref}}{V_m} - 1\right)\right\} \\ &\times \exp\left\{j\frac{1}{2}R_{mc}\cos\theta_m \frac{k_u^2(k_x^2 + k_u^2)}{k_x^3}\left(\frac{V_{m\_ref}}{V_m} - 1\right)^2\right\} \end{aligned} \tag{5.111}$$

Ignoring the higher-order terms of k_u/k_x, Equation (5.111) reduces to

$$\begin{aligned} S_H(k_x, k_u; R_{mc}, \eta_{pc}) &\approx \sigma_p \exp\left\{-jk_u V_m \eta_{pc}\right\} \exp\left\{-jk_u R_{mc} \sin\theta_m \frac{V_{m\_ref}}{V_m}\right\} \\ &\times \exp\left\{-j\left[R_{mc}\cos\theta_m - R_{mc\_ref}\cos\theta_{m\_ref}\right]k_x\right\} \\ &\times \exp\left\{j\frac{1}{2}R_{mc}\cos\theta_m \frac{k_u^2}{k_x}\left(\frac{V_{m\_ref}^2}{V_m^2} - 1\right)\right\} \end{aligned} \tag{5.112}$$

The last term in Equation (5.112) shows the residual cross-coupling phase induced by the range variant V_m. At the reference range, this residual phase degrades to zero and the targets can be well focused. At other ranges, however, it brings about the following effects:

1. The targets are defocused in the azimuth direction, which becomes worse as the targets move away from the reference range (because the residual phase is a quadratic term of k_u).
2. The targets have uncorrected RCM and are defocused in the range direction, since the power series of $1/k_x$ include linear, quadratic and every higher-order term. Among them, the linear term leads to uncorrected RCM, which is expressed by

$$\Delta R_{RCM}(k_u) = \frac{1}{2} R_{mc} \cos\theta_m \frac{k_u^2}{k_{x0}^2} \left(\frac{V_{m\_ref}^2}{V_m^2} - 1 \right) \tag{5.113}$$

The quadratic term leads to range defocusing. Though the higher-order terms also cause image quality degradation, they are usually small enough to be ignored.

When the residual phase is slowly varying with k_x, we can only keep the constant of the power series of $1/k_x$, that is, $\exp\{j R_{mc} \cos\theta_m (k_u^2/k_{x0})[(V_{m\_ref}^2/V_m^2) - 1]/2\}$. Because only the baseband of k_x can be observed in digital signal processing, k_{x0} is defined as the median of k_x, the phase term corresponding to which will be preserved in the time domain as a constant value. Accordingly, the compensation filter in the range-Doppler domain is given by

$$H_D(k_u; R_{mc}) = \exp\left\{ j \frac{1}{2} R_{mc} \cos\theta_m \frac{k_u^2}{k_{x0}} \left(1 - \frac{V_{m\_ref}^2}{V_m^2} \right) \right\} \exp\left\{ j k_u R_{mc} \sin\theta_m \frac{V_{m\_ref}}{V_m} \right\} \tag{5.114}$$

where the second term ensures that the target is located at the central position of its synthetic aperture time.

In the ω-k algorithm presented above, the cross-coupling between the range and azimuth is corrected by assuming that V_m is range invariant, whereas the azimuth compression accommodates the range variance of V_m to a certain extent. Therefore, for bistatic SAR with a significantly variant equivalent velocity, the ω-k algorithm is less precise than the RD and CS algorithms in theory.

5.3.2.4 Simulations and Conclusions

In this section, we show some simulation results by following the parameters in DLR-ONERA bistatic experiments described in [57] (as listed in Table 5.1), allowing one to understand the above algorithms better.

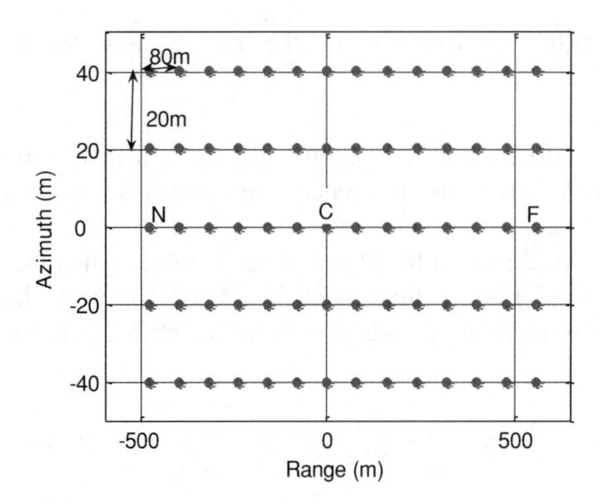

Figure 5.13 The simulated targets in the scenario. (© 2008 IEEE. Reprinted with permission from X. Qiu, D. Hu, C. Ding, "An Omega-K Algorithm With Phase Error Compensation for Bistatic SAR of a Translational Invariant Case", *IEEE Transaction on Geoscience and Remote Sensing*, **46**,8, pp. 2224–2231, 2008.)

The scene center is set to be the origin of the coordinates. The transmitter and receiver are relatively near the scenario while the baseline has a length of 1654.78 m, so that the bistatic angle is 29° in the across-track direction, which can be considered as a long baseline bistatic configuration. The simulated targets in the scenario are shown in Figure 5.13, where the height of all the targets is zero, the near-range target (N) is placed at −480 m in the range direction (the x axis), the far-range target (F) is placed at 560 m and the range swath is 1040 m. According to this range swath, the transmitting and receiving beam widths in the range direction are 10.16° and 15.77°, respectively, which can be regarded as a quite wide range swath. Supposing that the transmitting antenna has a larger beam footprint, the synthetic aperture time in the simulator is mainly determined by the receiving antenna.

Figure 5.14 demonstrates how the targets of the middle row locate in the focused image processed by the RD, CS and ω-k algorithms introduced above. Because the synthetic aperture time in this simulation is determined by the receiving antenna (which is the nonsquint case), the targets, which are focused at the central time of the synthetic aperture, are distributed in a horizontal line (instead of a diagonal line) along the range direction. Figure 5.15 illustrates the impulse responses of the near-range target N using the three algorithms, which reveals that these three algorithms have almost the same performance under these given simulation parameters. The range impulse responses are all nearly perfect. What should be clarified is that, since the range frequency axis is transformed in the ω-k algorithm, the range space interval between the adjacent pixels is changed, and thus the range profile of the ω-k algorithm here is different in scale from those of the other two algorithms. The azimuth impulse responses have a little dissymmetry between the left and right sidelobes due to the hyperbolic approximation error. Correspondingly, the image qualities of the targets at

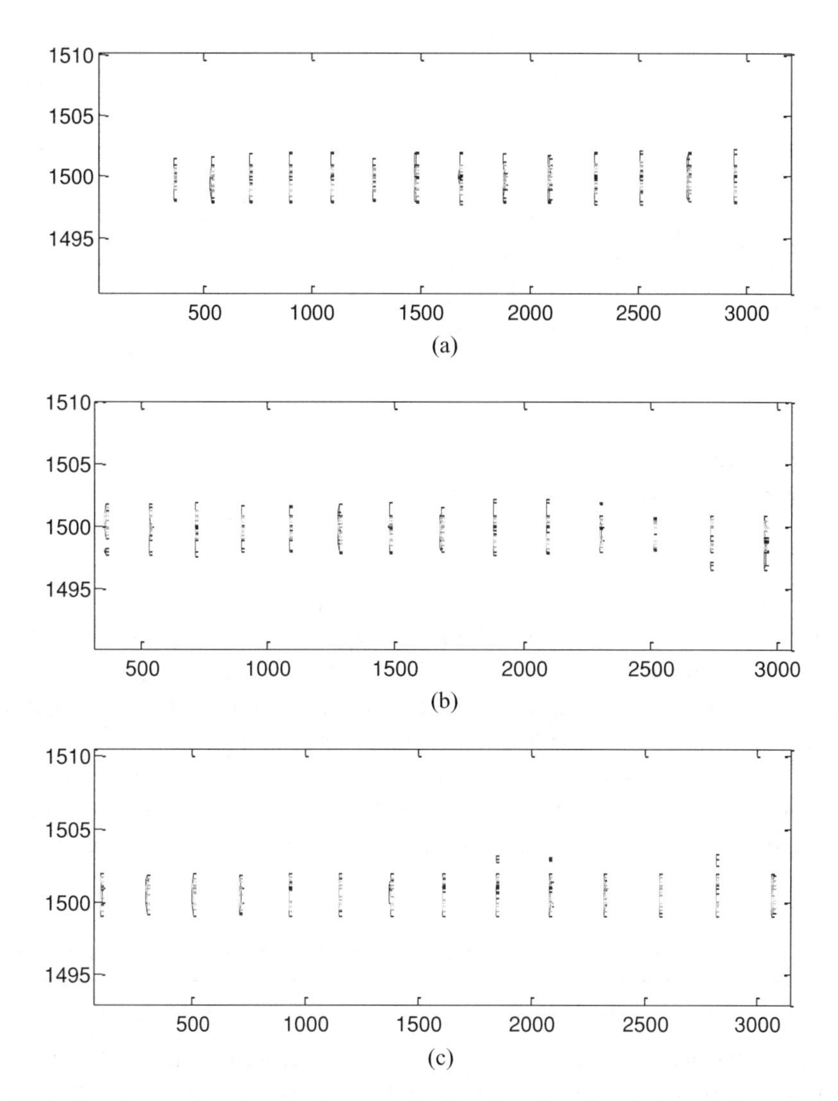

Figure 5.14 Imaging results using quasi-monostatic algorithms based on the hyperbolic approximation: (a) RD algorithm, (b) CS algorithm, (c) ω-k algorithm

the near, middle and far ranges are listed in Table 5.2, which also shows that the three algorithms have similar qualities under the given simulation parameters. Moreover, as the range variance of the equivalent velocity is left out of the consideration in SRC and cubic phase compensation, the image qualities at the edges of the range swath are slightly degraded compared with those at the scene center.

5.3.3 Advanced Hyperbolic Approximation

As discussed in Section 5.3.2, the hyperbolic approximation is more appropriate for the long baseline case as compared with the baseline middle-point approximation.

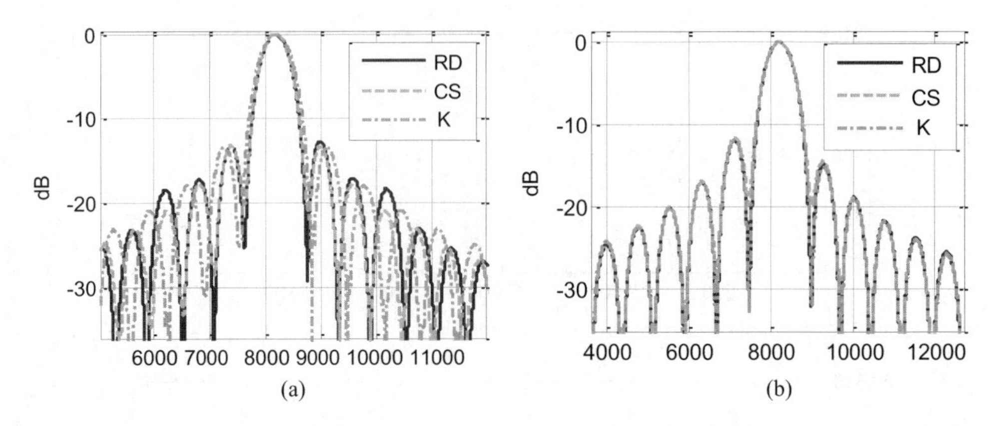

Figure 5.15 Impulse responses of the near-range target processed by different algorithms: (a) range impulse responses, (b) azimuth impulse responses

However, since the hyperbolic equation has three equivalent variables, only the constant, linear and quadratic terms of the flat-top hyperbola are exactly equalized; others are not. As shown in Figure 5.9, the cubic terms are partly equalized. Hence, it causes sidelobe dissymmetry in the azimuth direction, which confines its applicability and it is not suitable for wide aperture and large squint cases. [15] proposed an advanced hyperbolic equation by introducing an additional variable to the conventional hyperbolic equation and re-estimating the three variables, in order to enhance signal modeling precision for spaceborne SAR at higher orbits. In this section, we will

Table 5.2 Image qualities processed by different algorithms based on the hyperbolic approximation

		Range qualities				Azimuth qualities			
	Target	Resolution (m)	Widen ratio	PSLR (dB)	ISLR (dB)	Resolution (m)	Widen ratio	PSLR (dB)	ISLR (dB)
RD	N	0.266	0.886	−12.71	−9.41	0.192	0.900	−11.93	−10.05
	C	0.266	0.886	−13.04	−9.92	0.189	0.886	−12.45	−10.05
	F	0.267	0.891	−12.65	−9.25	0.191	0.893	−12.99	−10.24
CS	N	0.267	0.891	−13.11	−9.97	0.194	0.910	−11.70	−9.78
	C	0.265	0.882	−13.19	−10.09	0.189	0.886	−12.48	−10.07
	F	0.266	0.886	−13.01	−9.84	0.191	0.897	−12.63	−9.87
ω−k	N	0.266	0.886	−13.39	−10.18	0.193	0.904	−11.94	−10.13
	C	0.265	0.883	−13.24	−10.04	0.190	0.890	−12.45	−10.05
	F	0.266	0.887	−13.29	−10.14	0.191	0.896	−12.99	−10.21

Note: The image qualities of each target are obtained by selecting a 32 × 32 chip centered on the target, upsampling by a factor of 512 and estimating along the peak direction. The range resolution here shows the resolution in the bistatic range dimension and corresponds to the bistatic range divided by 2.

extend this advanced hyperbolic approximation to bistatic SAR and then establish the quasi-monostatic algorithms accordingly.

5.3.3.1 Principle of the Advanced Hyperbolic Approximation

The bistatic imaging geometry and range equation are the same as those given in Section 5.3.2. The hyperbolic approximation in (5.25) is replaced by the advanced hyperbolic approximation with an additional linear term, which is

$$R_{bi}(\eta) \approx 2R_{mono\_adv}(\eta) = 2\left[\sqrt{R_{mc}^2 + V_m^2(\eta-\eta_{pc})^2 - 2R_{mc}V_m(\eta-\eta_{pc})\sin(\theta_m)} + l_m(\eta-\eta_{pc})\right] \tag{5.115}$$

where the variables R_{mc}, V_m, θ_m and l_m are waiting to be resolved: R_{mc}, V_m and θ_m have the same definitions as given in Section 5.3.2, while l_m represents the additional linear coefficient.

Similarly, taking the Taylor series expansion of $R_{mono\_adv}(\eta)$ about $\eta = \eta_{pc}$, we arrive at

$$R_{mono\_adv}(\xi) = R_{mc} - (V_m\sin\theta_m - l_m)\xi + \frac{V_m^2\cos^2\theta_m}{2R_{mc}}\xi^2 + \frac{V_m^3\sin\theta_m\cos^2\theta_m\xi^3}{2R_{mc}^2} + \frac{V_m^4\cos^2\theta_m(5\sin^2\theta_m-1)\xi^4}{8R_{mc}^3} + o\left(\frac{V_m^4\xi^4}{R_{mc}^3}\right) \tag{5.116}$$

As the advanced hyperbolic approximation has one more variable than the conventional one, we can equate the first four lower-order terms and establish a set of equations as follows:

$$R_{Tc} + R_{Rc} = 2R_{mc} \tag{5.117}$$

$$V_T\sin\theta_T + V_R\sin\theta_R = 2V_m\sin\theta_m - 2l_m \tag{5.118}$$

$$\frac{V_T^2\cos^2\theta_T}{2R_{Tc}} + \frac{V_R^2\cos^2\theta_R}{2R_{Rc}} = 2\frac{V_m^2\cos^2\theta_m}{2R_{mc}} \tag{5.119}$$

$$\frac{V_T^3\sin\theta_T\cos^2\theta_T}{2R_{Tc}^2} + \frac{V_R^3\sin\theta_R\cos^2\theta_R}{2R_{Rc}^2} = 2\frac{V_m^3\sin\theta_m\cos^2\theta_m}{2R_{mc}^2} \tag{5.120}$$

Thus, the four variables can be resolved as

$$R_{mc} = (R_{Tc} + R_{Rc})/2 \tag{5.121}$$

$$V_m = \sqrt{V_{//m}^2 + V_{\perp m}^2} \tag{5.122}$$

$$\theta_m = \arctan(V_{//m}/V_{\perp m}) \tag{5.123}$$

$$l_m = V_{//m} - (V_{//T} + V_{//R})/2 \tag{5.124}$$

where $V_{//T}$, $V_{//R}$, a_T and a_R have the same definitions as those in Section 5.3.2, and

$$V_{\perp m} = V_m \cos\theta_m = \sqrt{\frac{a_T + a_R}{2} R_{mc}} \tag{5.125}$$

$$\begin{aligned} V_{//m} &= V_m \sin\theta_m = F_T V_{//T} + F_R V_{//R},\ F_T = \frac{a_T}{a_T + a_R}\frac{R_{mc}}{R_{Tc}}, \\ F_R &= \frac{a_R}{a_T + a_R}\frac{R_{mc}}{R_{Rc}} \end{aligned} \tag{5.126}$$

It can be seen that the rotational component of equivalent velocity in the advanced hyperbolic approximation is consistent with that in the conventional one, that is, the rotational velocity at the equivalent monostatic slant range caused by the average of transmitted and received centripetal forces. However, the translational component in the advanced hyperbolic approximation is modified compared with that in the conventional one: the latter one is the average of transmitted and received translational velocities without weighting and the former one is the average weighted by F_T and F_R. Furthermore, the additional linear coefficient compensates for the difference between the translational velocities in advanced and conventional hyperbolic approximations, so as to equalize both the linear and cubic terms exactly.

Note that the equivalent variables V_m, θ_m and l_m in the advanced hyperbolic approximation are range variant, just as the case in the hyperbolic approximation, which should be considered in the imaging process.

In view of the above derivation, the advanced hyperbolic approximation can completely equalize the flat-top hyperbola by constant, linear, quadratic and cubic terms. As to the approximation, using the Taylor series expansion of the transmitted range in Equation (5.26), the maximum phase induced by the higher-order terms within the synthetic aperture time is

$$\phi_{Terro4} = -\frac{2\pi}{\lambda} O\left(\frac{V_T^4 T_{syn}^4}{R_{Tc}^3}\right) = -\frac{2\pi}{\lambda} O\left(\frac{V_T^4}{R_{Tc}^3}\left(\frac{\lambda R_{Tc}}{D_a V_T}\right)^4\right) = O\left(\frac{\lambda^3 R_{Tc}}{D_a^4}\right) \tag{5.127}$$

This shows that the maximum phase is directly proportional to λ^3 and R_{Tc}, but inversely proportional to D_a^4. Hence, the advanced hyperbolic approximation is also quite appropriate for airborne SAR with high frequency.

Moreover, the main error of the advanced hyperbolic approximation is mainly attributed to the residual fourth-order term, which is given by

$$R_{err4}(\xi) = \frac{V_T^4\xi^4\cos^2\theta_T(5\sin^2\theta_T - 1)}{8R_{Tc}^3} + \frac{V_R^4\xi^4\cos^2\theta_R(5\sin^2\theta_R - 1)}{8R_{Rc}^3} - 2\frac{V_m^4\xi^4\cos^2\theta_m(5\sin^2\theta_m - 1)}{8R_{mc}^3} \tag{5.128}$$

Using (5.117) to (5.120), the above equation can be simplified to

$$R_{err4}(\xi) = \frac{a_T a_R}{2(a_T + a_R)}\left(\frac{V_{//T}}{R_{Tc}} + \frac{V_{//R}}{R_{Rc}}\right)^2 - \frac{1}{8(R_{Tc} + R_{Rc})}\left(a_T\sqrt{\frac{R_{Rc}}{R_{Tc}}} - a_R\sqrt{\frac{R_{Tc}}{R_{Rc}}}\right)^2 \tag{5.129}$$

Using the simulation parameters in Table 5.1, Figure 5.16(a) compares the fourth- and higher-order terms between the actual bistatic range and the advanced hyperbolic approximation, and Figure 5.16(b) shows the approximation error. These two figures verify that the advanced hyperbolic approximation completely equalizes the actual bistatic range up to the cubic term and partly equalizes the higher-order terms at the same time. In contrast with Figure 5.9(b), the approximation error here is reduced by two orders of magnitude, attributed to the introduction of the additional linear term and the re-estimation of the other variables. Thus, the ideal azimuth profile with symmetric sidelobes is obtained, as in Figure 5.16(c).

5.3.3.2 Variables Calculation

As derived in Section 5.3.2.2, the bistatic range parameters (including R_{Tc}, R_{Rc}, θ_T and θ_R) in each range gate can be obtained by using the estimated Doppler parameters (including f_{dc} and f_r) and the bistatic imaging geometry. Then, in terms of (5.121) to (5.124), the equivalent range parameters (including R_{mc}, V_m, θ_m and l_m) in the advanced hyperbolic approximation can be calculated.

Moreover, the relationship between the Doppler parameters and the equivalent range parameters can be given by

$$f_{dc} = \frac{2}{\lambda}(V_m\sin\theta_m - l_m) \tag{5.130}$$

$$f_r = -\frac{2V_m^2\cos^2\theta_m}{\lambda R_{mc}} \tag{5.131}$$

$$f_{r2} = \frac{3V_m^3\sin\theta_m\cos^2\theta_m}{\lambda R_{mc}^2} \tag{5.132}$$

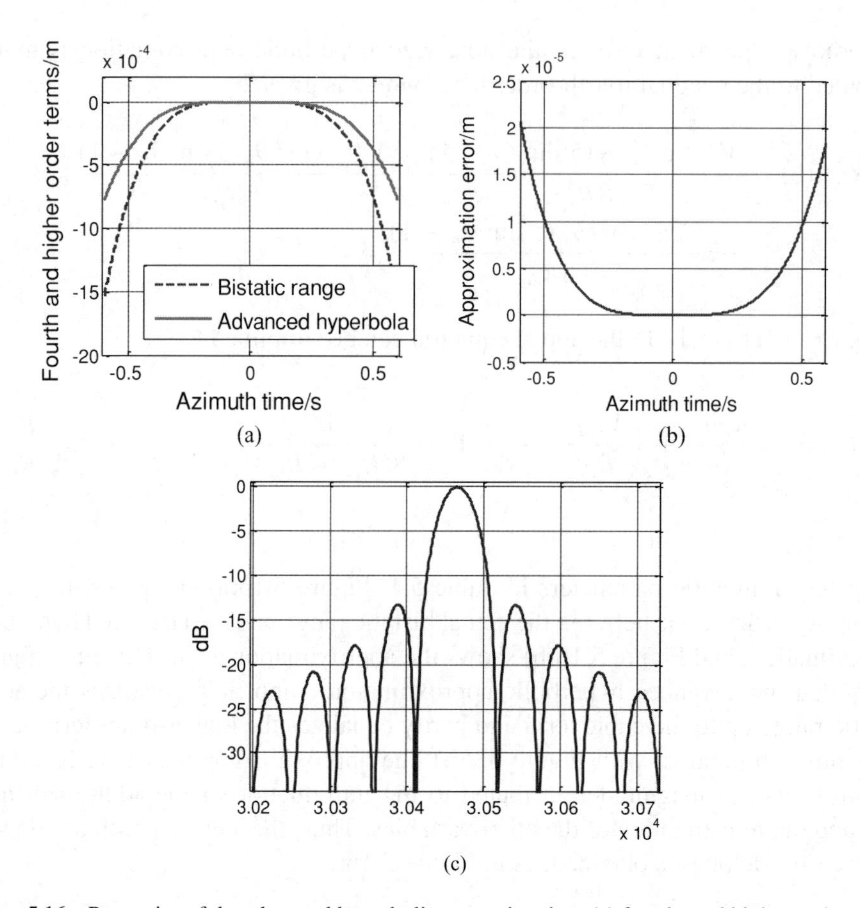

Figure 5.16 Properties of the advanced hyperbolic approximation: (a) fourth- and higher-order terms, (b) approximation error, (c) azimuth profile

where f_{2r} denotes the quadratic FM rate of the azimuth signal, that is, the third derivative of the range equation. Accordingly, V_m, θ_m and l_m are in the same expressions of f_{dc}, f_r and f_{2r}, as those in [58]:

$$V_m = \sqrt{\left(\frac{2R_{mc}f_{2r}}{3f_r}\right)^2 - \frac{\lambda R_{mc}f_r}{2}} \tag{5.133}$$

$$\theta_m = \arcsin\left(\frac{2R_{mc}f_{2r}}{3V_m f_r}\right) \tag{5.134}$$

$$l_m = -\frac{\lambda f_{dc}}{2} + \frac{2R_{mc}f_{2r}}{3f_r} \tag{5.135}$$

5.3.3.3 Quasi-Monostatic Algorithms Based on the Advanced Hyperbolic Approximation

Though an additional linear term is introduced, the two-dimensional signal spectrum based on the advanced hyperbolic approximation can still be analytically derived, which, moreover, is shown in a similar form to its counterpart based on the conventional hyperbolic approximation. Therefore, the well-known algorithms, such as the RD, CS, NLCS and ω-k algorithms, can be continued only by using some modifications. In the following text, the analytical spectrum will be deduced first and then the modified algorithms will be developed with some specific explanations for implementation.

By means of the advanced hyperbolic approximation, the received signal of target (R_{mc}, η_{pc}) in the two-dimensional time domain can be written as

$$\begin{aligned} s(t, \eta; R_{mc}, \eta_{pc}) \approx \sigma_p \text{rect}\left[\frac{t - 2R_{m\_adv}(\eta)/c}{T}\right] \exp\left\{j\pi b[t - 2R_{m\_adv}(\eta)/c]^2\right\} \\ \times w_a(\eta - \eta_{pc}) \exp\left\{-j\frac{4\pi R_{m\_adv}(\eta)}{\lambda}\right\} \end{aligned} \tag{5.136}$$

where $w_a()$ is the antenna beam pattern in the azimuth direction.

Using the POSP to evaluate the range Fourier transform, the signal in the range frequency and azimuth time domain is

$$\begin{aligned} s(f, \eta; R_{mc}, \eta_{pc}) = \sigma_p \text{rect}\left(\frac{f}{|bT|}\right) \exp\left\{-j\pi\frac{f^2}{b}\right\} w_a(\eta - \eta_{pc}) \\ \times \exp\left\{-j\frac{4\pi(f_0 + f)R_{m\_adv}(\eta)}{c}\right\} \end{aligned} \tag{5.137}$$

Using the POSP to evaluate the azimuth Fourier transform, the relationship between η and f_η at the stationary point can be written as

$$f_\eta = -\frac{2(f_0 + f)}{c}\left[\frac{V_m(V_m\eta - R_{mc}\sin\theta_m)}{\sqrt{(R_{mc}\cos\theta_m)^2 + (V_m\eta - R_{mc}\sin\theta_m)^2}} + l_m\right] \tag{5.138}$$

and

$$\eta = -\frac{R_{mc}\cos\theta_m}{V_m}\left[\frac{\left(\frac{cf_\eta}{2V_m(f_0 + f)} + \frac{l_m}{V_m}\right)}{\sqrt{1 - \left(\frac{cf_\eta}{2V_m(f_0 + f)} + \frac{l_m}{V_m}\right)^2}}\right] + \frac{R_{mc}\sin\theta_m}{V_m} \tag{5.139}$$

Both Equations (5.138) and (5.139) give the one-to-one correspondence between η and f_η, which is in favor of spectrum derivation. Then the two-dimensional spectrum can be given by

$$\begin{aligned} S(f, f_\eta; R_{mc}, \eta_{pc}) = {} & \sigma_p \text{rect}\left(\frac{f}{|bT|}\right) w_a(f_\eta - f_{dc}) \\ & \times \exp\left\{-j\frac{4\pi R_{mc}\sin\theta_m (f_0 + f) l_m}{cV_m}\right\} \\ & \times \exp\left\{-j2\pi f_\eta\left(\frac{R_{mc}\sin\theta_m}{V_m} + \eta_{pc}\right)\right\}\exp\left\{-j\pi\frac{f^2}{b}\right\} \\ & \times \exp\left\{-j\frac{4\pi R_{mc}\cos\theta_m}{\lambda}\sqrt{D^2(f_\eta) + 2E(f_\eta)\left(\frac{f}{f_0}\right) + F\left(\frac{f}{f_0}\right)^2}\right\} \end{aligned} \tag{5.140}$$

where

$$D(f_\eta) = \sqrt{1 - \left(\frac{\lambda f_\eta}{2V_m} + \frac{l_m}{V_m}\right)^2} \tag{5.141}$$

$$E(f_\eta) = 1 - \frac{l_m}{V_m}\left(\frac{\lambda f_\eta}{2V_m} + \frac{l_m}{V_m}\right) \tag{5.142}$$

$$F = 1 - \left(\frac{l_m}{V_m}\right)^2 \tag{5.143}$$

By expanding the square root term in the last phase expression around $f = 0$ and keeping terms up to f^3, we have

$$\begin{aligned} S(f, f_\eta; R_{mc}, \eta_{pc}) \approx {} & \sigma_p \text{rect}\left(\frac{f}{|bT|}\right) w_a(f_\eta - f_{dc})\exp\left\{-j\pi\frac{f^2}{b}\right\} \\ & \times \exp\left\{-j2\pi f_\eta\left(\frac{R_{mc}\sin\theta_m}{V_m} + \eta_{pc}\right)\right\} \\ & \times \exp\left\{-j\frac{4\pi}{\lambda}\left[R_{mc}\cos\theta_m D(f_\eta) + \frac{R_{mc}\sin\theta_m l_m}{V_m}\right]\right\} \text{ (zeroth-order term)} \\ & \times \exp\left\{-j\frac{4\pi f}{c}\left[\frac{R_{mc}\cos\theta_m E(f_\eta)}{D(f_\eta)} + \frac{R_{mc}\sin\theta_m l_m}{V_m}\right]\right\} \text{ (linear term)} \end{aligned}$$

$$\times \exp\left\{ j\pi f^2 \left[\frac{cR_{mc}\cos\theta_m f_\eta^2}{2V_m^2 f_0^3 D^3(f_\eta)} \right] \right\} \text{(quadratic term)}$$
$$\times \exp\left\{ -j\pi f^3 \left[\frac{cR_{mc}\cos\theta_m f_\eta^2 E(f_\eta)}{2V_m^2 f_0^4 D^5(f_\eta)} \right] \right\} \text{(cubic term)} \tag{5.144}$$

Each phase term in (5.144) is interpreted as follows:

1. The first phase term represents the phase spectrum relating to the transmitted chirp pulse.
2. The second phase term, solely dependent on f_η, causes an azimuth shift in the azimuth time domain, which locates the target at the closest approach time, $\eta_{pc} + R_{mc}\sin\theta_m / V_m$.
3. The zeroth-order term denotes the azimuth modulation.
4. The linear term, due to the cross-coupling between the range and azimuth, denotes the RCM in the Doppler domain.
5. The quadratic and cubic terms, due to the cross-coupling as well, represent an additional range modulation besides that of the transmitted chirp pulse. They are significant especially when the squint angle is large.

RD Algorithm Based on the Advanced Hyperbolic Approximation

In terms of the two-dimensional spectrum in (5.144), the range compression and the cubic phase compensation filters can be designed with reference to the target at the reference range. The range compression filter is given by

$$H_{rc\&src}(f, f_\eta) = \exp\left\{ j\pi \frac{f^2}{b_r(f_\eta; R_{mc\_ref})} \right\} \tag{5.145}$$

where

$$b_r(f_\eta; R_{mc}) = \frac{b}{1 - b\left[\dfrac{cR_{mc}\cos\theta_m f_\eta^2}{2V_m^2 f_0^3 D^3(f_\eta)} \right]} \tag{5.146}$$

The cubic phase compensation filter is given by

$$H_{3err}(f, f_\eta) = \exp\left\{ j\pi f^3 \left[\frac{cR_{mc}\cos\theta_m f_\eta^2 E(f_\eta)}{2V_m^2 f_0^4 D^5(f_\eta)} \right] \right\} \tag{5.147}$$

After multiplying by $H_{rc\&src}(f, f_\eta)$ and $H_{3err}(f, f_\eta)$, and taking the range inverse Fourier transform, the signal turns to

$$\begin{aligned} s(t, f_\eta; R_{mc}, \eta_{pc}) = \sigma_p \text{sinc}\left\{b\left[t - \frac{2R_{mc}}{c}\left(\frac{\cos\theta_m E(f_\eta)}{D(f_\eta)} + \frac{\sin\theta_m l_m}{V_m}\right)\right]\right\} w_a(f_\eta - f_{dc}) \\ \times \exp\left\{-j2\pi f_\eta\left(\frac{R_{mc}\sin\theta_m}{V_m} + \eta_{pc}\right)\right\} \\ \times \exp\left\{-j\frac{4\pi}{\lambda}\left[R_{mc}\cos\theta_m D(f_\eta) + \frac{R_{mc}\sin\theta_m l_m}{V_m}\right]\right\} \end{aligned} \tag{5.148}$$

It can be seen that the amount of RCM to be corrected should be as follows:

$$\Delta R_{RCM}(f_\eta; R_{mc}) = R_{mc}\left[\frac{\cos\theta_m E(f_\eta)}{D(f_\eta)} + \frac{\sin\theta_m l_m}{V_m} - 1\right] \tag{5.149}$$

Afterwards, the azimuth compression filter can be multiplied in each range gate, which is

$$H_a(f_\eta; R_{mc}) = \exp\left\{j\frac{4\pi R_{mc}}{\lambda}\left[\cos\theta_m D(f_\eta) + \frac{\sin\theta_m l_m}{V_m} - 1\right] + j2\pi f_\eta \frac{R_{mc}\sin\theta_m}{V_m}\right\} \tag{5.150}$$

Finally, the target is located at $t = 2R_{mc}/c$ in the range direction and $\eta = \eta_{pc}$ in the azimuth direction.

Similarly, the range variance of the four equivalent variables can be accommodated in RCMC and azimuth compression operations, which are the advantages of the RD algorithm. However, the range variance cannot be accommodated in range compression and cubic phase compensation operations, which are the shortcomings of this RD algorithm.

CS Algorithm Based on the Advanced Hyperbolic Approximation

To facilitate the chirp scaling operation, consider that the cubic phase in (5.144) has been compensated with reference to the reference range, which has the same expression as (5.147). Then, taking the range inverse Fourier transform to (5.144), we arrive at

$$\begin{aligned} S(t, f_\eta; R_{mc}, \eta_{pc}) \approx \sigma_p \text{rect}\left\{ \frac{t - \frac{2R_{mc}}{c}\left[\frac{\cos\theta_m E(f_\eta)}{D(f_\eta)} + \frac{l_m \sin\theta_m}{V_m}\right]}{T} \right\} w_a(f_\eta - f_{dc}) \\ \times \exp\left\{ j\pi b_r(f_\eta, R_{mc}) \left[t - \frac{2R_{mc}}{c}\left(\frac{\cos\theta_m E(f_\eta)}{D(f_\eta)} + \frac{l_m \sin\theta_m}{V_m}\right)\right]^2 \right\} \\ \times \exp\left\{ -j2\pi f_\eta \left(\frac{R_{mc}\sin\theta_m}{V_m} + \eta_{pc} \right) \right\} \\ \times \exp\left\{ -j\frac{4\pi}{\lambda}\left[R_{mc}\cos\theta_m D(f_\eta) + \frac{R_{mc} l_m \sin\theta_m}{V_m} \right] \right\} \end{aligned} \tag{5.151}$$

At the RCMC stage, ignoring the range variance of $b_r(f_\eta, R_{mc})$ means that

$$b_r(f_\eta, R_{mc}) \approx b_r(f_\eta, R_{mc\_ref}) = b_{r\_ref}(f_\eta) \tag{5.152}$$

Furthermore, assuming the equivalent variables V_m, θ_m and l_m are range invariant, the differential RCM represented by the time delay can be written as follows:

$$t_m(f_\eta, R_{mc}) - t_{m\_ref}(f_\eta) \approx [C_{s\_ref}(f_\eta) + 1](2R_{mc}/c - 2R_{mc\_ref}/c) \tag{5.153}$$

where $t_m(f_\eta, R_{mc})$ denotes the time delay due to the RCM:

$$t_m(f_\eta, R_{mc}) = \frac{2R_{mc}}{c}\left[\frac{\cos\theta_m E(f_\eta)}{D(f_\eta)} + \frac{l_m \sin\theta_m}{V_m}\right] \tag{5.154}$$

and $t_{m\_ref}(f_\eta) = t_m(f_\eta, R_{mc\_ref})$. In addition,

$$C_{s\_ref}(f_\eta) = \left[\frac{\cos\theta_{m\_ref} E(f_\eta; R_{mc\_ref})}{D(f_\eta; R_{mc\_ref})} + \frac{l_{m\_ref}\sin\theta_{m\_ref}}{V_{m\_ref}}\right] - 1 \tag{5.155}$$

Accordingly, the linear CS function, used for differential RCMC, is established as follows:

$$H_{cs}(t, f_\eta) = \exp\left\{ j\pi b_{r\_ref}(f_\eta) C_{s\_ref}(f_\eta)\left[t - t_{m\_ref}(f_\eta)\right]^2 \right\} \tag{5.156}$$

By multiplying with the CS function, the signal turns into

$$
\begin{aligned}
S_{\eta cs}(t, f_\eta; R_{mc}, \eta_{pc}) \approx\ & \sigma_p \mathrm{rect}\left\{\left[t - \frac{2R_{mc}}{c}\left(\frac{\cos\theta_m E(f_\eta)}{D(f_\eta)} + \frac{l_m \sin\theta_m}{V_m}\right)\right]\Big/ T\right\} w_a(f_\eta - f_{dc}) \\
& \times \exp\left\{j\pi b_{r\_ref}(f_\eta)[1 + C_{s\_ref}(f_\eta)][t - t'_m(f_\eta, R_{mc})]^2\right\} \\
& \times \exp\left\{j\frac{4\pi}{c^2}\pi b_{r\_ref}(f_\eta) C_{s\_ref}(f_\eta)\left[1 + C_{s\_ref}(f_\eta)\right]\left(R_{mc} - R_{mc\_ref}\right)^2\right\} \\
& \times \exp\left\{-j2\pi f_\eta\left(\frac{R_{mc}\sin\theta_m}{V_m} + \eta_{pc}\right)\right\} \\
& \times \exp\left\{-j\frac{4\pi}{\lambda}\left[R_{mc}\cos\theta_m D(f_\eta) + \frac{R_{mc} l_m \sin\theta_m}{V_m}\right]\right\}
\end{aligned}
\tag{5.157}
$$

where $t'_m(f_\eta, R_{mc})$ denotes the desired RCM trajectory after CS operation, which is

$$
t'_m(f_\eta, R_{mc}) = t_{m\_ref}(f_\eta) + \frac{2}{c}(R_{mc} - R_{mc\_ref}) \tag{5.158}
$$

It can be seen that all the RCM trajectories are parallel with those of the reference target.

In the two-dimensional frequency domain, the spectrum becomes

$$
\begin{aligned}
S_{cs}(f, f_\eta; R_{mc}, \eta_{pc}) \approx\ & \sigma_p \exp\left\{-j\frac{\pi f^2}{b_{r\_ref}(f_\eta)[1 + C_{s\_ref}(f_\eta)]}\right\} \exp\left\{-j2\pi f t'_m(f_\eta, R_{mc})\right\} \\
& \times \exp\left\{j\frac{4\pi}{c^2}\pi b_{r\_ref}(f_\eta) C_{s\_ref}(f_\eta)\left[1 + C_{s\_ref}(f_\eta)\right]\left(R_{mc} - R_{mc\_ref}\right)^2\right\} \\
& \times \exp\left\{-j2\pi f_\eta\left(\frac{R_{mc}\sin\theta_m}{V_m} + \eta_{pc}\right)\right\} \\
& \times \exp\left\{-j\frac{4\pi}{\lambda}\left[R_{mc}\cos\theta_m D(f_\eta) + \frac{R_{mc} l_m \sin\theta_m}{V_m}\right]\right\}
\end{aligned}
\tag{5.159}
$$

Therefore, the range matched filter for range compression and bulk RCMC should be designed by

$$
H_r(f, f_\eta) = \exp\left\{j\frac{\pi f^2}{b_{r\_ref}(f_\eta)[1 + C_{s\_ref}(f_\eta)]}\right\} \exp\left\{-j2\pi f\left[t_{m\_ref}(f_{\eta\_ref}) - t_{m\_ref}(f_\eta)\right]\right\} \tag{5.160}
$$

where f_{η_ref} is the Doppler centroid of the reference target

$$f_{\eta\_ref} = \frac{2}{\lambda}(V_{m\_ref} \sin\theta_{m\_ref} - l_{m\_ref}) \tag{5.161}$$

After range processing, the target is located at $\tau = 2R_{mc}/c$ in the range direction. Finally, the azimuth matched filter should be designed by

$$\begin{aligned} H_a(f_\eta; R_{mc}) = \exp&\left\{-j\frac{4\pi}{c^2}\pi b_{r\_ref}(f_\eta)C_{s\_ref}(f_\eta)\left[1 + C_{s\_ref}(f_\eta)\right]\left(R_{mc} - R_{mc\_ref}\right)^2\right\} \\ &\times \exp\left\{-j\frac{4\pi}{\lambda}\left[R_{mc}\cos\theta_m D(f_\eta) + \frac{R_{mc}l_m \sin\theta_m}{V_m}\right]\right\} \\ &\times \exp\left\{j2\pi f_\eta \frac{R_{mc}\sin\theta_m}{V_m}\right\} \end{aligned} \tag{5.162}$$

The first exponential term responds for the residual phase induced by the CS operation and the second exponential term responds for the azimuth modulation.

NLCS Algorithm Based on theAdvanced Hyperbolic Approximation

As discussed above, the CS algorithm cannot handle the range variance in range compression. Thus, at the edges of the swath, the imaging performance degrades as the swath extends. Since the signal expressions and properties of the advanced hyperbolic approximation are similar to those of the hyperbolic approximation, the NLCS algorithm can also be derived based on the advanced hyperbolic approximation.

Above all, in order to enhance the range compression performance, the range FM rate is modified by

$$b_r(f_\eta; R_{mc}) - b_{r\_ref}(f_\eta) \approx k_s(f_\eta)[t_m(f_\eta; R_{mc}) - t_{m\_ref}(f_\eta)] \tag{5.163}$$

where $k_s(f_\eta)$ is the variation slope with respect to $t_m(f_\eta; R_{mc})$. This expression is more precise than that in (5.152), and thus makes the range compression operation in the NLCS algorithm more precise than that in the RD or CS algorithms. Moreover, the required shift for differential RCMC is assumed linear with range, which is given by

$$t_m(f_\eta; R_{mc}) - t_{m\_ref}(f_\eta) \approx \alpha(f_\eta)[t_m(f_{\eta\_ref}; R_{mc}) - t_{m\_ref}(f_{\eta\_ref})] \tag{5.164}$$

where $f_{\eta\_ref} = 2(V_{m\_ref} \sin\theta_d - l_{m\_ref})/\lambda$ is the reference azimuth frequency for RCMC. Accordingly, the new RCM trajectory after the scaling operation, $t_{m,new}(f_\eta; R_{mc})$, is given by

$$t_{m,new}(f_\eta; R_{mc}) = t_{m\_ref}(f_\eta) + [t_m(f_{\eta\_ref}; R_{mc}) - t_{m\_ref}(f_{\eta\_ref})] \tag{5.165}$$

This means that the distance in RCM trajectories between a given range and the reference range is independent of f_η, that is, the differential RCMs have been corrected. Therefore, the bulk RCM trajectories can be efficiently corrected afterwards by phase multiplying in the range frequency domain.

Referring to (5.146) and (5.154), the derivation of $k_s(f_\eta)$ and $\alpha(f_\eta)$ are given in this paragraph. Using Taylor series expansion up to the linear term, the left side of (5.163) can be written by

$$b_r(f_\eta; R_{mc}) - b_{r\_ref}(f_\eta) \approx \left.\frac{db_r(f_\eta; R_{mc})}{dR_{mc}}\right|_{R_{mc\_ref}} (R_{mc} - R_{mc\_ref}) \tag{5.166}$$

$$\begin{aligned}\frac{db_r(f_\eta; R_{mc})}{dR_{mc}} &= \frac{\partial b_r(f_\eta; R_{mc})}{\partial R_{mc}} + \frac{\partial b_r(f_\eta; R_{mc})}{\partial V_m}\frac{dV_m}{dR_{mc}} \\ &\quad + \frac{\partial b_r(f_\eta; R_{mc})}{\partial \theta_m}\frac{d\theta_m}{dR_{mc}} + \frac{\partial b_r(f_\eta; R_{mc})}{\partial l_m}\frac{dl_m}{dR_{mc}}\end{aligned} \tag{5.167}$$

Similarly, the right side of (5.163) can be written by

$$t_m(f_\eta; R_{mc}) - t_{m\_ref}(f_\eta) \approx \left.\frac{dt_m(f_\eta; R_{mc})}{dR_{mc}}\right|_{R_{mc\_ref}} (R_{mc} - R_{mc\_ref}) \tag{5.168}$$

$$\begin{aligned}\frac{dt_m(f_\eta; R_{mc})}{dR_{mc}} &= \frac{\partial t_m(f_\eta; R_{mc})}{\partial R_{mc}} + \frac{\partial t_m(f_\eta; R_{mc})}{\partial V_m}\frac{dV_m}{dR_{mc}} \\ &\quad + \frac{\partial t_m(f_\eta; R_{mc})}{\partial \theta_m}\frac{d\theta_m}{dR_{mc}} + \frac{\partial t_m(f_\eta; R_{mc})}{\partial l_m}\frac{dl_m}{dR_{mc}}\end{aligned} \tag{5.169}$$

Since the range variance of $b_r(f_\eta; R_{mc})$ and $t_m(f_\eta; R_{mc})$ is mainly due to R_{mc}, we have

$$\begin{aligned}k_s(f_\eta) &= \frac{\partial b_r(f_\eta; R_{mc})/\partial R_{mc}\big|_{R_{mc\_ref}}}{\partial t_m(f_\eta; R_{mc})/\partial R_{mc}\big|_{R_{mc\_ref}}} \\ &= \frac{b_{r\_ref}^2(f_\eta)}{f_0}\left\{\left(\frac{\lambda f_\eta}{2V_{m\_ref}}\right)^2 \Big/ \left[D_{ref}^2(f_\eta)E_{ref}(f_\eta) + \frac{l_{m\_ref}\tan\theta_{m\_ref}D_{ref}^3(f_\eta)}{V_{m\_ref}}\right]\right\}\end{aligned} \tag{5.170}$$

By substituting f_{η_ref} for f_η in (5.168) and (5.169), the right side of (5.164) is also obtained and then $\alpha(f_\eta)$ can be resolved:

$$\alpha(f_\eta) = \left[\frac{E_{ref}(f_\eta)}{D_{ref}(f_\eta)} + \frac{l_{m\_ref}}{V_{m\_ref}} \tan\theta_{m\_ref}\right] \Big/ \left[\frac{1}{\cos\theta_d} + \frac{l_{m\_ref}}{V_{m\_ref}}(\tan\theta_{m\_ref} - \tan\theta_d)\right] \tag{5.171}$$

Following the concept and derivation of the NLCS algorithm based on the hyperbolic approximation, the first step of the NLCS algorithm based on the advanced hyperbolic approximation is to multiply a cubic phase filter, which is

$$H_{Y3}(f, f_\eta) = \exp\left\{j\frac{2\pi}{3}Y(f_\eta)f^3\right\} \tag{5.172}$$

After multiplication, the signal in (5.144) turns to

$$\begin{aligned}
S_{Y_3}(f, f_\eta; R_{mc}, \eta_{pc}) \approx \sigma_p & \exp\left\{-j2\pi f_\eta\left(\frac{R_{mc}\sin\theta_m}{V_m} + \eta_{pc}\right)\right\} \\
& \times \exp\left\{-j\frac{4\pi}{\lambda}\left[R_{mc}\cos\theta_m D(f_\eta) + \frac{R_{mc}\sin\theta_m l_m}{V_m}\right]\right\} \quad \text{(zeroth-order term)} \\
& \times \exp\left\{-j\frac{4\pi f}{c}\left[\frac{R_{mc}\cos\theta_m E(f_\eta)}{D(f_\eta)} + \frac{R_{mc}\sin\theta_m l_m}{V_m}\right]\right\} \quad \text{(linear term)} \\
& \times \exp\left\{-j\pi\frac{f^2}{b_r(f_\eta; R_{mc})}\right\} \quad \text{(quadratic term)} \\
& \times \exp\left\{j\frac{2\pi}{3}Y_3(f_\eta)f^3\right\} \quad \text{(cubic term)}
\end{aligned} \tag{5.173}$$

The cubic phase coefficient of the signal $Y_3(f_\eta)$ is combined with the cubic phase filter, where the range variance can be ignored:

$$Y_3(f_\eta) = Y(f_\eta) - \frac{3cR_{m,c\_ref}\cos\theta_{m\_ref} f_\eta^2 E_{ref}(f_\eta)}{4V_{m\_ref}^2 f_0^4 D_{ref}^5(f_\eta)} \tag{5.174}$$

In general bistatic SAR, the cubic phase term can hardly affect the stationary point, compared with the quadratic phase term. Thus, after the range inverse Fourier transform, the signal in the range-Doppler domain is

$$\begin{aligned}
S_{Y_3}(t, f_\eta; R_{mc}, \eta_{pc}) = \sigma_p & \exp\left\{-j2\pi f_\eta \left(\frac{R_{mc}\sin\theta_m}{V_m} + \eta_{pc}\right)\right\} \\
& \times \exp\left\{-j\frac{4\pi}{\lambda}\left[R_{mc}\cos\theta_m D(f_\eta) + \frac{R_{mc}\sin\theta_m l_m}{V_m}\right]\right\} \\
& \times \exp\{j\pi b_r(f_\eta; R_{mc})[t - t_m(f_\eta, R_{mc})]^2\} \\
& \times \exp\left\{j\frac{2\pi}{3}Y_3(f_\eta)b^3_{r_ref}(f_\eta)[t - t_m(f_\eta, R_{mc})]^3\right\}
\end{aligned} \tag{5.175}$$

The range variance of the cubic phase is also ignored by using $b_r(f_\eta; R_{m,c}) \approx b_{r\_ref}(f_\eta)$.

Afterwards, the NLCS function, centered on the RCM trajectory of the reference range and having quadratic coefficient $p(f_\eta)$ and cubic coefficient $q(f_\eta)$, is introduced:

$$H_{NLCS}(t, f_\eta) = \exp\{j\pi p(f_\eta)[t - t_{m\_ref}(f_\eta)]^2 + j\frac{2\pi}{3}q(f_\eta)[t - t_{m\_ref}(f_\eta)]^3\} \tag{5.176}$$

After multiplying (5.175) by (5.176), the range phase becomes

$$\begin{aligned}
\Theta_{rg}(t, f_\eta; R_{mc}) = \pi b_r(f_\eta; R_{mc})&[t - t_m(f_\eta; R_{mc})]^2 + \pi p(f_\eta)[t - t_{m_ref}(f_\eta)]^2 \\
& + \frac{2\pi}{3}Y_3(f_\eta)b^3_{r_ref}(f_\eta)[t - t_m(f_\eta; R_{mc})]^3 \\
& + \frac{2\pi}{3}q(f_\eta)[t - t_{m_ref}(f_\eta)]^3
\end{aligned} \tag{5.177}$$

To equalize the range FM rate and correct the differential RCM at the same time, two conditions should be satisfied, which jointly determine the three variables $Y_3(f_\eta)$, $p(f_\eta)$ and $q(f_\eta)$:

1. The first condition is that the zero frequency position of (5.177) should be at $t_{m,new}(f_\eta; R_{mc})$:

$$f(t_{m,new}(f_\eta; R_{mc}), f_\eta; R_{mc}) = \frac{1}{2\pi}\left.\frac{d\Theta_{rg}(t, f_\eta; R_{mc})}{dt}\right|_{t_{m,new}(f_\eta; R_{mc})} = 0 \tag{5.178}$$

By means of (5.163), (5.164) and (5.165), the above expression results in

$$\begin{aligned}[b_{r\_ref}(f_\eta) + k_s(f_\eta)\alpha(f_\eta)\Delta t_m][1 - \alpha(f_\eta)]\Delta t_m + p(f_\eta)\Delta t_m \\ + Y_3(f_\eta)b^3_{r\_ref}(f_\eta)[1 - \alpha(f_\eta)]^2\Delta t_m^2 + q(f_\eta)\Delta t_m^2 = 0\end{aligned} \quad (5.179)$$

where $\Delta t_m = t_m(f_{\eta\_ref}; R_{m,c}) - t_{m\_ref}(f_{\eta\_ref})$. Setting both the linear and the quadratic coefficients of Δt_m to zero, we can establish two equations:

$$\begin{cases} b_{r\_ref}(f_\eta)[1 - \alpha(f_\eta)] + p(f_\eta) = 0 & (1) \\ k_s(f_\eta)\alpha(f_\eta)[1 - \alpha(f_\eta)] + Y_3(f_\eta)b^3_{r\_ref}(f_\eta)[1 - \alpha(f_\eta)]^2 + q(f_\eta) = 0 & (2) \end{cases} \quad (5.180)$$

2. The second condition is that the range FM rate at $t_{m,new}(f_\eta; R_{mc})$ should be constant:

$$\left.\frac{df(t, f_\eta; R_{mc})}{dt}\right|_{t_{m,new}(f_\eta; R_{mc})} = b_{const}(f_\eta) \quad (5.181)$$

Similarly, by means of (5.163), (5.164) and (5.165), the above expression results in

$$\begin{aligned}b_{r\_ref}(f_\eta) + k_s(f_\eta)\alpha(f_\eta)\Delta t_m + p(f_\eta) + 2Y_3(f_\eta)b^3_{r\_ref}(f_\eta)[1 - \alpha(f_\eta)]\Delta t_m \\ + 2q(f_\eta)\Delta t_m = b_{const}(f_\eta)\end{aligned} \quad (5.182)$$

Setting the linear coefficient of Δt_m to zero, we can establish the third equation:

$$k_s(f_\eta)\alpha(f_\eta) + 2Y_3(f_\eta)b^3_{r\_ref}(f_\eta)[1 - \alpha(f_\eta)] + 2q(f_\eta) = 0 \quad (3) \quad (5.183)$$

Combining Equations (1), (2) and (3), we have

$$Y_3(f_\eta) = \frac{k_s(f_\eta)[\alpha(f_\eta) - 0.5]}{b^3_{r\_ref}(f_\eta)[\alpha(f_\eta) - 1]} \quad (5.184)$$

$$p(f_\eta) = b_{r\_ref}(f_\eta)[\alpha(f_\eta) - 1] \quad (5.185)$$

$$q(f_\eta) = \frac{k_s(f_\eta)[\alpha(f_\eta) - 1]}{2} \quad (5.186)$$

$$b_{const}(f_\eta) = b_{r\_ref}(f_\eta)\alpha(f_\eta) \quad (5.187)$$

The denominator in (5.184) implies that $|\alpha(f_\eta) - 1| \gg 0$ should be satisfied in order to avoid the effect of a cubic phase term on the stationary point; thus, the reference azimuth frequency should be chosen outside the Doppler bandwidth.

Substituting (5.184), (5.185), (5.186) and (5.187) into (5.177), the range phase changes to

$$\begin{aligned}\Theta_{rg}(t, f_\eta; R_{mc}) = &\frac{\pi k_s(f_\eta)\alpha^2(f_\eta)}{3[\alpha(f_\eta)-1]}[t - t_{m,new}(f_\eta; R_{mc})]^3 \\ &+\pi b_{const}(f_\eta)[t - t_{m,new}(f_\eta; R_{mc})]^2 + \Theta_0(f_\eta; R_{mc})\end{aligned} \tag{5.188}$$

where

$$\Theta_0(f_\eta; R_{mc}) = \pi b_{r\_ref}(f_\eta)\alpha(f_\eta)[\alpha(f_\eta)-1]\Delta t_m^2 + \frac{\pi}{3}k_s(f_\eta)\alpha^2(f_\eta)[\alpha(f_\eta)-1]\Delta t_m^3 \tag{5.189}$$

After range Fourier transform, the range phase in the range time domain is

$$\begin{aligned}\Theta_{rg}(f, f_\eta; R_{mc}) = &\frac{\pi k_s(f_\eta)\alpha^2(f_\eta) f^3}{3[\alpha(f_\eta)-1]b_{const}^3(f_\eta)} - \frac{\pi f^2}{b_{const}(f_\eta)} \\ &-2\pi f t_{m,new}(f_\eta; R_{mc}) + \Theta_0(f_\eta; R_{mc})\end{aligned} \tag{5.190}$$

Accordingly, the range filter for range compression, cubic phase compensation and RCMC is given by

$$\begin{aligned}H_r(f, f_\eta) = \exp\Bigg\{ &j\frac{\pi f^2}{b_{const}(f_\eta)} - j\frac{\pi k_s(f_\eta)\alpha^2(f_\eta) f^3}{3[\alpha(f_\eta)-1]b_{const}^3(f_\eta)} \\ &+j2\pi f\left[t_{m\_ref}(f_\eta) - \frac{2R_{mc\_ref}}{c}\right]\Bigg\}\end{aligned} \tag{5.191}$$

The azimuth filter is given by

$$H_a(f_\eta; R_{mc}) = \exp\left\{ j2\pi f_\eta \frac{R_{mc}\sin\theta_m}{V_m} + j\frac{4\pi}{\lambda}\left[R_{mc}\cos\theta_m D(f_\eta) + \frac{R_{mc}\sin\theta_m l_m}{V_m}\right]\right\} \tag{5.192}$$

Finally, the target is set at $t = 2R_{mc\_ref}/c + \Delta t_m$ in the range direction and $\eta = \eta_{pc}$ in the azimuth direction.

ω-k Algorithm Based on the Advanced Hyperbolic Approximation

As a comparable algorithm, the ω-k algorithm will be deduced in this section. As a result of the additional linear coefficient, the cross-coupling phase between the range and azimuth is accompanied by a linear phase with respect to k. Therefore, how to accommodate the range variance of V_m after Stolt interpolation needs further consideration. Alternatively, the invariance region is used here to ensure the imaging performance.

Let $2\pi(f + f_0)/c = k$ and $2\pi f_\eta / V_{m\_ref} = k_u$; the signal spectrum in terms of f and f_η, shown in (5.140), are transformed to that in terms of wavenumbers (spatial frequencies) k and k_u:

$$\begin{aligned} S(k, k_u; R_{mc}, \eta_{pc}) \approx \sigma_p \exp & \left\{ -j \left(k_u \frac{V_{m\_ref}}{V_m} + 2k \frac{l_m}{V_m} \right) R_{mc} \sin\theta_m - j k_u V_m \eta_{pc}] \right\} \\ & \times \exp\left\{ -j\pi \frac{f^2}{b} \right\} \\ & \times \exp\left\{ -j R_{mc} \cos\theta_m \sqrt{(2k)^2 - \left(k_u \frac{V_{m\_ref}}{V_m} + 2k \frac{l_m}{V_m} \right)^2} \right\} \end{aligned} \tag{5.193}$$

The reference function multiplication (RFM), applied in the two-dimensional wavenumber domain, is given by

$$\begin{aligned} H_{ref}(k, k_u; R_{mc\_ref}) = \exp & \left\{ j\pi \frac{f^2}{b} \right\} \exp\left\{ j \left(k_u + 2k \frac{l_{m\_ref}}{V_{m\_ref}} \right) R_{mc\_ref} \sin\theta_{m\_ref} \right\} \\ & \times \exp\left\{ j R_{mc\_ref} \cos\theta_{m\_ref} \sqrt{(2k)^2 - \left(k_u + 2k \frac{l_{m\_ref}}{V_{m\_ref}} \right)^2} \right\} \end{aligned} \tag{5.194}$$

If the following approximations are made, which are $V_m \approx V_{m\_ref}$ and $l_m / V_m \approx l_{m\_ref} / V_{m\_ref}$, then the phase remaining in the signal spectrum after RFM operation turns out to be as follows:

$$\begin{aligned} \Theta_{RFM}(k, k_u; R_{mc}, \eta_{pc}) = & -k_u V_m \eta_{pc} - k_u (R_{mc} \sin\theta_m - R_{mc\_ref} \sin\theta_{m\_ref}) \\ & -2k \frac{l_{m\_ref}}{V_{m\_ref}} (R_{mc} \sin\theta_m - R_{mc\_ref} \sin\theta_{m\_ref}) \\ & -(R_{mc} \cos\theta_m - R_{mc\_ref} \cos\theta_{m\_ref}) \\ & \times \sqrt{(2k)^2 - \left(k_u + 2k \frac{l_{m\_ref}}{V_{m\_ref}} \right)^2} \end{aligned} \tag{5.195}$$

Therefore, the Stolt interpolation based on the advanced hyperbolic approximation can be

$$k_x = \sqrt{(2k)^2 - [k_u + 2k(l_{m\_ref}/V_{m\_ref})]^2} \tag{5.196}$$

Then, the phase remaining in the signal spectrum after RFM turns out to be as follows:

$$\begin{aligned} \Theta_{RFM}(k, k_u; R_{mc}, \eta_{pc}) = &-k_u(V_m \eta_{pc} + R_{mc}\sin\theta_m - R_{mc\_ref}\sin\theta_{m\_ref}) \\ &-k_x(R_{mc}\cos\theta_m - R_{mc\_ref}\cos\theta_{m\_ref}) \\ &-\Gamma(k_x, k_u)(R_{mc}\sin\theta_m - R_{mc\_ref}\sin\theta_{m\_ref}) \end{aligned} \tag{5.197}$$

where

$$\Gamma(k_x, k_u) = \frac{-k_u + \sqrt{\left(k_u^2 + k_x^2\right) V_{m\_ref}^2 \Big/ l_{m\_ref}^2 - k_x^2}}{1 - V_{m\_ref}^2 \Big/ l_{m\_ref}^2} \tag{5.198}$$

Since V_m and l_m/V_m are considered to be range invariant in this algorithm, the targets other than the reference range will not be delicately focused. Besides, $\Gamma(k_x, k_u)$ contains coupling between k_u and k_x, and $R_{mc}\sin\theta_m - R_{mc\_ref}\sin\theta_{m\_ref}$ may not be zero in the range other than the reference range. These will aggravate the defocusing effect of the nonreference targets. Therefore, the ω-k algorithm based on the advanced hyperbolic approximation introduced here will not be better than the ω-k algorithm based on the hyperbolic approximation.

5.3.3.4 Simulations and Conclusions

In order to verify the performance of advanced hyperbolic approximation, the simulation results following the bistatic SAR geometry and parameters given in Figure 5.13 and Table 5.1 are shown.

Figure 5.17 demonstrates the compressed contours of the targets in the middle line along the range direction, respectively, using the RD, CS and ω-k algorithms based on the advanced hyperbolic approximation. For the RD and CS algorithms, the azimuth compression stage is arranged after range compression and RCMC stages, where the range variance of the azimuth signal can be accommodated conveniently, and thus the linear azimuth phase shown in the first exponential term in (5.144) can be corrected. As a result, the targets are located at the beam center crossing time, which behave as a straight line since a composite beam pattern is determined by the nonsquint receiving antenna in this simulation. For the ω-k algorithm, the range variance of V_m and l_m are

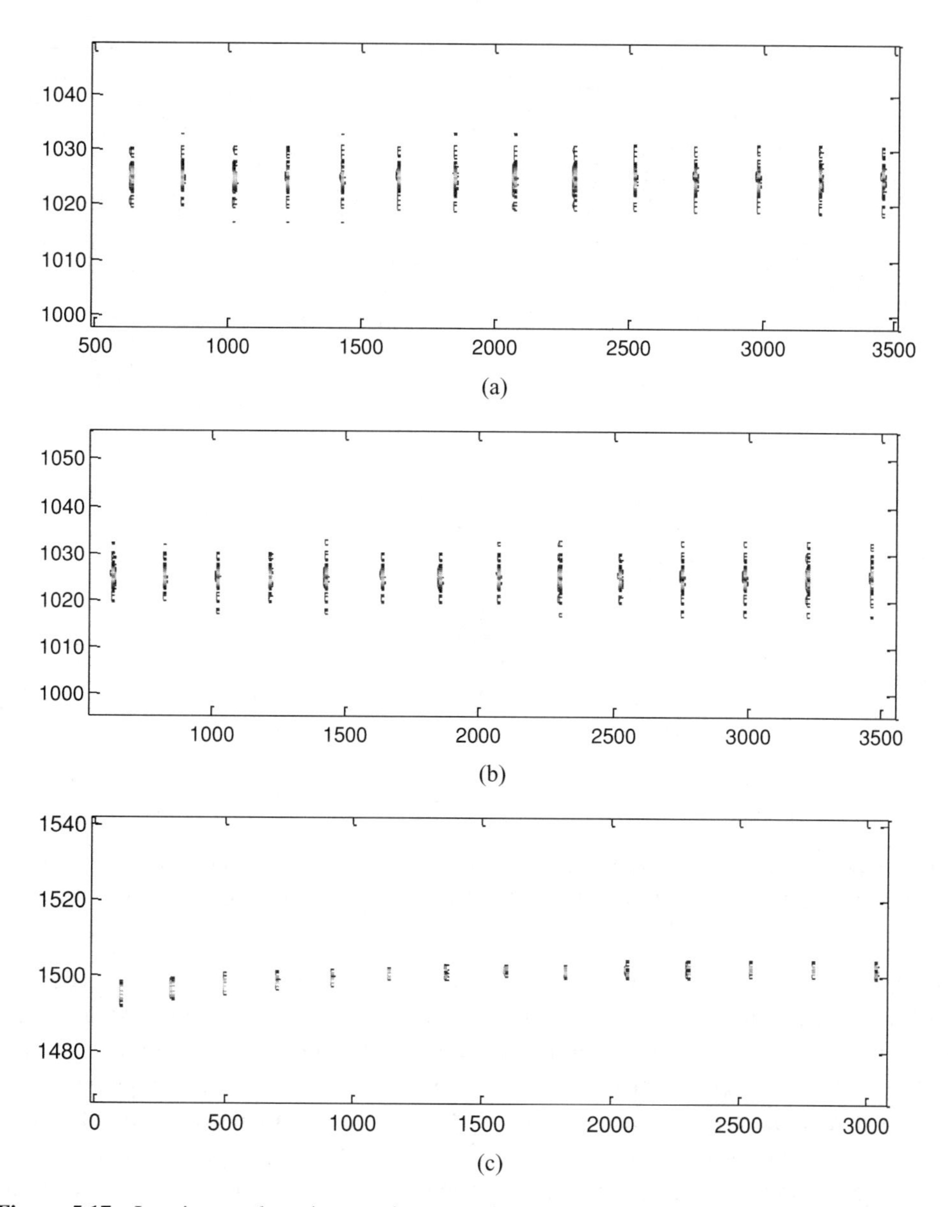

Figure 5.17 Imaging results using quasi-monostatic algorithms based on the advanced hyperbolic approximation: (a) RD algorithm, (b) CS algorithm, (c) ω-k algorithm

not considered in the Stolt interpolation operation as the target cannot be precisely located at the beam center. That is the reason why the targets at a horizontal line in the scene are distributed in a curved line in the image.

Figure 5.18 shows the range and azimuth impulse responses of the near-range target N using different algorithms. The range pulse responses of the RD and the CS

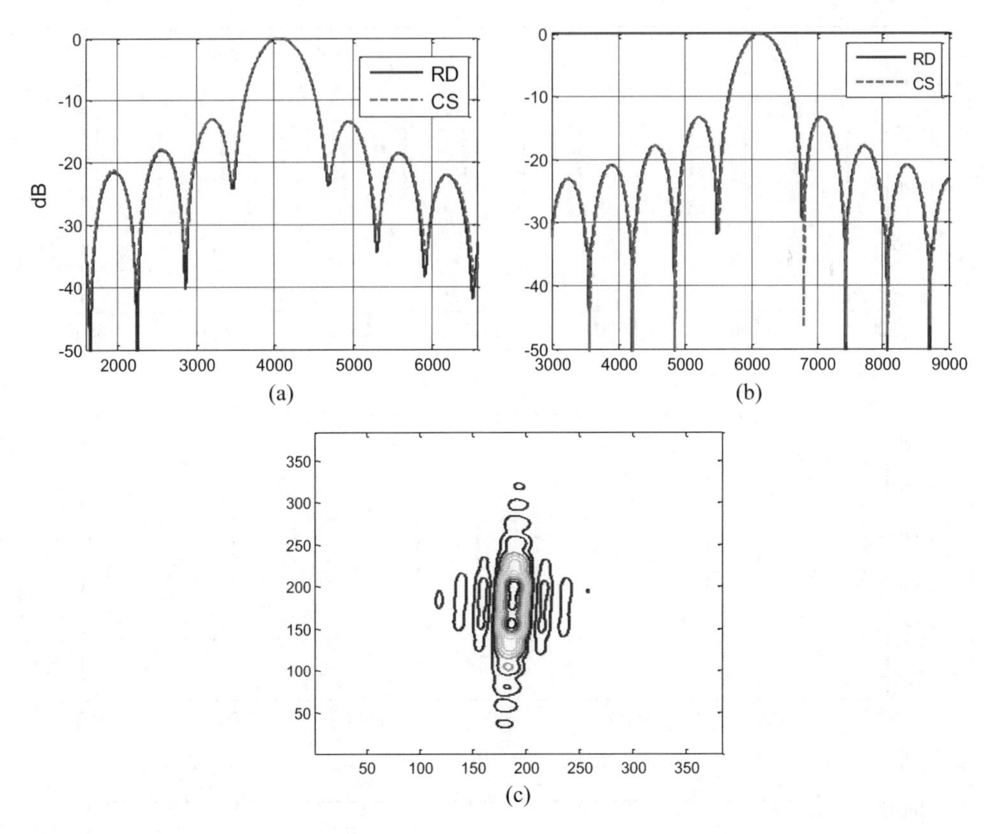

Figure 5.18 Impulse responses of the near-range target using different algorithms: (a) range impulse responses of RD and CS, (b) azimuth impulse responses of RD and CS, (c) magnitude contour of ω-k

algorithms are very similar to each other, as shown in Figure 5.18(a), and are as perfect as those given by the algorithms based on the hyperbolic approximation. Meanwhile, the azimuth dissymmetry between the left and right sidelobes disappears, as shown in Figure 5.18(b), because the advanced hyperbolic approximation compensates for the residual third-order terms ignored by the hyperbolic approximation. The two-dimensional contour of the ω-k algorithm in Figure 5.18(c) implies that the image quality deteriorates at the near range. This means that the ω-k algorithm based on the advanced hyperbolic approximation cannot handle the range variance very well, as the RD and the CS algorithms do.

The image qualities of the targets at near, central and far ranges are listed in Table 5.3. For the reference target, the three algorithms have almost theoretical qualities. For the targets at the edge of the swath, the RD and CS algorithms show similar qualities, where the range resolution slightly degrades since the range variances are ignored in SRC and higher-order range phase compensation; the ω-k algorithm severely degrades in azimuth performance because the range variance is ignored in the Stolt interpolation operation.

Table 5.3 Image qualities by different algorithms based on advanced hyperbolic approximation

		Range qualities				Azimuth qualities			
	Target	Resolution (m)	Widen ratio	PSLR (dB)	ISLR (dB)	Resolution (m)	Widen ratio	PSLR (dB)	ISLR (dB)
RD	N	0.266	0.887	−13.07	−10.84	0.195	0.898	−13.41	−10.71
	C	0.265	0.884	−13.43	−10.94	0.193	0.887	−13.11	−10.55
	F	0.265	0.884	−13.01	−10.80	0.192	0.884	−13.19	−10.56
CS	N	0.269	0.895	−13.13	−10.79	0.196	0.900	−13.41	−10.85
	C	0.266	0.887	−13.43	−10.93	0.194	0.888	−13.12	−10.55
	F	0.266	0.887	−12.96	−10.82	0.193	0.886	−13.12	−10.55
ω-k	N	0.291	0.970	−12.60	−9.96	0.625	2.92	−0.244	3.328
	C	0.265	0.884	−13.70	−11.64	0.194	0.888	−13.33	−10.86
	F	0.266	0.885	−13.13	−10.16	0.188	0.881	−10.82	−8.276

Note: The image qualities of each target are obtained by selecting a 32×32 chip centered on the target, upsampling by a factor of 512 and estimating along the peak direction. The range resolution here shows the resolution in the bistatic range dimension and corresponds to the bistatic range divided by 2.

5.4 Imaging Algorithms Based on Analytical Explicit Spectrums

All the imaging algorithms introduced in Section 5.3 start with simplifying the bistatic range equation in order to find a simple and analytical solution to the stationary point and then to derive the two-dimensional BPTRS to develop the bistatic SAR algorithms following the monostatic ones. These algorithms bring in an approximation error in the time domain and thus the precision of the two-dimensional BPTRS is restricted by the precision of the bistatic range history approximation; meanwhile, though the approximation error analysis is evident in the time domain, it is quite complicated in the two-dimensional frequency domain. Therefore, how the approximation errors affect the imaging quality in quantity is hard to measure and consequently hard to compensate. In order to overcome the restriction of the range history approximation and enhance the capability of controlling the approximation error, many researchers attempted to find the explicit BPTRS of bistatic SAR via other methods. Among them, some expressions of the spectrum are too complex, which makes it hard to develop algorithms based on these spectrums further [14]. By now, some explicit BPTRS are preferred and have been successful in developing imaging algorithms, including the LBF method proposed by Loffeld [3, 16, 17, 59–62], the MSR method proposed by Neo [18–20] and the IDW method by Zhang [2, 22]. These three methods will be described one by one in the following context.

5.4.1 *Imaging Algorithm Based on LBF*

The key concept of the LBF method in developing the imaging algorithm is to take the Taylor series expansion of the transmitted and received phases around their own stationary points, respectively, up to the quadratic terms, and then to obtain the approximate explicit BPTRS.

As illustrated in Figure 5.8, ignoring the antenna pattern and other weighting factors, the received bistatic SAR signal after range compression of target $P(x_p, y_p, z_p)$ can be written as

$$s_P(t,\eta) = \sigma_p \text{sinc}\left\{\frac{t - [R_T(\eta) + R_R(\eta)]/c}{1/B}\right\} \exp\left\{-j\frac{2\pi}{\lambda}[R_T(\eta) + R_R(\eta)]\right\} \tag{5.199}$$

The range Fourier transform turns the signal into

$$S_P^t(f,\eta) = \sigma_p \exp\left\{-j\frac{2\pi(f+f_0)}{c}[R_T(\eta) + R_R(\eta)]\right\} \tag{5.200}$$

After the azimuth Fourier transform, the two-dimensional spectrum is given by

$$\begin{aligned} S_P(f, f_\eta) &= \sigma_p \int_\eta \exp\left\{-j\frac{2\pi(f+f_0)}{c}[R_T(\eta) + R_R(\eta)] - j2\pi f_\eta \eta\right\} d\eta \\ &= \sigma_p \int_\eta \exp\left\{-j\phi_T(f, f_\eta, \eta) - j\phi_R(f, f_\eta, \eta)\right\} d\eta \end{aligned} \tag{5.201}$$

where

$$\phi_T(f, f_\eta, \eta) = \frac{2\pi(f+f_0)}{c} R_T(\eta) + \mu_T 2\pi f_\eta \eta \tag{5.202}$$

$$\phi_R(f, f_\eta, \eta) = \frac{2\pi(f+f_0)}{c} R_R(\eta) + \mu_R 2\pi f_\eta \eta \tag{5.203}$$

$R_T(\eta)$ and $R_R(\eta)$ are already shown in (5.22) and (5.23). The stationary point of ϕ_T and ϕ_R are denoted by $\tilde{\eta}_T$ and $\tilde{\eta}_R$, that is, $\partial\phi_T/\partial\eta|_{\tilde{\eta}_T} = 0$ and $\partial\phi_R/\partial\eta|_{\tilde{\eta}_R} = 0$. Here, μ_T and μ_R are the weighting factors that represent the different contributions of the transmitting and the receiving phases to the azimuth modulation. They are confined by

$$\mu_T + \mu_R = 1 \tag{5.204}$$

Taking the Taylor series expansion of ϕ_T and ϕ_R around $\tilde{\eta}_T$ and $\tilde{\eta}_R$, respectively, and keeping terms up to the second order, we have

$$S_P(f, f_\eta) = \sigma_p \exp\left\{-j\phi_T(f, f_\eta, \tilde{\eta}_T) - j\phi_R(f, f_\eta, \tilde{\eta}_R)\right\}$$
$$\times \int_\eta \exp\left\{-j \left.\frac{\partial^2\phi_T}{\partial\eta^2}\right|_{\tilde{\eta}_T} (\eta - \tilde{\eta}_T)^2 - j \left.\frac{\partial^2\phi_R}{\partial\eta^2}\right|_{\tilde{\eta}_R} (\eta - \tilde{\eta}_R)^2\right\} d\eta \tag{5.205}$$

It can be seen that the phase in the integration is a quadratic polynomial of η. Accordingly, its stationary point can be calculated by

$$\tilde{\eta}_b = \frac{\ddot{\phi}_T\tilde{\eta}_T + \ddot{\phi}_R\tilde{\eta}_R}{\ddot{\phi}_T + \ddot{\phi}_R} \tag{5.206}$$

where

$$\ddot{\phi}_T = \left.\frac{\partial^2\phi_T}{\partial\eta^2}\right|_{\tilde{\eta}_T} \quad \text{and} \quad \ddot{\phi}_R = \left.\frac{\partial^2\phi_R}{\partial\eta^2}\right|_{\tilde{\eta}_R}$$

Therefore, (5.205) can be simplified to

$$S_P(f, f_\eta) = \sigma_p \exp\left\{-j\phi_T(f, f_\eta, \tilde{\eta}_T) - j\phi_R(f, f_\eta, \tilde{\eta}_R)\right\}$$
$$\times \exp\left\{-j\frac{\ddot{\phi}_T\ddot{\phi}_R}{\ddot{\phi}_T + \ddot{\phi}_R}(\tilde{\eta}_T - \tilde{\eta}_R)^2\right\} \tag{5.207}$$

Using a sequence of derivations and resolutions, we can find the following results:

$$\tilde{\eta}_Y = \eta_{Y0} - \frac{R_{Y,0}}{V_Y^2}\frac{\mu_Y c f_\eta}{F_Y^{1/2}(f, f_\eta)} \tag{5.208}$$

$$\phi_Y(f, f_\eta, \tilde{\eta}_Y) = \frac{2\pi}{c}R_{Y0}F_Y^{1/2}(f, f_\eta) + \mu_Y 2\pi f_\eta \eta_{T0} \tag{5.209}$$

$$\ddot{\phi}_Y = \frac{2\pi(f + f_0)}{c}\frac{V_Y^2}{R_{Y0}}\frac{F_Y^{3/2}(f, f_\eta)}{(f + f_0)^3} \tag{5.210}$$

$$F_Y(f, f_\eta) = (f + f_0)^2 - \left(\frac{\mu_Y c f_\eta}{V_Y}\right)^2 \tag{5.211}$$

where the subscript "Y" can denote T or R, R_{Y0} is the closest range and η_{Y0} is the azimuth time corresponding to R_{Y0}. The relationships between R_{Y0}, η_{Y0} and the variables defined in (5.22) and (5.23) are as follows:

$$\eta_{Y0} = \eta_{pc} + \frac{R_{Yc}\sin\theta_Y}{V_Y} \tag{5.212}$$

$$R_{Y0} = R_{Yc}\cos\theta_Y \tag{5.213}$$

As a result, if an explicit expression of the spectrum, which is denoted in (5.207), is desired, the weighting factors μ_T and μ_R should be resolved. In the original LBF, both the transmitting phase and receiving phase are equally weighted, which means $\mu_T = \mu_R = 0.5$. Further study reveals that the equally weighted LBF will suffer from a significant degradation if the transmitter and the receiver have very different distances to the scenario, or have very different velocities relative to the scenario, or have very different squint angles. Extended LBFs with unequal weighting factors are proposed in [60], [63] and [64], which enhance the precision of the LBF spectrum. In [63] and [64], the weighting factors are fixed according to the TBP of the Doppler FM rate caused by the transmitter and receiver, respectively. In [60], a better method of optimizing the weighting factors are proposed, which will be introduced in the following.

From the above derivations we know that if $\tilde{\eta}_T = \tilde{\eta}_R = \tilde{\eta}_b$, then $\tilde{\eta}_b$ is exactly the bistatic phase stationary point that satisfies $\left.\frac{\partial(\phi_T + \phi_R)}{\partial\eta}\right|_{\tilde{\eta}_b} = 0$. This is the case we expect. It is because the analytical solution of $\left.\frac{\partial(\phi_T + \phi_R)}{\partial\eta}\right|_{\tilde{\eta}_b} = 0$ is hard to obtain that the above LBF method was thought out. Therefore, in order to obtain the proper weighting factors, we try to make $\tilde{\eta}_T$ and $\tilde{\eta}_R$ as close as to $\tilde{\eta}_b$. Then we get

$$\begin{aligned}\eta_{T0} - \eta_{R0} - \frac{R_{T0}}{V_T^2}\frac{\mu_T c f_\eta}{\sqrt{(f+f_0)^2 - \left(\mu_T c f_\eta / V_T\right)^2}} \\ + \frac{R_{R0}}{V_R^2}\frac{\mu_R c f_\eta}{\sqrt{(f+f_0)^2 - \left(\mu_R c f_\eta / V_R\right)^2}} \to 0\end{aligned} \tag{5.214}$$

Expanding the square root terms in (5.214) to their Taylor series, we have

$$\begin{aligned}\eta_{T0} - \eta_{R0} - \left(\frac{R_{T0}}{V_T^2}\frac{\mu_T c f_\eta}{f+f_0} - \frac{R_{R0}}{V_R^2}\frac{\mu_R c f_\eta}{f+f_0}\right) - \left\{\frac{R_{T0}}{V_T^2} O\left[\left(\frac{\mu_T c f_\eta / V_T}{f+f_0}\right)^2\right]\right. \\ \left. - \frac{R_{R0}}{V_R^2} O\left[\left(\frac{\mu_R c f_\eta / V_R}{f+f_0}\right)^2\right]\right\} \to 0\end{aligned} \tag{5.215}$$

If all the higher-order terms are taken out of consideration, then the following equation is obtained:

$$\eta_{T0} - \eta_{R0} = \frac{R_{T0}}{V_T^2}\frac{\mu_T c f_\eta}{f + f_0} - \frac{R_{R0}}{V_R^2}\frac{\mu_R c f_\eta}{f + f_0} \tag{5.216}$$

Together with the equation $\mu_T + \mu_R = 1$, the weighting factors can be resolved, which are

$$\begin{cases} \mu_T = \dfrac{V_T^2/R_{T0}}{V_T^2/R_{T0} + V_R^2/R_{R0}}\left[1 + (\eta_{T0} - \eta_{R0})\dfrac{V_R^2}{R_{R0}}\dfrac{f + f_0}{c f_\eta}\right] \\ \mu_R = \dfrac{V_R^2/R_{R0}}{V_T^2/R_{T0} + V_R^2/R_{R0}}\left[1 - (\eta_{T0} - \eta_{R0})\dfrac{V_T^2}{R_{T0}}\dfrac{f + f_0}{c f_\eta}\right] \end{cases} \tag{5.217}$$

It can be seen that μ_T and μ_R are functions of f and f_η, so the spectrum in (5.207) is quite complex. However, when $\eta_{T0} - \eta_{R0} = 0$, the weighting factors can be independent of f and f_η, based on which a relatively simpler spectrum can be obtained. Moreover, if $\eta_{T0} - \eta_{R0} = 0$ and $V_T^2/R_{T0} = V_R^2/R_{R0}$, then $\mu_T = \mu_R = 0.5$, which reduces to the basic LBF.

Let us firstly consider the basic LBF for the translational invariant configuration. With $\mu_T = \mu_R = 0.5$, $V_T = V_R = V$ and substituting (5.208) to (5.217) into (5.207), the LBF spectrum is as follows:

$$\begin{aligned} S_P(f, f_\eta) = \sigma_p \exp&\left\{-j\frac{2\pi}{c}(R_{T0} + R_{R0})F^{1/2}(f, f_\eta) - j2\pi f_\eta \frac{\eta_{T0} + \eta_{R0}}{2}\right\} \\ \times \exp&\left\{-j\frac{2\pi}{c}\frac{F^{3/2}(f, f_\eta)}{(f + f_0)^2}\frac{1}{R_{T0} + R_{R0}}\right. \\ &\times\left.\left[\eta_{T0} - \eta_{R0} - \frac{R_{T0} - R_{R0}}{V^2}\frac{c f_\eta/2}{F^{1/2}(f, f_\eta)}\right]^2\right\} \end{aligned} \tag{5.218}$$

where

$$F(f, f_\eta) = (f + f_0)^2 - \left(\frac{c f_\eta}{2V}\right)^2 \tag{5.219}$$

It can be seen that the first phase term in (5.218) is similar to the phase spectrum of monostatic SAR, so it is usually named the quasi-monostatic term, while the second phase term in (5.218) is called the bistatic deformation term.

Moreover, in the tandem configuration, $R_{T0} = R_{R0} = R_0$; then Equation (5.218) becomes

$$S_{P,tandem}(f, f_\eta) = \sigma_p \exp\left\{-j\frac{4\pi}{c}R_0 F^{1/2}(f, f_\eta) - j2\pi f_\eta \frac{\eta_{T0} + \eta_{R0}}{2}\right\} \times \exp\left\{-j\frac{2\pi}{c}\frac{F^{3/2}(f, f_\eta)}{(f + f_0)^2}\frac{1}{2R_0}(\eta_{T0} - \eta_{R0})^2\right\} \tag{5.220}$$

where it can be found that the quasi-monostatic term is exactly the phase spectrum of some monostatic SAR signal. As a result, we can consider that the tandem bistatic SAR raw data can be transformed into some monostatic SAR raw data through a range-variant filter, which is

$$H_{BD,tandem}(f, f_\eta; R_0) = \exp\left\{j\frac{2\pi}{c}\frac{F^{3/2}(f, f_\eta)}{(f + f_0)^2}\frac{1}{2R_0}(\eta_{T0} - \eta_{R0})^2\right\} \tag{5.221}$$

Here, $H_{BD,tandem}(f, f_\eta; R_0)$ can be considered as the approximate filter of "Rocca's smile"; it is range variant, which is consistent with the conclusion of the DMO method. For more explanation about the relationship between LBF and DMO, the reader can refer to [20].

Based on the BPTRS obtained through LBF, a simple but lack of precision way for bistatic SAR focusing is ignoring the range variance of the bistatic deformation phase term and firstly compensating the bistatic deformation phase term according to that of the reference range, and then using quasi-monostatic SAR imaging algorithms to handle the quasi-monostatic phase term. Imaging algorithms with higher precision for the translational invariant case have been proposed based on LBF; one can refer to [62] to [65].

5.4.2 *Imaging Algorithm Based on MSR*

The method of series reversion (MSR) proposed by Neo is another approach used to obtain an approximate BPTRS. The principle of MSR can be explained as follows: if y can be written as the polynomial series of x with no constant term, that is,

$$y = a_1 x + a_2 x^2 + a_3 x^3 + \cdots \tag{5.222}$$

then x can also be written as the polynomial series of y, that is,

$$x = A_1 y + A_2 y^2 + A_3 y^3 + \cdots \tag{5.223}$$

Substituting (5.223) into (5.222), we have

$$\begin{aligned} y &= a_1(A_1 y + A_2 y^2 + A_3 y^3 + \cdots) + a_2(A_1 y + A_2 y^2 + A_3 y^3 + \cdots)^2 \\ &\quad + a_3(A_1 y + A_2 y^2 + A_3 y^3 + \cdots)^3 + \cdots \\ &= a_1 A_1 y + (a_1 A_2 + a_2 A_1^2) y^2 + (a_1 A_3 + 2a_2 A_1 A_2 + a_3 A_1^3) y^3 + \cdots \end{aligned} \tag{5.224}$$

Equating the corresponding coefficients, we have

$$\begin{aligned} A_1 &= a_1^{-1} \\ A_2 &= -a_2 a_1^{-3} \\ A_3 &= a_1^{-5}(2a_2^2 - a_1 a_3) \\ &\vdots \end{aligned} \tag{5.225}$$

Inspired by the principle of MSR, the bistatic range history is represented as the polynomial of η. Because of such a range history equation, when POSP is applied to find the stationary point $\tilde{\eta}_b$, an equation where f_η is expressed by polynomial series of $\tilde{\eta}_b$ is obtained. Then an analytical solution of $\tilde{\eta}_b$, which is represented by the polynomial of f_η, can be achieved. In addition, the two-dimensional analytical BPTRS can be deduced and the precision of the BPTRS can be controlled by adjusting the kept order of the polynomial.

According to the above idea, the bistatic range, R_{bi}, is expanded to its Taylor series around $\eta_{p,c}$. Let $\bar{\eta} = \eta - \eta_{p,c}$; we then have[1]

$$R_{bi}(\bar{\eta}) = R_{bi\_cen} + k_1\bar{\eta} + k_2\bar{\eta}^2 + k_3\bar{\eta}^3 + k_4\bar{\eta}^4 + \cdots = k_1\bar{\eta} + R_{bi1}(\bar{\eta}) \tag{5.226}$$

where

$$k_1 = \frac{dR_T(\bar{\eta})}{d\bar{\eta}}\Big|_{\bar{\eta}=0} + \frac{dR_R(\bar{\eta})}{d\bar{\eta}}\Big|_{\bar{\eta}=0} = -V_T \sin\theta_T - V_R \sin\theta_R \tag{5.227}$$

$$k_2 = \frac{1}{2!}\left[\frac{d^2R_T(\bar{\eta})}{d\bar{\eta}^2}\Big|_{\bar{\eta}=0} + \frac{d^2R_R(\bar{\eta})}{d\bar{\eta}^2}\Big|_{\bar{\eta}=0}\right] = \frac{V_T^2\cos\theta_T}{R_{Tc}} + \frac{V_R^2\cos\theta_R}{R_{Rc}} \tag{5.228}$$

$$k_3 = \frac{1}{3!}\left[\frac{d^3R_T(\bar{\eta})}{d\bar{\eta}^3}\Big|_{\bar{\eta}=0} + \frac{d^3R_R(\bar{\eta})}{d\bar{\eta}^3}\Big|_{\bar{\eta}=0}\right] = \frac{3V_T^3\cos^2\theta_T\sin\theta_T}{R_{Tc}^2} + \frac{3V_R^3\cos^2\theta_R\sin\theta_R}{R_{Rc}^2} \tag{5.229}$$

$$\begin{aligned} k_4 &= \frac{1}{4!}\left[\frac{d^4R_T(\bar{\eta})}{d\bar{\eta}^4}\Big|_{\bar{\eta}=0} + \frac{d^4R_R(\bar{\eta})}{d\bar{\eta}^4}\Big|_{\bar{\eta}=0}\right] \\ &= \frac{3V_T^4\cos^2\theta_T(4\sin^2\theta_T - \cos^2\theta_T)}{R_{Tc}^3} + \frac{3V_R^4\cos^2\theta_R(4\sin^2\theta_R - \cos^2\theta_R)}{R_{Rc}^3} \end{aligned} \tag{5.230}$$

[1]Formula 5.222–5.225: from Yew Lam Neo; Wong, F.; Cumming, I.G. "A Two-Dimensional Spectrum for Bistatic SAR Processing Using Series Reversion", *IEEE Geoscience and Remote Sensing Letters*, **4**(1), pp. 93–96, 2007.

On condition that $k_1 = 0$, the signal of target P after range compression is given by

$$s_{1p}(t, \bar{\eta}) = \sigma_p \ \text{sinc}\left[t - \frac{R_{bi1}(\bar{\eta})}{c}\right] \exp\left\{-j\frac{2\pi}{\lambda} R_{bi1}(\bar{\eta})\right\} \tag{5.231}$$

Accordingly, the two-dimensional spectrum is given by

$$\begin{aligned} S_{1p}(f, f_{\bar{\eta}}) &= \sigma_p \int_{\bar{\eta}} \exp\left\{-j\frac{2\pi(f + f_0)}{c} R_{bi1}(\bar{\eta}) - j2\pi f_{\bar{\eta}}\bar{\eta}\right\} d\bar{\eta} \\ &\approx \sigma_p \exp\left\{-j\frac{2\pi(f + f_0)}{c} R_{bi1}(\tilde{\bar{\eta}}_b) - j2\pi f_{\bar{\eta}}\tilde{\bar{\eta}}_b\right\} \end{aligned} \tag{5.232}$$

The stationary point $\tilde{\bar{\eta}}_b$ satisfies

$$-f_\eta = (f + f_0/c) \cdot \left.\frac{\partial R_{bi1}(\bar{\eta})}{\partial \bar{\eta}}\right|_{\tilde{\bar{\eta}}_b}$$

that is,

$$-\left(\frac{c}{f + f_0}\right) f_{\bar{\eta}} = 2k_2\tilde{\bar{\eta}}_b + 3k_3\tilde{\bar{\eta}}_b^2 + 4k_4\tilde{\bar{\eta}}_b^3 + \cdots \tag{5.233}$$

Using MSR, the stationary point $\tilde{\bar{\eta}}_b$ can be represented as the polynomial of $f_{\bar{\eta}}$, which is

$$\tilde{\bar{\eta}}_b(f_{\bar{\eta}}) = A_1\left(-\frac{c}{f + f_0} f_{\bar{\eta}}\right) + A_2\left(-\frac{c}{f + f_0} f_{\bar{\eta}}\right)^2 + A_3\left(-\frac{c}{f + f_0} f_{\bar{\eta}}\right)^3 + \cdots \tag{5.234}$$

where

$$\begin{aligned} A_1 &= 1/2k_2 \\ A_2 &= -\frac{3k_3}{8k_2^3} \\ A_3 &= \frac{9k_3^2 - 4k_2k_4}{16k_2^5} \\ &\vdots \end{aligned} \tag{5.235}$$

Substituting (5.234) and (5.233) into (5.232), the two-dimensional spectrum becomes

$$\begin{aligned} S_{1p}(f, f_{\bar{\eta}}) \approx \sigma_p \exp\left\{-j\frac{2\pi(f+f_0)}{c}R_{bi\_cen}\right\} \\ \times \exp\left\{j2\pi\frac{c}{4k_2(f_0+f)}f_{\bar{\eta}}^2\right\} \\ \times \exp\left\{j2\pi\frac{c^2k_3}{8k_2^3(f+f_0)^2}f_{\bar{\eta}}^3\right\} \\ \times \exp\left\{j2\pi\frac{c^3(9k_3^2-4k_2k_4)}{64k_2^5(f+f_0)^3}f_{\bar{\eta}}^4\right\} \end{aligned} \tag{5.236}$$

Consider the case when $k_1 \neq 0$; the signal of target P after range compression is given by

$$\begin{aligned} s_p(t, \bar{\eta}) &= \sigma_p \text{sinc}\left(t - \frac{R_{bi1}(\bar{\eta}) + k_1\bar{\eta}}{c}\right)\exp\left\{-j\frac{2\pi}{\lambda}[R_{bi1}(\bar{\eta}) + k_1\bar{\eta}]\right\} \\ &= s_{1p}\left(t - \frac{k_1\bar{\eta}}{c}, \bar{\eta}\right)\exp\left\{-j\frac{2\pi f_0}{c}k_1\bar{\eta}\right\} \end{aligned} \tag{5.237}$$

In terms of the following Fourier transform properties

$$\begin{aligned} g(t, \eta) &\leftrightarrow G(f, f_\eta) \\ g(t, \eta)\exp\{-j2\pi f_k \eta\} &\leftrightarrow G(f, f_\eta + f_k) \\ g(t - \kappa\eta, \eta) &\leftrightarrow G(f, f_\eta + \kappa f) \end{aligned} \tag{5.238}$$

we arrive at

$$S_p(f, f_{\bar{\eta}}) = S_{1p}\left(f, f_{\bar{\eta}} + (f_0 + f)\frac{k_1}{c}\right) \tag{5.239}$$

Therefore, the BPTRS is given by

$$\begin{aligned} S_p(f, f_{\bar{\eta}}) \approx \sigma_p \exp\left\{-j\frac{2\pi(f+f_0)}{c}R_{bi\_cen}\right\} \\ \times \exp\left\{j2\pi\frac{c}{4k_2(f_0+f)}\left[f_{\bar{\eta}}+(f+f_0)\frac{k_1}{c}\right]^2\right\} \\ \times \exp\left\{j2\pi\frac{c^2k_3}{8k_2^3(f+f_0)^2}\left[f_{\bar{\eta}}+(f+f_0)\frac{k_1}{c}\right]^3\right\} \\ \times \exp\left\{j2\pi\frac{c^3(9k_3^2-4k_2k_4)}{64k_2^5(f+f_0)^3}\left[f_{\bar{\eta}}+(f+f_0)\frac{k_1}{c}\right]^4\right\} \end{aligned} \quad (5.240)$$

The above approximate two-dimensional BPTRS keeps the polynomial expression to the fourth order. Moreover, its precision can be enhanced by increasing the kept order. It should be noticed that the coefficients k_1, k_2 and so on are all varying with the target position (while in a translational invariant configuration they are range variant). This increases the difficulty of bistatic SAR focusing based on the above BPTRS. Some imaging algorithms have been developed based on (5.240) by using some approximations in which the processing filters have complex analytical expressions, which can be seen in [19], [58], [66] and [67].

5.4.3 Imaging Algorithm Based on IDW

Unlike the previous two methods, the method of instantaneous Doppler wavenumber (IDW) uses the free space wave equation and the concept of wavenumber in order to derive the BPTRS directly in the wavenumber domain.

For the purpose of allowing one to understand the IDW method easily, the wavenumber concept is presented first. Actually, the term "wavenumber" has been mentioned in Chapter 2 when the ω-k algorithm was introduced; herein its definition is further specified. The "wavenumber" here represents the angular wavenumber, which defines the number of wavelengths per 2π of unit distance, that is, the number of 2π in the phase displacement per unit distance. For example, f_c is the carrier frequency of the electromagnetic wave and v is the wave propagating velocity, the magnitude of the wavenumber is $K = 2\pi f_c/v$ and the direction of the wavenumber is the same as the wave propagating direction. If the wave propagates along $\boldsymbol{R}$, the signal expression in the time-spatial domain is given by $\exp\{j2\pi f_c t - j\boldsymbol{K}\cdot\boldsymbol{R}\}$. As the time domain signal can be transformed and analyzed in the frequency domain, the spatial domain signal can be analyzed in the corresponding wavenumber domain. The difference is that the time domain is one dimensional and so is the frequency domain, but the spatial domain is multidimensional and thus the wavenumber is directional.

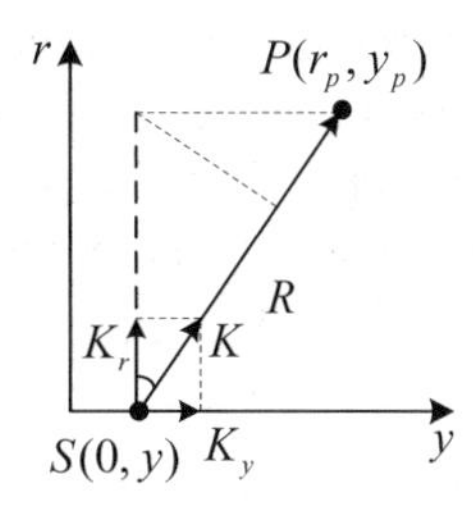

Figure 5.19 Wavenumber analysis of monostatic SAR

Let us first derive the monostatic SAR point target spectrum using the IDW method, so as to better understand this method.

As shown in Figure 5.19, in the slant range plane denoted by (r, y), SAR is located at $\boldsymbol{S}:(0,0)$ when the azimuth time is zero and is located at $\boldsymbol{S}:(0,y)$ when the time is η and the target P is located at $\boldsymbol{P}:(r_p, y_p)$. The transmitted signal reaches the target and is then scattered back to be received by the radar. Let the carrier frequency of the transmitted signal be $f_0 + f$; then the received signal after demodulation is given by $s_p(y) = \exp\{-j(\boldsymbol{K}_T \bullet \boldsymbol{R}_T + \boldsymbol{K}_R \bullet \boldsymbol{R}_R)\}$, where $\boldsymbol{R}_T = \boldsymbol{P} - \boldsymbol{S}$, $\boldsymbol{R}_R = \boldsymbol{S} - \boldsymbol{P}$, $K_T = K_R = 2\pi(f_0 + f)/c$ in magnitude and the direction of $\boldsymbol{K}_T$ and $\boldsymbol{K}_R$ are identical to that of $\boldsymbol{R}_T$ and $\boldsymbol{R}_R$, respectively. With no effect on analysis, $\boldsymbol{K}_T$ and $\boldsymbol{K}_R$ can be uniformed by $\boldsymbol{K}$, the direction of which is the same with $\boldsymbol{K}_T$; $\boldsymbol{R}_T$ and $\boldsymbol{R}_R$ can be uniformed by $\boldsymbol{R}$, the direction of which is the same with $\boldsymbol{R}_T$. As a result, the received signal turns to $s_p(y) = \exp\{-j2\boldsymbol{K} \bullet \boldsymbol{R}\}$. Though the direction of $\boldsymbol{K}$ and $\boldsymbol{R}$ change continuously as the radar moves, they are identical to each other all along. Therefore, $s_p(y) = \exp\{-j2KR\}$, where $R = r_p \cos\theta + (y_p - y)\sin\theta$, and then we arrive at

$$s_p(y) = \exp\{-j2Kr_p \cos\theta - j2K(y_p - y)\sin\theta\} \tag{5.241}$$

Here, θ is a function of y.

Similar to the Fourier transform from the time domain into the frequency domain, the above signal can be transformed into the wavenumber domain corresponding to y, which is given by

$$S_p(K_y) = \int_y \exp\{-j2Kr_p \cos\theta - j2K(y_p - y)\sin\theta\} \exp\{-jK_y y\} dy \tag{5.242}$$

In a similar way to how we get the SAR spectrum in the frequency domain, the POSP is used here to resolve the above integrand. According to POSP, the stationary point $\tilde{y}$ satisfies the following equation:

$$-2K \left.\frac{dR}{dy}\right|_{\tilde{y}} = K_y \tag{5.243}$$

This means that the stationary point in the azimuth distance (i.e., y) for a given wavenumber is the azimuth position, where the instantaneous wavenumber of the received signal equals the given wavenumber. Suppose that for a given instantaneous wavenumber K_{yA} the corresponding stationary point is y_A and corresponding angle is θ_A. Consider the relationship between K_{yA}, y_A and θ_A. We then have

$$K_{yA} = \frac{2\pi f_{\eta A}}{V} = \frac{2\pi}{V}\frac{2V\sin\theta_A}{\lambda} = \frac{2\pi(f+f_0)}{c}2\sin\theta_A = 2K\sin\theta_A \tag{5.244}$$

Thus,

$$\sin\theta_A = \frac{K_{yA}}{2K} \tag{5.245}$$

Using (5.244), the solution to the integrand (5.242) can be written as

$$\begin{aligned} S_p(K_{yA}) &= \exp\{-j2Kr_p\cos\theta_A - jK_{yA}(y_p - y_A)\}\exp\{-jK_{yA}y_A\} \\ &= \exp\{-j2Kr_p\sqrt{1-(K_{yA}/2K)^2} - jK_{yA}y_p\} \end{aligned} \tag{5.246}$$

Supposing the transmitted signal is a wide-band signal and the range compression is already done, according to (5.244) the received signal in the wavenumber domain can be rewritten as

$$S_p(K, K_y) = \exp\{-jr_p\sqrt{(2K)^2 - K_y^2} - jK_y y_p\} \tag{5.247}$$

According to the relationship between the frequency and wavenumber, the received signal in the frequency domain is obtained, which is

$$S_p(f, f_\eta) = \exp\left\{-jr_p\sqrt{\left[\frac{2\pi(f+f_0)}{c/2}\right]^2 - \left[\frac{2\pi f_\eta}{V}\right]^2} - j\frac{2\pi f_\eta}{V}y_p\right\} \tag{5.248}$$

This is simply the two-dimensional spectrum expression of monostatic SAR that we know from Chapter 2.

In the same way, let us now derive the BPTRS in both the wavenumber and two-dimensional frequency domain. We firstly start with the tandem configuration. We suppose that when the azimuth time is zero, the middle point of the transmitting antennas and the receiving antennas is at (0, 0). At a given time η, the

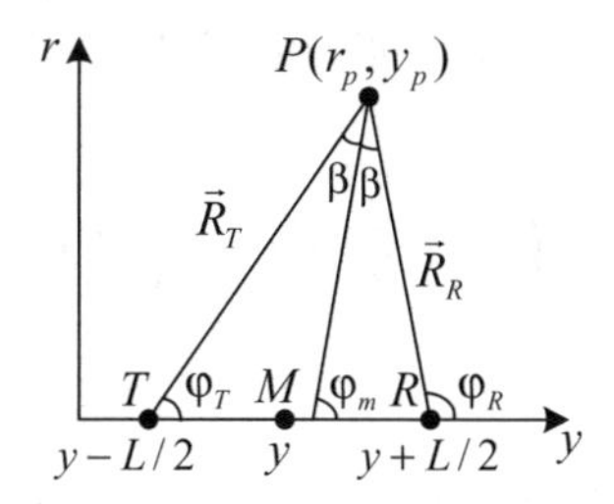

Figure 5.20 Wavenumber analysis of bistatic SAR of the tandem configuration

middle point of the transmitting and receiving antennas, M, is located at $(0, y)$ (as shown in Figure 5.20). At this place, the received signal of target P can be written as

$$\begin{aligned} s_{p,tandem}(y) &= \exp\{-jK(R_T + R_R)\} \\ &= \exp\{-jKr_p(\sin\varphi_T + \sin\varphi_R) \\ &\quad -jK[(y_p - y + L/2)\cos\varphi_T + (y_p - y - L/2)\cos\varphi_R]\} \end{aligned} \tag{5.249}$$

According to the geometrical relationship in Figure 5.20, we have

$$\begin{cases} \varphi_m = (\varphi_T + \varphi_R)/2 \\ \beta = (\varphi_R - \varphi_T)/2 \end{cases} \tag{5.250}$$

where β is half of the bistatic angle and φ_m is the included angle between the bistatic angular bisector and the y axis. Then, in terms of the trigonometric functions, (5.249) can be simplified to

$$\begin{aligned} s_{p,tandem}(y) = \exp\bigg\{ &-j2K\cos\beta\sin\varphi_m\left(r_p + \frac{L}{2}\tan\beta\right) \\ &-j2K\cos\beta\cos\varphi_m(y_p - y)\bigg\} \end{aligned} \tag{5.251}$$

In the tandem mode, the instantaneous Doppler frequency is given by

$$f_\eta = \frac{V}{\lambda}(\cos\varphi_T + \cos\varphi_R) = \frac{2V}{\lambda}\cos\varphi_m\cos\beta \tag{5.252}$$

Thus, the instantaneous Doppler wavenumber is given by

$$K_y = \frac{2\pi f_\eta}{V} = \frac{4\pi}{\lambda}\cos\varphi_m\cos\beta = 2K\cos\beta\cos\varphi_m \tag{5.253}$$

Using the POSP and (5.253), the signal spectrum in the wavenumber domain is written as

$$S_{p,tandem}(K, K_y) = \exp\left\{-j\left(r_p + \frac{L}{2}\tan\beta\right)\sqrt{(2K\cos\beta)^2 - K_y^2} - jK_y y_p\right\} \tag{5.254}$$

It can be seen that the signal spectrum in the wavenumber domain of the tandem bistatic SAR highly resembles that of the monostatic SAR. However, since β is variant with range r_p, the wavenumber cannot be completely independent of the range in the first phase term, which is different from monostatic SAR. Moreover, β is changing while the transmitter and the receiver are moving, so it is also azimuth variant and hence is variant with K_y. To make the spectrum explicit and make the development of algorithms based on this spectrum easier, the azimuth variance of β is usually ignored, which makes the spectrum an approximated one. If the range variance of β is also ignored, the following filter applied in the two-dimensional wavenumber domain can transform the bistatic SAR raw data into the equivalent monostatic SAR (denoted as M in Figure 5.20) raw data.

$$H_L(K, K_y) = \exp\left\{j\frac{L}{2}\tan\beta_{ref}\sqrt{(2K\cos\beta_{ref})^2 - K_y^2}\right\} \tag{5.255}$$

This can be considered as another approximated implementation of DMO in the wavenumber domain. The approximated equivalent monostatic SAR has a new wavenumber variable, which is $K_m = K\cos\beta_{ref}$.

Let us now consider the general translational invariant configuration. As shown in Figure 5.21, at a given moment, the middle point of the transmitting and receiving antennas is located at (x_M, y_M, z_M), the transmitting and receiving antennas are located at $(x_M - L_x/2, y_M - L_y/2, z_M - L_z/2)$ and $(x_M + L_x/2, y_M + L_y/2, z_M + L_z/2)$, respectively, where

$$\begin{cases} L_x = x_R - x_T \\ L_y = y_R - y_T \\ L_z = z_R - z_T \end{cases} \tag{5.256}$$

We also suppose that $y_M = 0$ when the azimuth time is zero. Then, at a given moment, the received signal from target P is $s_{p,\mathrm{TI}}(y_M) = \exp\{-jK(R_T + R_R)\}$ and

$$\begin{cases} R_T = R_{T0}\cos\theta_T + (y_p - y_T)\sin\theta_T = \left(R_{M0} - \dfrac{\Delta R_0}{2}\right)\cos\theta_T + \left(y_p - y_M + \dfrac{L_y}{2}\right)\sin\theta_T \\ R_R = R_{R0}\cos\theta_R + (y_p - y_R)\sin\theta_R = \left(R_{M0} + \dfrac{\Delta R_0}{2}\right)\cos\theta_R + \left(y_p - y_M - \dfrac{L_y}{2}\right)\sin\theta_R \end{cases} \tag{5.257}$$

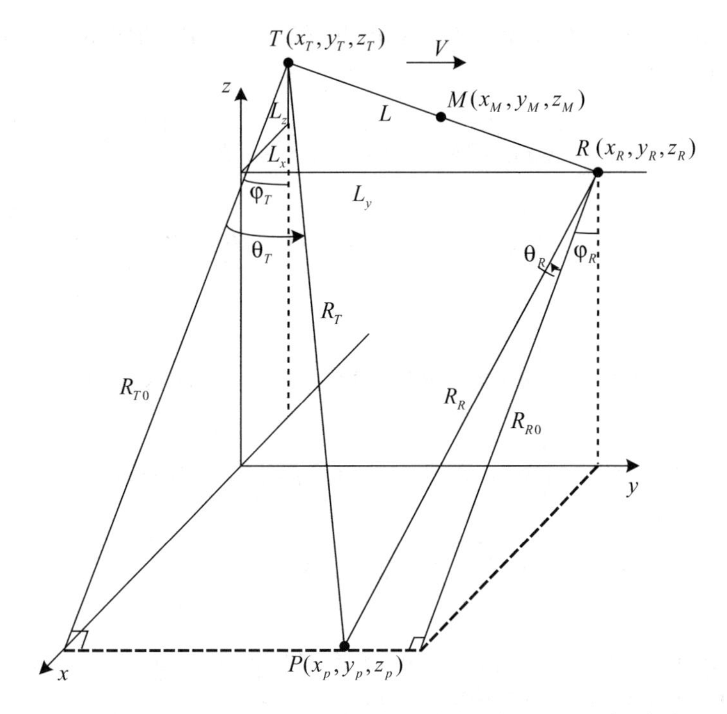

Figure 5.21 Wavenumber analysis of translational invariant bistatic SAR

where $\Delta R_0 = R_{R0} - R_{T0}$, $R_{M0} = (R_{T0} + R_{R0})/2$, $\sin\theta_T = (y_p - y_T)/R_T$ and $\sin\theta_R = (y_p - y_R)/R_R$. Let

$$\begin{cases} \beta = \dfrac{\theta_T - \theta_R}{2} \\ \theta_m = \dfrac{\theta_T + \theta_R}{2} \end{cases} \tag{5.258}$$

Then $s_{p,\mathrm{TI}}(y_M)$ can be expressed as

$$\begin{aligned} s_{p,\mathrm{TI}}(y_M) = \exp\Bigg\{ & -j2K\cos\beta\cos\theta_m\left(R_{M0} + \frac{L_y}{2}\tan\beta\right) \\ & -j2K\cos\beta\sin\theta_m\left(y_p - y_M + \frac{\Delta R_0}{2}\tan\beta\right)\Bigg\} \end{aligned} \tag{5.259}$$

Similarly, the instantaneous Doppler wavenumber is

$$K_y = \frac{2\pi f_\eta}{V} = \frac{4\pi}{\lambda}\sin\theta_m\cos\beta = 2K\cos\beta\sin\theta_m \tag{5.260}$$

so we have $\sin\theta_m = K_y/(2K\cos\beta)$. Using the POSP, the BPTRS in the wavenumber domain is given by

$$S_{p,\mathrm{TI}}(K, K_y) = \exp\left\{-jR_{M0}\sqrt{(2K\cos\beta)^2 - K_y^2} - jK_y y_p\right\} \times \exp\left\{-j\frac{L_y}{2}\tan\beta\sqrt{(2K\cos\beta)^2 - K_y^2}\right\}\exp\left\{-jK_y\frac{\Delta R_0}{2}\tan\beta\right\} \tag{5.261}$$

Hence, the following filter can be employed to focus the reference target:

$$H_{ref}(K, K_y) = \exp\left\{j\left(R_{M0\_ref} + \frac{L_y}{2}\tan\beta_{ref}\right)\sqrt{(2K\cos\beta_{ref})^2 - K_y^2}\right\} \times \exp\left\{jK_y\frac{\Delta R_{0\_ref}}{2}\tan\beta_{ref}\right\} \tag{5.262}$$

Subsequently, if $\cos\beta$ in the square root of (5.261) is assumed to be range invariant, the following wavenumber transformation,

$$K_r = \sqrt{(2K\cos\beta_{ref})^2 - K_y^2} \tag{5.263}$$

can fulfill the job of decoupling, and finally the focused image can be obtained by the two-dimensional inverse Fourier transform following that.

What should be kept in mind is that β_{ref} is variant K_y and K, so we should calculate β_{ref} for each (K, K_y) so as to perform the wavenumber transformation. For the calculation of β_{ref}, a method has been given in [22]; here we provide a numerical method that is very convenient to use.

From (5.260), we have $2\cos\beta_{ref}\sin\theta_{m,ref} = K_y/K$, so β_{ref} is actually dependent on K_y/K. For the reference target $(x_{ref}, y_{ref}, z_{ref})$, we can calculate the value of $2\cos\beta_{ref}\sin\theta_{m,ref}$ (i.e., K_y/K) for each azimuth time and also the value of $\cos\beta_{ref}$ for each azimuth time; then we get a curve of $\cos\beta_{ref}$ versus K_y/K. Then, for each (K, K_y), we just need to calculate the value of K_y/K and then get the corresponding $\cos\beta_{ref}$ on this curve through one-dimensional interpolation.

If the range variance of R_{M0}, ΔR_0 and β are considered and assumed to meet the linear variance model, then

$$\Delta R_0 \approx \Delta R_{0\_ref} + \alpha(R_{bi0} - R_{bi0\_ref}) \tag{5.264}$$

$$R_{M0} \approx R_{M0\_ref} + \rho(R_{bi0} - R_{bi0\_ref}) \tag{5.265}$$

$$\beta \approx \beta_{ref} + \mu(R_{bi0} - R_{bi0\_ref}) \tag{5.266}$$

Table 5.4 Image qualities processed by the algorithm based on IDW

	Range qualities				Azimuth qualities			
Target	Resolution (m)	Widen ratio	PSLR (dB)	ISLR (dB)	Resolution (m)	Widen ratio	PSLR (dB)	ISLR (dB)
N	0.268	0.895	−13.01	−9.65	0.386	1.810	−3.02	−4.02
C	0.266	0.886	−13.57	−9.92	0.190	0.890	−13.24	−9.79
F	0.267	0.889	−13.32	−9.98	0.193	0.902	−11.94	−9.26

more accurate algorithms can be derived based on the BPTRS of IDW. Interested readers can refer to [22] for more details.

Here, some simulation results are given using the above algorithm without considering the range variance of R_{M0}, ΔR_0 and β. The simulation parameters are the same as those listed in Table 5.1, and the simulated targets are the same as those in Figure 5.13. The image qualities for the near/center/far targets are shown in Table 5.4 and the focusing results are shown in Figure 5.22. It can be seen that the center target is well focused while the near and far targets are not well focused because the range variances of R_{M0}, ΔR_0 and β are neglected. Besides, the imaging results for the center raw of targets in Figure 5.13 are shown in Figure 5.23. It can be seen that it is not a straight line in the image. This is also because the range variances of ΔR_0 and β are neglected, so when using (5.262) as the reference filter, the targets not at the reference range will have the following term left:

$$\exp\left\{-jK_y\left[\frac{\Delta R_0}{2}\tan\beta - \frac{\Delta R_{0\_ref}}{2}\tan\beta_{ref}\right]\right\}$$

This term causes the different azimuth shifts for targets at different ranges.

5.5 Imaging Algorithms Based on Accurate Implicit Spectrums

The algorithms introduced in the former section were all based on the analytical explicit BPTRS, which is usually an approximated one. As the bases of such algorithms are not totally precise, some researchers try to develop algorithms based on a precise BPTRS to avoid the errors caused by the approximation of BPTRS. However, the precise BPTRS cannot be explicit and can only be expressed by implicit forms. Ender first developed algorithms based on this idea [28–31, 68–70], in which an accurate but implicit BPTRS was given and the numerical way of calculating this spectrum was provided. Based on the implicit BPTRS, ω-k-like algorithms based on phase decomposition and wavenumber domain transformation can be developed. Giroux proposed an analytical way to decompose the phase spectrum based on some approximations to the bistatic SAR geometry [32,71]. Ender provided a numerical

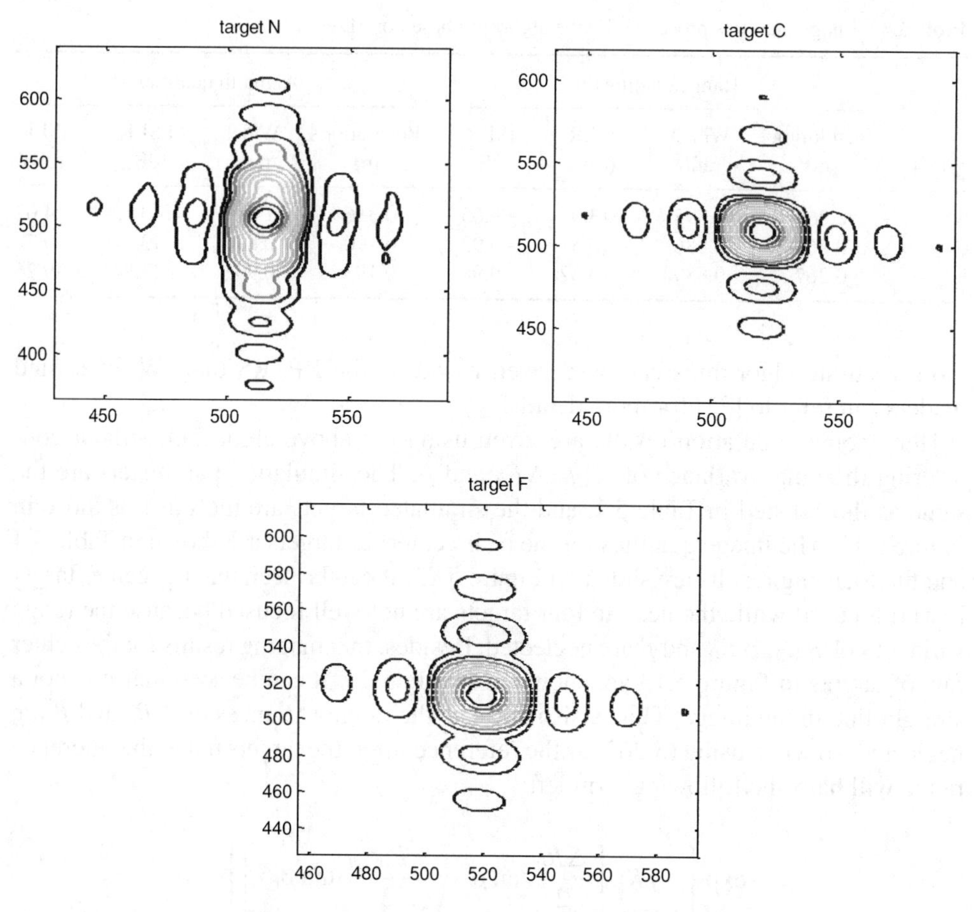

Figure 5.22 Imaging results of targets N, C and F

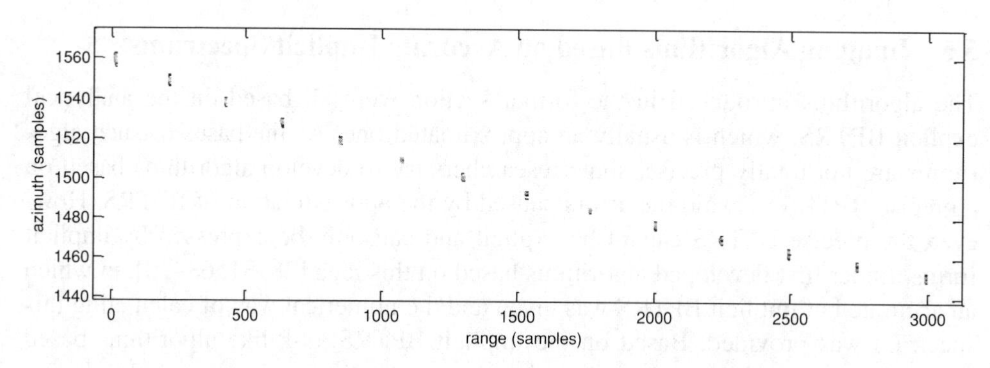

Figure 5.23 Imaging result of the center raw of simulated targets

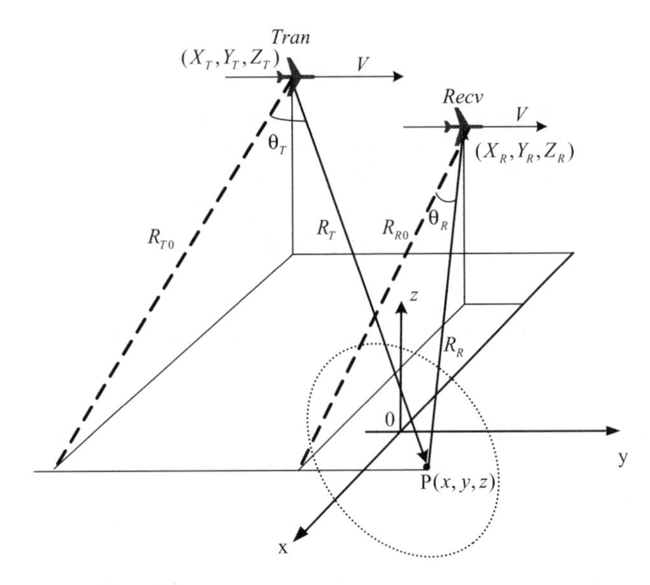

Figure 5.24 Bistatic SAR geometry of the translational invariant configuration

way of phase decomposition that has a better performance in two-dimensional decoupling [31]. Qiu proposed another analytical way of phase decomposition that can achieve the high precision achieved by the numerical way of Ender [34]. Furthermore, Qiu analyzed the residual coupling phase after decomposition and wavenumber transformation, and made some compensation to them, which further enhanced the performance of the algorithm in wide swath imaging. In the following text, algorithms based on implicit BPTRS using different phase decomposition methods will be introduced.

5.5.1 *Implicit BPTRS*

For the convenience of narration, the geometry of translational invariant bistatic SAR is shown in Figure 5.24. The notation used here is the same as in Figure 5.8. As the height of the scene is a function of x, y, that is, $z = z(x, y)$, the received signal after the range Fourier transform can be expressed as follows:

$$s(f, \eta) = \iint dxdy\sigma(x, y)\exp\{j2\pi ft_{dly}\}\exp\left\{-j\frac{2\pi(f_0 + f)}{c}R_{bi}(\eta; x, y)\right\} \tag{5.267}$$

where

$$R_{bi}(\eta; x, y) = R_T(\eta; x, y) + R_R(\eta; x, y) \tag{5.268}$$

$$R_T(\eta; x, y) = \sqrt{R_{T0}^2 + (Y_T + V\eta - y)^2} \tag{5.269}$$

$$R_R(\eta; x, y) = \sqrt{R_{R0}^2 + (Y_R + V\eta - y)^2} \tag{5.270}$$

Let $k = 2\pi(f_0 + f)/c$ and apply the Fourier transform to η; then we get

$$S(f, f_\eta) = \exp\{j2\pi f t_{dly}\} \iint dxdy\sigma(x, y) \int_\eta \exp\{-jkR_{bi}(\eta; x, y) - j2\pi f_\eta \eta\}d\eta \tag{5.271}$$

Let $u = V\eta - y$ and $k_u = 2\pi f_\eta / V$, and ignore the constant factors caused by variable substitution; we arrive at

$$\begin{aligned} S(k, k_u) = {} & \exp\{j2\pi f t_{dly}\} \iint dxdy\sigma(x, y) \exp\{-jk_u y\} \\ & \times \int_u \exp\left\{-jk\left[\sqrt{R_{T0}^2 + (Y_T + u)^2} + \sqrt{R_{R0}^2 + (Y_R + u)^2}\right] - jk_u u\right\}du \end{aligned} \tag{5.272}$$

Let

$$\Phi(u) = k\left[\sqrt{R_{T,0}^2 + (Y_T + u)^2} + \sqrt{R_{R,0}^2 + (Y_R + u)^2}\right] + k_u u \tag{5.273}$$

According to POSP, the stationary point $\tilde{u}$ should satisfy the following equation:

$$\frac{\tilde{u} + Y_T}{\sqrt{R_{T,0}^2 + (Y_T + \tilde{u})^2}} + \frac{\tilde{u} + Y_R}{\sqrt{R_{R,0}^2 + (Y_R + \tilde{u})^2}} = -\frac{k_u}{k} \tag{5.274}$$

If we use the notation of $C = k_u / k$ then $\tilde{u}$ can be expressed as a function of $C$and x, that is, $\tilde{u} = \tilde{u}(C, x)$. Thus, the phase of (5.272) at the stationary point will be

$$\Phi(\tilde{u}) = k[R_{bi}(C, x) + C\tilde{u}(C, x)] \tag{5.275}$$

Hence, the implicit BPTRS can be expressed as follows:

$$S(k, k_u) = \exp\{j2\pi f t_{dly}\} \times \iint dxdy\sigma(x, y)\exp\{-jk_u y\}\exp\{-jk[R_{bi}(C, x) + C\tilde{u}(C, x)]\} \tag{5.276}$$

5.5.2 Decomposition of the Phase Spectrum

As shown in Chapter 2, in monostatic SAR with a straight flying path, the phase of a monostatic SAR spectrum can be expressed as the linear combination of a set of wavenumber domain bases. Transformation from the original wavenumber variables to these new wavenumber bases can fulfill the job of two-dimensional decoupling and then the two-dimensional inverse Fourier transform can result in a focused image. Following the same idea, we here try to decompose the phase of BPTRS into the linear combination of proper wavenumber domain bases.

5.5.2.1 Method of Giroux

The decomposition method proposed by Giroux is introduced in the following text. First, let

$$\begin{cases} r = \left(\sqrt{(x - X_T)^2 + (Z_T - z)^2} + \sqrt{(x - X_R)^2 + (Z_R - z)^2}\right)/2 \\ r_I = \left(\sqrt{(x - X_T)^2 + (Z_T - z)^2} - \sqrt{(x - X_R)^2 + (Z_R - z)^2}\right)/2 \end{cases} \tag{5.277}$$

where r and r_I are functions of x. If r is considered to be the active variable, then r_I can be considered as a function of r. Therefore we use the following notation:

$$\begin{aligned} G(C, r) &= R_{bi}(C, r) + Cu^*(C, r) \\ &= \sqrt{(r + r_I)^2 + (Y_T + u^*)^2} + \sqrt{(r - r_I)^2 + (Y_R + u^*)^2} + Cu^*(C, r) \end{aligned} \tag{5.278}$$

Expand $G(C, r)$ around

$$r = r_{ref} = \left(\sqrt{(x_{ref} - X_T)^2 + (Z_T - z)^2} + \sqrt{(x_{ref} - X_R)^2 + (Z_R - z)^2}\right)/2$$

to its Taylor series and keep to the first-order term; then

$$G(C,r) \approx G(C,r_{ref}) + \left(\frac{(r_{ref}+r_{I_{ref}})\left(1+\frac{dr_I}{dr}|_{r_{ref}}\right)}{\sqrt{(r_{ref}+r_{I_{ref}})^2+[Y_T+u^*(C,r_{ref})]^2}} + \frac{(r_{ref}-r_{I_{ref}})\left(1-\frac{dr_I}{dr}|_{r_{ref}}\right)}{\sqrt{(r_{ref}-r_{I_{ref}})^2+[Y_R+u^*(C,r_{ref})]^2}} \right)(r-r_{ref}) \quad (5.279)$$

where the first-order term contains the coupling. The method of Giroux is to make the following variable substitution:

$$k_G = k\left(\frac{r_{ref}+r_{I_{ref}}}{\sqrt{(r_{ref}+r_{I_{ref}})^2+[Y_T+u^*(C,r_{ref})]^2}} + \frac{r_{ref}-r_{I_{ref}}}{\sqrt{(r_{ref}-r_{I_{ref}})^2+[Y_R+u^*(C,r_{ref})]^2}} \right) \quad (5.280)$$

In order to make the phase linear with the new range wavenumber variable k_G and the range variable r, and meanwhile to make the phase independent with azimuth wavenumber variable k_u (which is similar to the Stolt conversion in the monostatic ω-k algorithm). However, as $dr_I/dr|_{r_{ref}} \neq 0$, the above method cannot decouple the phase completely. Therefore, the residual phase error of the method is not very small and the precision of this method cannot be very high.

Let us take a second glance at K_G. From the bistatic geometry, we have

$$\frac{r_{ref}+r_{I_{ref}}}{\sqrt{(r_{ref}+r_{I_{ref}})^2+[Y_T+u^*(C,r_{ref})]^2}} = \cos\theta_{T,ref}$$
$$\text{and} \quad \frac{r_{ref}-r_{I_{ref}}}{\sqrt{(r_{ref}-r_{I_{ref}})^2+[Y_R+u^*(C,r_{ref})]^2}} = \cos\theta_{R,ref} \quad (5.281)$$

Therefore,

$$k_G = k(\cos\theta_{T,ref}+\cos\theta_{R,ref}) = 2k\cos\theta_{m,ref}\cos\beta_{ref} = \sqrt{(2k\cos\beta_{ref})^2-k_u^2} \quad (5.282)$$

which is exactly the same as K_r in the method of IDW.

5.5.2.2 Method of Ender

The decomposition method proposed by Ender is to make the phase as follows:

$$\Phi(u^*) = k[h(C)q(x) + g(C)] \tag{5.283}$$

where $q(x)$ is independent with frequency and $h(C)$ and $g(C)$ are independent with range. As $G(C, x) = R_{bi}(C, x) + Cu^*(C, x)$ is a curved surface on a two-dimensional plane denoted by (C, x), it is quite difficult to get the above decomposition. The approach proposed by Ender is through a numerical way as follows:

1. Choose a certain C to be noted as C_0 (usually the median of $C$is chosen).
2. Calculate the value of $G(C, x)$ at C_0 and let

$$q(x) = G(C_0, x) = R_{bi}(C_0, x) + C_0 u^*(C_0, x) \tag{5.284}$$

   This means that $h(C_0) = 1$ and $g(C_0) = 0$.
3. Calculate the value of $G(C, x)$ at another C and find a curve where $G(C, x)$ versus $q(x)$; then the constant and linear coefficients ($g(C)$ and $h(C)$, respectively) can be obtained by a linear fit to the curve.

Based on the above decomposition, a focused image can be obtained using the two-dimensional inverse Fourier transform after compensating the near-range phase and making the substitution of $(k, k_u) \rightarrow [kh(C), k_u]$, that is,

$$\begin{aligned} f[q(x), y] = \iint & S[hk(C), k_u] \exp\{-j2\pi f t_{dly}\} \exp\{jkg(C)\} \\ & \times \exp\{jk_u y\} \exp\{jkh(C)q(x)\} dk_u d\{kh(C)\} \end{aligned} \tag{5.285}$$

To make the application of the method clearer, the flow of the wavenumber algorithm based on Ender's decomposition is described in detail as follows:

1. Choose a certain series of control points (x_i, C_j), $i = 0, \ldots, N-1$, $j = 0, \ldots, M-1$, and calculate the value of $u^*(C_j, x_i)$ according to (5.274) through a numerical approach. Here, N and M are the numbers of the control points on x and C. As the stationary point is slowly variant with x and C, N and M need not be large; $N = M = 32$ is enough.
2. Calculate the value of $G(C_0, x_i)$ (i.e., $q(x_i)$), $i = 0, \ldots, N-1$ according to Equation (5.278), and meanwhile calculate $G(C_j, x_i)$ to find the curve of $G(C_j, x_i) \sim q(x_i)$; then find $h(C_j)$ and $g(C_j)$ through a linear fit.
3. Compensate the near-range phase in the two-dimensional frequency domain, which means multiplying the signal by $\exp\{-j2\pi f t_{dly}\}$.
4. Compensate the constant phase by multiplying the signal by $\exp\{jkg(C)\}$, where $g(C)$ can be obtained through interpolation by knowing $g(C_j)$.

5. Perform the wavenumber domain conversion, that is, convert the (k, k_u) plane to the (k_q, k_u) plane, where

$$k_q = kh(C) \tag{5.286}$$

6. Multiply the signal by $\exp\{jk_q q(x_{near})\}$ to make the targets remain at their proper locations.
7. Perform the two-dimensional inverse Fourier transform to (k_q, k_u) and a focused image $f[q(x) - q(x_{near}), y]$ is obtained.

5.5.2.3 Method of Qiu

The residual phase error of the decomposition method proposed by Ender is relatively small, but $G(C_j, x_i)$ at several control points was calculated through the numerical way, which makes the method not very efficient. Is there any other approach that does not need a number of control points and still achieves a comparable small residual error? The decomposition method proposed by the authors themselves is such a method. For the sake of simplification, the approach is called the method of Qiu.

Firstly, $G(C, x)$ is rewritten as follows:

$$\begin{aligned} G(C, x) &= R_{bi}(C, x) + Cu^*(C, x) \\ &= \sqrt{(x - X_T)^2 + (Z_T - z)^2 + (Y_T + u^*)^2} \\ &\quad + \sqrt{(x - X_R)^2 + (Z_R - z)^2 + (Y_R + u^*)^2} + Cu^*(C, x) \end{aligned} \tag{5.287}$$

Extending the above equation around x_{ref} to its Taylor series, we have

$$\begin{aligned} G(C, x) \approx & \left(\frac{x_{ref} - X_T}{\sqrt{(x_{ref} - X_T)^2 + (Z_T - z)^2 + [Y_T + u^*(C, x_{ref})]^2}} \right. \\ & \left. + \frac{x_{ref} - X_R}{\sqrt{(x_{ref} - X_R)^2 + (Z_R - z)^2 + [Y_R + u^*(C, x_{ref})]^2}} \right) (x - x_{ref}) \\ & + \frac{\partial u}{\partial x}\Big|_{x_{ref}} \left(\frac{Y_T + u^*(C, x_{ref})}{\sqrt{(x_{ref} - X_T)^2 + (Z_T - z)^2 + [Y_T + u^*(C, x_{ref})]^2}} \right. \\ & \left. + \frac{Y_R + u^*(C, x_{ref})}{\sqrt{(x_{ref} - X_R)^2 + (Z_R - z)^2 + [Y_R + u^*(C, x_{ref})]^2}} + C \right) (x - x_{ref}) \end{aligned} \tag{5.288}$$

According to Equation (5.274), the second parentheses in the above equation is equal to zero. Therefore, the decomposition of Qiu is to make the following substitution:

$$k_Q = k\left(\frac{x_{ref} - X_T}{\sqrt{(x_{ref} - X_T)^2 + (Z_T - z)^2 + [Y_T + u^*(C, x_{ref})]^2}} + \frac{x_{ref} - X_R}{\sqrt{(x_{ref} - X_R)^2 + (Z_R - z)^2 + [Y_R + u^*(C, x_{ref})]^2}}\right) \tag{5.289}$$

The above equation is quite similar to Equation (5.280) in form; however, the decomposition method of Qiu is more precise than the method of Giroux. This is because, in the method of Giroux, the r_I is changing with r. Only in the case where $X_T = X_R$ and $Z_T = Z_R$ (i.e., the tandem configuration), do we have $r_I \equiv 0$, which means that the Giroux method can achieve the same precision as the method of Qiu. At the tandem configuration, k_G and k_Q are linear to each other, showing that the Giroux method has the same precision as the Qiu method.

Let us have a second glance at k_Q. According to the geometry in Figure 5.24, Equation (5.289) can be written as follows:

$$k_Q = k\left(\cos\phi_{T,ref}\cos\theta_{T,ref} + \cos\phi_{R,ref}\cos\theta_{R,ref}\right) \tag{5.290}$$

In the tandem configuration, $\phi_{T,ref} = \phi_{R,ref} = \phi_{T,R,ref}$, so $k_Q = k\cos\phi_{T,R,ref}(\cos\theta_{T,ref} + \cos\theta_{R,ref})$, which is linear to k_G.

5.5.3 *The Residual Phase Error and Phase Error Compensation*

The linear decomposition described by (5.283) is the key step of the imaging algorithms based on phase decomposition in the wavenumber domain. However, in the elliptical cylindrical symmetry of the bistatic SAR geometry, a complete linear decomposition like Equation (5.283) does not exist. This also can be seen from other viewpoints. In the IDW method, the bistatic angle is 2β and β is changing with the range location (i.e., $\beta = \beta(x)$). The combined bistatic wavenumber is $k_{bi} = 2k\cos[\beta(x)]$, which is also changing with x. This implies that the phase of the bistatic spectrum cannot be completely decomposed linearly. In the hyperbolic approximation, the equivalent velocity is changing with range, which also implies that the complete linear decomposition does not exist and the residual error is unavoidable. In the following text, the residual phase error is considered.

If the residual phase error of linear decomposition is denoted by $E(C, x)$, the phase of the bistatic spectrum can be written as follows:

$$\Phi(u^*) = k[h(C)q(x) + g(C) + E(C, x)] \tag{5.291}$$

The amount of $E(C, x)$ can be obtained through the numerical approach. Here, we express $E(C, x)$ as a polynomial of $q(x)$, which is

$$E(C, x) = f(C)[q(x) - q_0]^2 + p(C)[q(x) - q_0]^3 + \cdots \tag{5.292}$$

where q_0 is the q at which the residual error is zero and $f(C)$, $g(C)$ can be calculated through a polynomial fit to $E(C, x)$. Thus, the residual phase error is

$$\Phi_{res}(k, C, q) = kf(C)[q(x) - q_0]^2 + kp(C)[q(x) - q_0]^3 + \cdots \tag{5.293}$$

After the range-invariant phase compensation and the wavenumber conversion from (k, k_u) to (k_q, k_u), the residual phase error turns out to be $\Phi_{res}(k_q, k_u, q)$. Let $\tilde{f}(k_q, k_u) = kf(C)$ and $\tilde{p}(k_q, k_u) = kp(C)$, and expand $\Phi_{res}(k_q, k_u, q)$ into the following series:

$$\begin{aligned}
\Phi_{res}(k_q, k_u, q) &\approx \tilde{f}(k_{q0}, k_u)(q - q_0)^2 + \tilde{p}(k_{q0}, k_u)(q - q_0)^3 && (1)\\
&+ \left[(q - q_0)^2 \frac{\partial \tilde{f}}{\partial k_q}|_{k_{q0}} + (q - q_0)^3 \frac{\partial \tilde{p}}{\partial k_q}|_{k_{q0}}\right](k_q - k_{q0}) && (2)\\
&+ \frac{1}{2}\left[(q - q_0)^2 \frac{\partial^2 \tilde{f}}{\partial k_q^2}|_{k_{q0}} + (q - q_0)^3 \frac{\partial^2 \tilde{p}}{\partial k_q^2}|_{k_{q0}}\right](k_q - k_{q0})^2 && (3)\\
&\approx \tilde{f}(k_{q0}, k_{u0})(q - q_0)^2 + \tilde{p}(k_{q0}, k_{u0})(q - q_0)^3 && (1.0)\\
&+ \left[(q - q_0)^2 \frac{\partial \tilde{f}}{\partial k_q}|_{k_{q0}} + (q - q_0)^3 \frac{\partial \tilde{p}}{\partial k_q}|_{k_{q0}}\right](k_q - k_{q0}) && (2)\\
&+ \frac{1}{2}\left[(q - q_0)^2 \frac{\partial^2 \tilde{f}}{\partial k_q^2}|_{k_{q0}} + (q - q_0)^3 \frac{\partial^2 \tilde{p}}{\partial k_q^2}|_{k_{q0}}\right](k_q - k_{q0})^2 && (3)\\
&+ \left[(q - q_0)^2 \frac{\partial \tilde{f}}{\partial k_u}|_{k_{q0},k_{u0}} + (q - q_0)^3 \frac{\partial \tilde{p}}{\partial k_u}|_{k_{q0},k_{u0}}\right](k_u - k_{u0}) && (1.1)\\
&+ \frac{1}{2}\left[(q - q_0)^2 \frac{\partial^2 \tilde{f}}{\partial k_u^2}|_{k_{q0},k_{u0}} + (q - q_0)^3 \frac{\partial^2 \tilde{p}}{\partial k_u^2}|_{k_{q0},k_{u0}}\right](k_u - k_{u0})^2 && (1.2)
\end{aligned} \tag{5.294}$$

where k_{q0}, k_{u0} are the median of k_q, k_u, respectively; then the impact of each term above can be clarified as follows:

1. Constant term. The term (1.0) only depends on q. It can be considered as a constant phase for each target, which will not have any influence on focusing, but only remains to the image plane. This phase contains the range information of a target, which is the phase interferometry needs.

2. Linear term of the range wavenumber. The term (2) is a linear term of $k_q - k_{q0}$, which is variant with q and k_u. The linear phase of $k_q - k_{q0}$ will cause a range shift in the q dimension after performing an inverse Fourier transform on k_q. Therefore, the location of a target at q will appear at $\tilde{q}(k_u, q)$ in the (k_u, q) domain, where $\tilde{q}(k_u, q)$ is as follows:

$$\tilde{q}(k_u, q) = q + (q - q_0)^2 \frac{\partial \tilde{f}}{\partial k_q}|_{k_{q0}} + (q - q_0)^3 \frac{\partial \tilde{p}}{\partial k_q}|_{k_{q0}} \tag{5.295}$$

This range shift will firstly cause a residual RCM as the shift is changing with k_u. Therefore, a residual RCM correction will be needed if the following condition is not satisfied:

$$|\tilde{q}(k_u, q) - \tilde{q}(k_{u0}, q)| \leq \Delta q/2 \tag{5.296}$$

where Δq is the resolution of q. The second influence of this range shift is a small displacement of the target location in the focused image, which should be taken into consideration when doing geocoding.

3. Quadratic term of the range wavenumber. The term (3) is a quadratic term of $k_q - k_{q0}$, which will cause the range to disperse and needs secondary range compression (SRC). Fortunately, this term is usually very small and can be ignored.
4. The residual azimuth phase. The residual azimuth phase includes the linear term of k_u (i.e., term 1.1), the quadratic term of k_u (i.e., term 1.2) and all the higher-order terms. The linear term causes the azimuth location shift. This makes the target which should appear at y shift to y_m, which is as follows:

$$y_m = y + (q - q_0)^2 \frac{\partial \tilde{f}}{\partial k_u}|_{k_{q0},k_{u0}} + (q - q_0)^3 \frac{\partial \tilde{p}}{\partial k_u}|_{k_{q0},k_{u0}} \tag{5.297}$$

It can be seen that the azimuth shift is a function of q, which causes image distortion. The quadratic term and the higher-order terms, which also depend on q, will cause image defocusing and should be compensated for delicate processing.

Based on the analysis above, here we provide the phase compensation method to extend the processing range swath of imaging algorithms based on phase decomposition:

1. The constant phase term. There is no need to compensate for the constant phase, which should remain for SAR image applications that need phase information (e.g., SAR interferometry).
2. Linear term of the range wavenumber. If (5.296) is satisfied for all q, then the effect of the linear term can be neglected; if not, the effect of the linear term (i.e., the residual RCM) should be corrected. Generally, the residual RCM can be

corrected through interpolation in the (q, k_u) domain according to the following equation:

$$q_{\Delta RCM} = \tilde{q}(k_u, q) - q = (q - q_0)^2 \frac{\partial \tilde{f}}{\partial k_q}|_{k_{q0}} + (q - q_0)^3 \frac{\partial \tilde{p}}{\partial k_q}|_{k_{q0}} \quad (5.298)$$

Here, the (q, k_u) domain can be obtained by inverse Fourier transform on k_q, and $\partial \tilde{f}/\partial k_q|_{k_{q0}}$ and $\partial \tilde{p}/\partial k_q|_{k_{q0}}$ can be obtained through a polynomial fit to $\tilde{f}(k_q, k_u)$ and $\tilde{p}(k_q, k_u)$, respectively.

3. Quadratic term of the range wavenumber. The effect of the quadratic residual phase term should be controlled. A simple way to control the defocusing effect is to estimate the value of term (3) in Equation (5.294) and limit it below $\pi/4$ through range swath segmentation.
4. The residual azimuth phase. The azimuth shift and azimuth defocusing can be corrected by multiplying the following function in the (q, k_u) domain:

$$H_{azerr}(k_u, q) = \exp\{j[\tilde{f}(k_{q0}, k_u)(q - q_0)^2 + \tilde{p}(k_{q0}, k_u)(q - q_0)^3]\} \quad (5.299)$$

5.5.4 Simulation Results

To illustrate the validity of the analysis above, some simulation results are given in the following text. The simulation parameters and the simulated scenario are the same as those in Section 5.3.2.4. The simulation results are as follows.

5.5.4.1 Imaging Results without Residual Phase Error Compensation

The imaging results without residual phase error compensation using the methods of Ender, Qiu and Giroux are shown in Figures 5.25 to 5.27 and in Table 5.5. Figure 5.25 shows the focusing results of target N by the three methods, where dispersion in the azimuth is noticeable in all focusing results and dispersion in the Giroux method result is worse than the other two results.

The azimuth and range profiles of targets N, C and F focused by the three methods are shown in Figure 5.26 and the focusing qualities are listed in Table 5.5. From these we can see that, using the method of Giroux, target C in the middle of the swath can be well focused, but the targets at the edge of the swath cannot be well focused, especially the near-range target N. When using the method of Qiu, the middle-range target, which is the reference target, is well focused; the near-range and far-range targets are defocused to some extent, though the defocusing is less than the results of Giroux. When using the method of Ender, the focusing results of target N and target F are slightly better than the results of Qiu, while the focusing result of target C is not as perfect as that of the Qiu method. This is because the phase decomposition strategy of Ender's method is to take the residual phase error for the whole scenario to its minimum, so it cannot guarantee that there is no phase error of the middle-range target C. So from the view of the focusing quality of the whole scene, the method

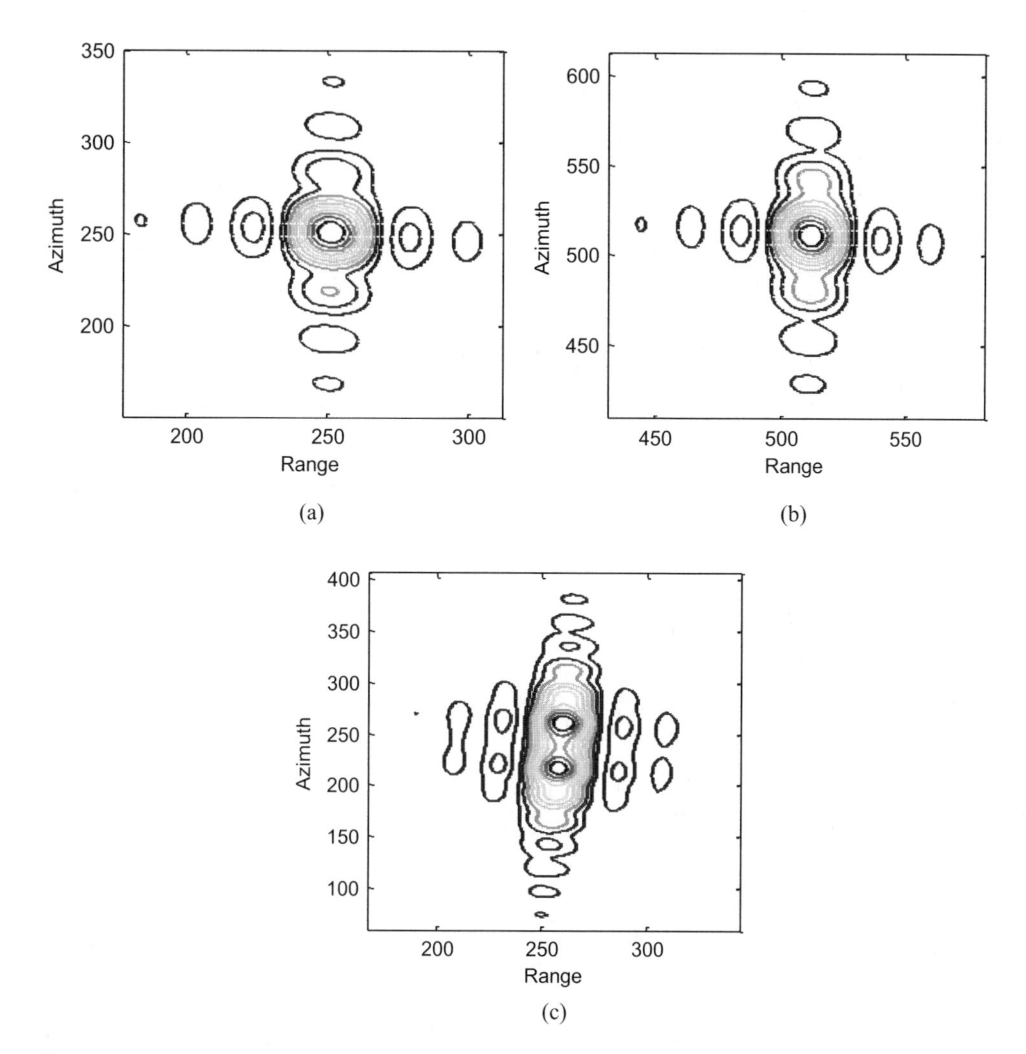

Figure 5.25 Imaging results of target N: (a) method of Ender, (b) method of Qiu, (c) method of Giroux

of Ender is a little better than the method of Qiu, while the method of Qiu has the advantage of making less numerical calculations than the method of Ender.

What should be clarified is that the range resolution in Table 5.5 means the resolution of the range axis of the focused image (i.e., q), so here the range resolutions of the three different methods are different. However, if the resolution of the ground range x is considered, the results of these three methods are coincident. The range axes of the focused images by the three methods are as follows:

$$\text{Ender: } q(x) = 1.351x + 1.603 \times 10^{-4}x^2 - 2.889 \times 10^{-8}x^3$$

$$\text{Qiu: } q_Q(x) = x + 1.192 \times 10^{-4}x^2 - 2.110 \times 10^{-8}x^3$$

$$\text{Giroux: } q_G(x) = 0.678x + 8.071 \times 10^{-4}x^2 - 1.510 \times 10^{-8}x^3$$

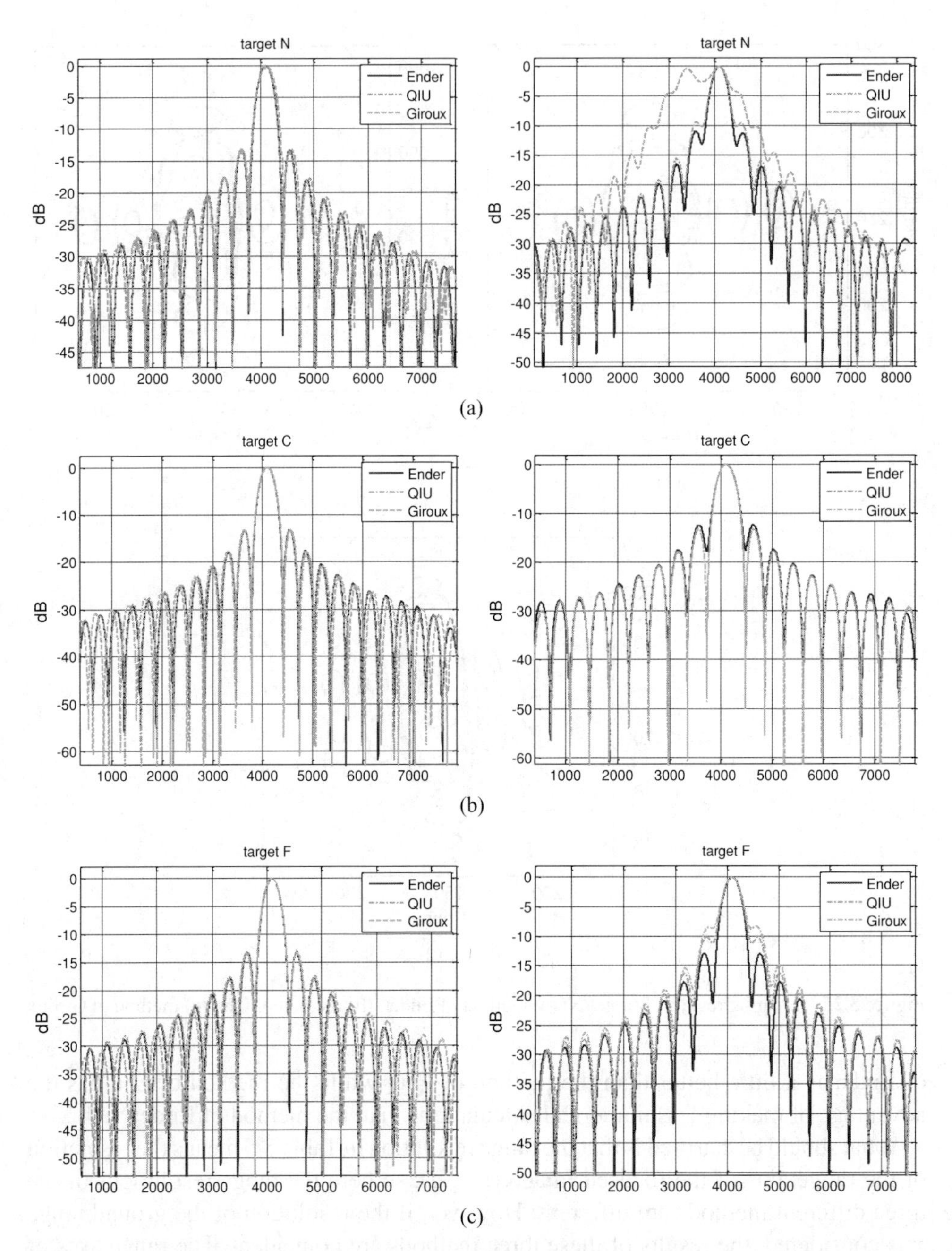

Figure 5.26 Azimuth and range profiles of targets N, C and F focused by the methods of Ender, Qiu and Giroux without phase error compensation (left column, range; right column, azimuth): (a) focusing results of target N, (b) focusing results of target C, (c) focusing results of target F

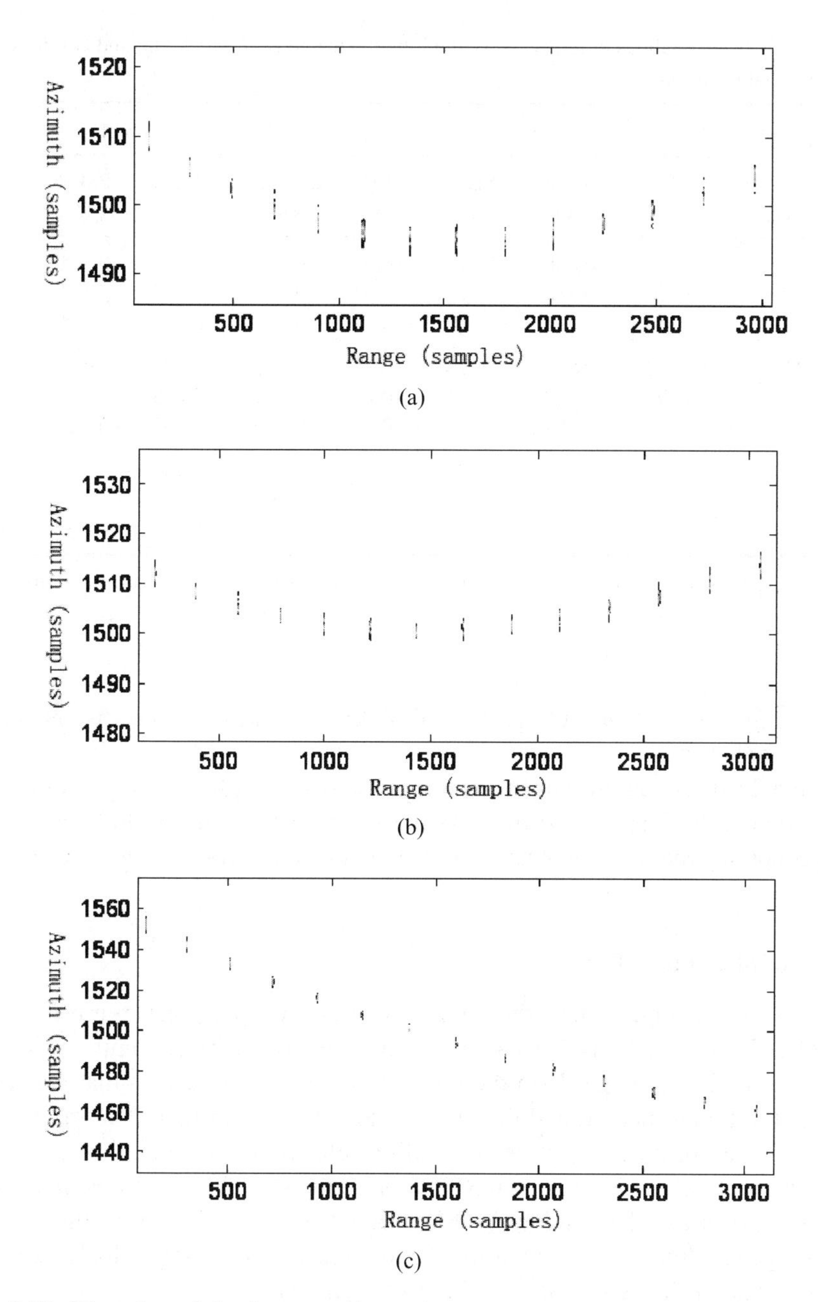

Figure 5.27 Distortion of the images without phase error compensation: (a) method of Ender, (b) method of Qiu, (c) method of Giroux

Table 5.5 Focusing qualities of targets N, C and F by the methods of Ender, Qiu and Giroux without phase error compensation

		Range				Azimuth			
Method	Target	Resolution (m)	Widen ratio	PSLR (dB)	ISLR (dB)	Resolution (m)	Widen ratio	PSLR (dB)	ISLR (dB)
Ender	N	0.532	0.888	−13.14	−10.03	0.196	0.918	−10.99	−8.19
	C	0.530	0.886	−13.17	−10.04	0.191	0.897	−12.39	−9.32
	F	0.530	0.886	−13.29	−10.15	0.191	0.897	−12.90	−9.87
Qiu	N	0.392	0.885	−13.14	−10.04	0.202	0.945	−9.42	−6.87
	C	0.392	0.887	−13.21	−10.01	0.190	0.889	−13.25	−10.11
	F	0.392	0.887	−13.25	−10.10	0.199	0.932	−10.13	−7.41
Giroux	N	0.298	0.993	−13.02	−10.31	0.672	3.147	−0.191	2.552
	C	0.266	0.886	−13.21	−10.03	0.190	0.889	−13.27	−10.13
	F	0.266	0.888	−13.13	−10.05	0.205	0.961	−8.35	−6.14

Note: The range resolution in this table means the resolution of the range axis of the focused image (i.e., q).

Therefore, $q(x) \approx 1.351 q_Q(x)$, $q_G(x) \approx 0.678 q_Q(x)$. These are consistent with the results in Table 5.5.

Figure 5.27 shows another phenomenon if the residual phase error is not compensated, which is the image distortion. As the image distortion should be avoided for better geocoding, the residual phase error calculation and compensation are necessary.

5.5.4.2 Residual Phase Error

According to Section 5.5.3, the decomposition residual phase errors for the Ender/Qiu/Giroux methods are calculated and exhibited here. Figure 5.28 shows the residual RCMs of target N in the three methods. It can be seen that the residual RCMs of the Ender method and the Qiu method are smaller than 0.06 m, which are smaller than half of the image range resolution cells of the Ender method and the Qiu method, respectively, so the residual RCM correction step can be omitted in these two methods. In contrast, the residual RCM of target N in the Giroux method is larger than a range resolution cell, which means that the imaging range block size of this method is smaller than those of the other two methods.

Figure 5.29 shows the quadratic term of k_q for target N. It can be seen that the residual phase errors caused by this term in the three methods for target N are all very small, which is only on the order of 0.01 rad. This is far smaller than $\pi/4$, so the residual secondary range compression is not needed in this simulation.

Figure 5.30 shows the azimuth shifts caused by the residual azimuth linear phase term. These are coincident with the image deformation in Figure 5.27. The azimuth

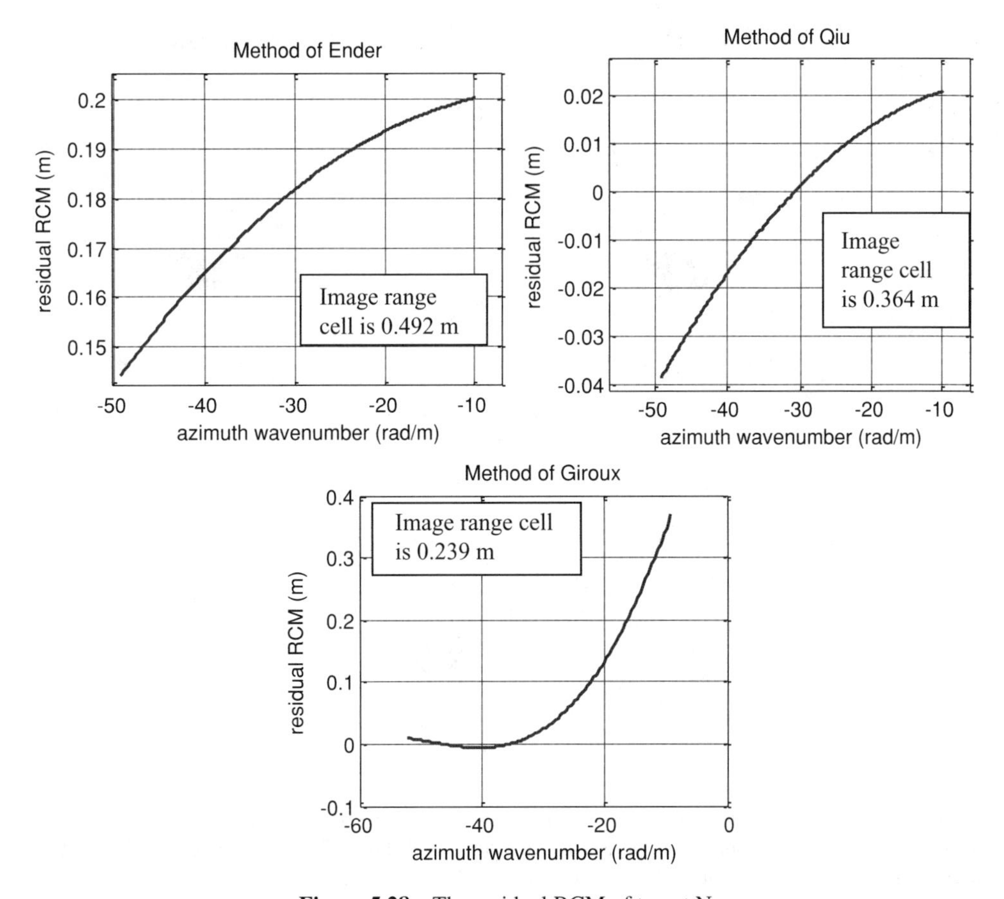

Figure 5.28 The residual RCM of target N

shifts for the middle range (i.e., the reference range used here) in the Qiu method and the Giroux method are zero, while the shift of the Ender method in the middle range is not zero. This is because the Ender method is based on a linear fit for the entire scenario and will not always make the middle-range fit error to be zero.

Figure 5.31 shows the azimuth residual phase errors, which are range variant. It can be seen that the residual phase error of the Ender method is the smallest, while the residual phase error of the Giroux method is the largest. These are coincident with the focusing results shown in Figure 5.26. Besides, another phenomenon seen in Figure 5.31 is that the phase errors of the Ender method and the Qiu method are nearly symmetrical at the near and far range, while in the method of Giroux, the phase errors at the near range and the far range have opposite sign. This is because the decomposition residual phase error of the Giroux method contains the linear term of range (see Equations (5.279) and (5.280)).

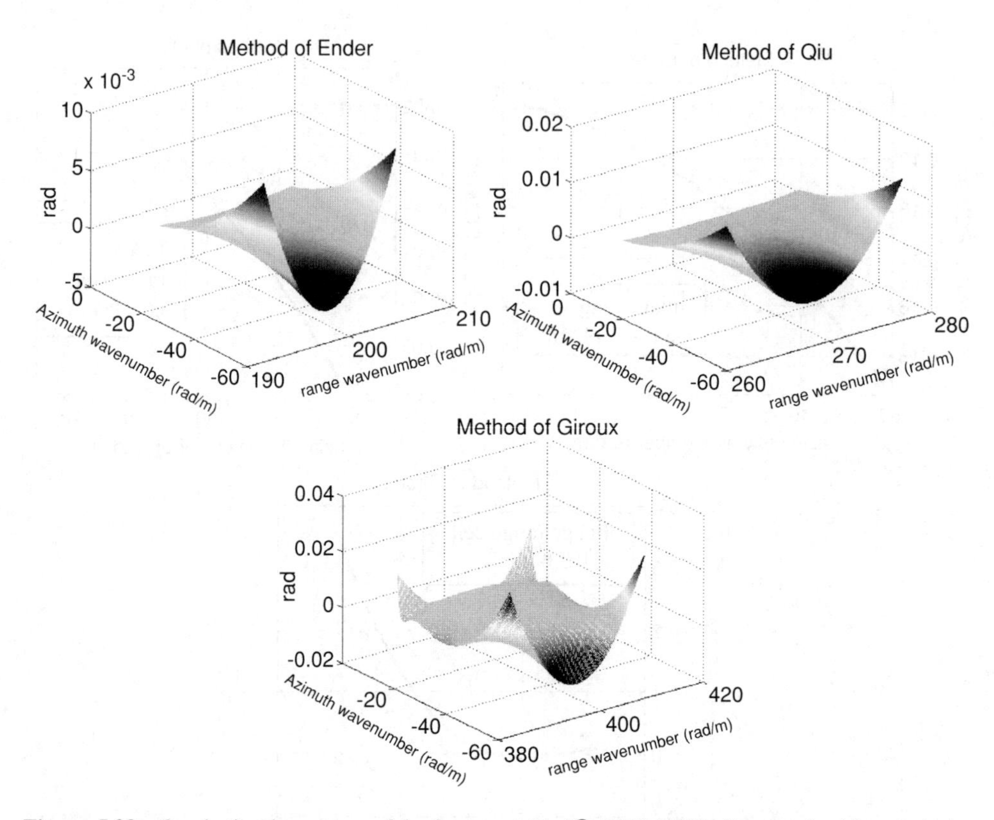

Figure 5.29 Quadratic phase term of k_q for target N. (© 2008 IEEE. Reprinted with permission from X. Qiu, D. Hu, C. Ding, "An Omega-K Algorithm With Phase Error Compensation for Bistatic SAR of a Translational Invariant Case", *IEEE Transaction on Geoscience and Remote Sensing*, **46**, 8, pp. 2224–2231, 2008.)

5.5.4.3 Imaging Result with Residual Phase Compensation

Compensating the residual phase error using the method introduced in Section 5.5.3, the imaging results are exhibited as follows. Figure 5.32 shows the azimuth profiles of target N focused by the three methods with residual phase compensation. It can be

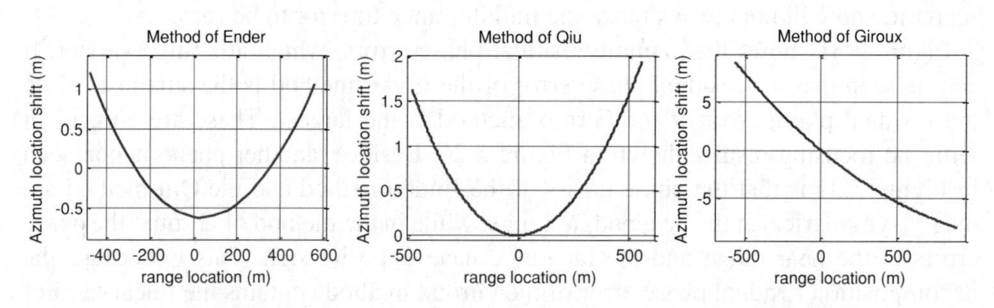

Figure 5.30 The range-dependent azimuth shifts of the three methods

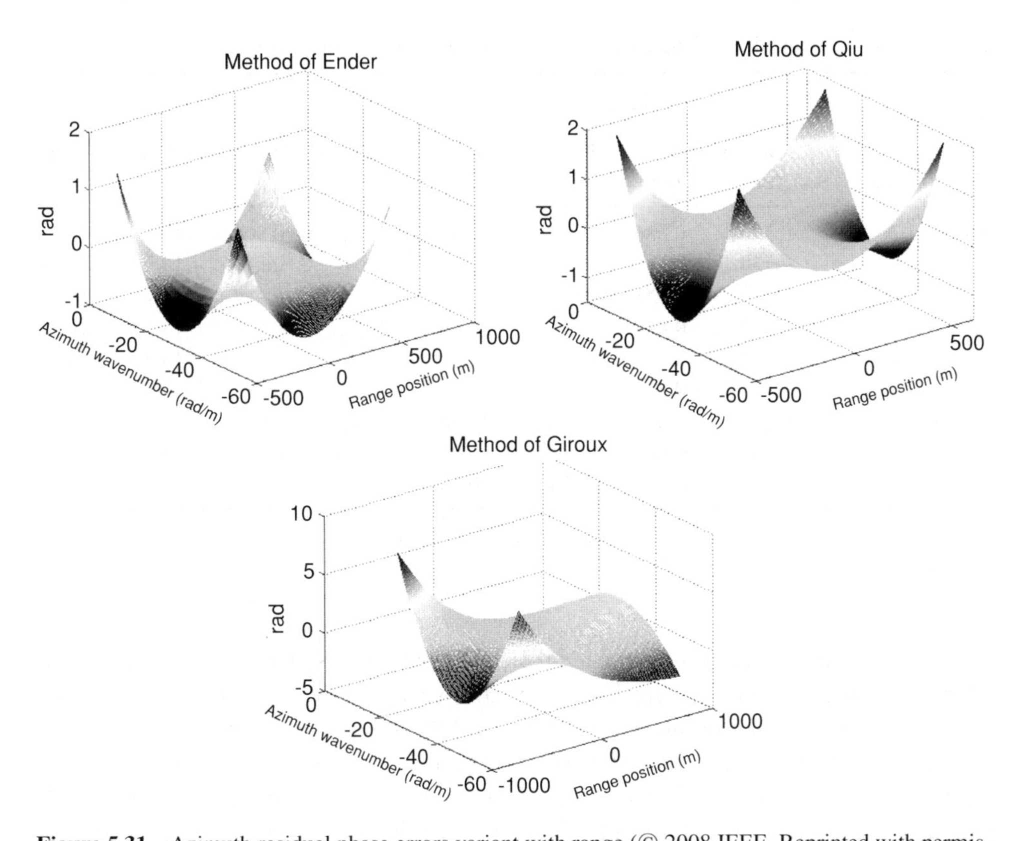

Figure 5.31 Azimuth residual phase errors variant with range.(© 2008 IEEE. Reprinted with permission from X. Qiu, D. Hu, C. Ding, "An Omega-K Algorithm With Phase Error Compensation for Bistatic SAR of a Translational Invariant Case", *IEEE Transaction on Geoscience and Remote Sensing*, **46**, 8, pp. 2224–2231, 2008.)

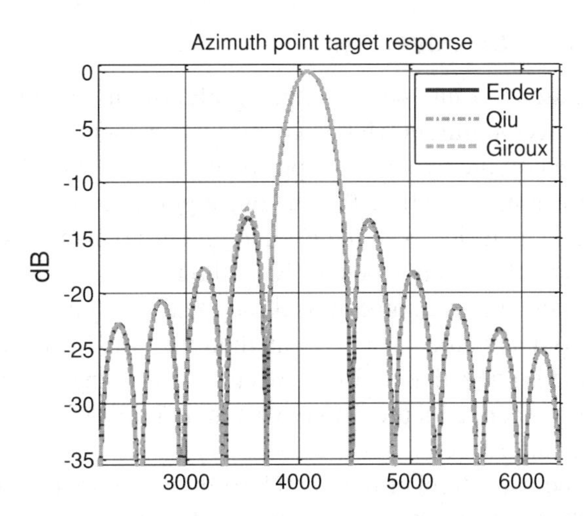

Figure 5.32 Azimuth profiles of target N after residual phase error compensation

Table 5.6 Imaging qualities of the three methods with phase error compensation

		Range				Azimuth			
Method	Target	Resolution (m)	Widen ratio	PSLR (dB)	ISLR (dB)	Resolution (m)	Widen ratio	PSLR (dB)	ISLR (dB)
Ender	N	0.530	0.886	−13.26	−10.06	0.192	0.900	−13.15	−10.15
	C	0.530	0.886	−13.23	−10.04	0.190	0.889	−13.25	−10.11
	F	0.531	0.887	−13.31	−10.14	0.191	0.894	−13.27	−10.22
Qiu	N	0.392	0.886	−13.29	−10.07	0.192	0.900	−13.17	−10.18
	C	0.392	0.886	−13.21	−10.01	0.190	0.889	−13.25	−10.11
	F	0.392	0.886	−13.36	−10.16	0.191	0.894	−13.24	−10.24
Giroux	N	0.268	0.892	−13.42	−10.20	0.193	0.902	−12.35	−10.04
	C	0.266	0.886	−13.21	−10.04	0.190	0.889	−13.21	−10.08
	F	0.266	0.886	−13.32	−10.18	0.191	0.894	−12.95	−10.15

**Note:* The range resolution in this table means the resolution of the range axis of the focused image (i.e., q).

seen that the profiles are close to the ideal profile. This can also be seen from Table 5.6, where the imaging qualities are listed. By comparing Table 5.6 with Table 5.5, we can see that the focusing qualities are enhanced by the residual phase compensation step, especially the azimuth qualities. Moreover, from Figure 5.33, we can see that the nonlinear image distortion is corrected through residual phase compensation, which validates the correctness of the phase error analysis and the phase error compensation.

5.6 Comparison of the Algorithms

In the previous sections, a number of imaging algorithms for bistatic SAR of the translational invariant configuration has been introduced, including the RD/CS/NLCS/ω-k algorithms based on the hyperbolic approximation and the advanced hyperbolic approximation of the range equation, the imaging algorithms based on LBF, MSR and IDW, and the imaging algorithms of the Ender/Qiu/Giroux methods with phase error compensation. Which algorithm is better? How should a proper algorithm be chosen for a special case? Here we give some hints through the comparison of the algorithms.

Literature [20] compares the basic LBF method with the MSR method and shows that the LBF method is equivalent to MSR, which keeps to the second-order term. As the MSR method can change its precision by changing the orders of its range history series, it is not compared here.

In this section, the wavenumber domain algorithms are firstly compared and then the wavenumber domain algorithms are compared with the range-Doppler domain algorithms.

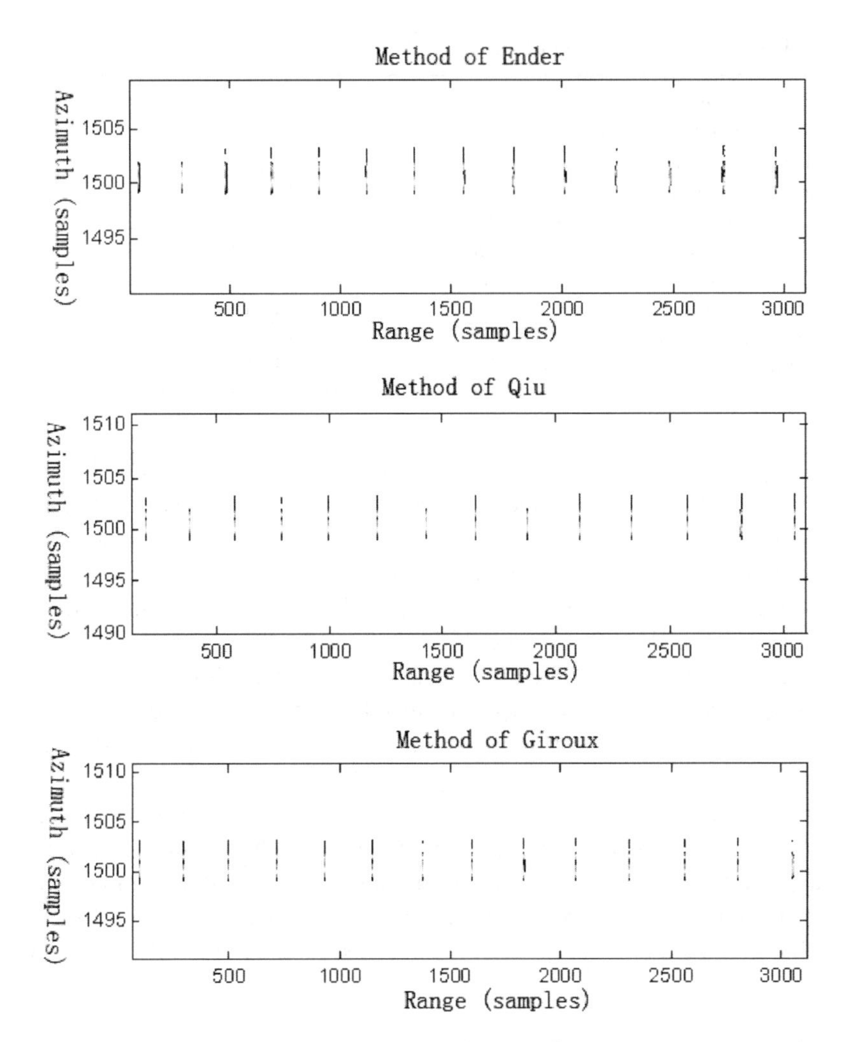

Figure 5.33 Imaging results for the simulated targets after phase error compensation

5.6.1 Comparison of the Wavenumber Domain Algorithms

Firstly, let us consider the wavenumber domain algorithms introduced in the previous sections, which are the ω-k algorithm based on the hyperbolic approximation and the advanced hyperbolic approximation, the ω-k algorithm based on the IDW method, the Ender algorithm, the Giroux algorithm and the Qiu algorithm with and without residual phase error compensation.

From the analysis in Section 5.3.3.3 and the simulation results in Sections 5.3.2.4 and 5.3.3.4, we know that the ω-k algorithm based on the hyperbolic approximation is better than the ω-k algorithm based on the advanced hyperbolic approximation. From the analysis in Section 5.5.2, we know that the Qiu and Ender methods have

Table 5.7 Simulation parameters

Center frequency	9.6 GHz
Transmitted bandwidth	1000 MHz
Transmitted pulse duration	1 μs
Range sampling rate	1100 MHz
Original position of transmitting antenna	(−4040.0, 1250, 6997.7) m
Original position of receiving antenna	(−6446.5, 0, 3873.5) m
Antenna size in azimuth direction (R)	0.35 m
Flying velocity (T&R)	200 m/s
Pulse repetition frequency	1371.43 Hz
Azimuth processing bandwidth	937.14 Hz

comparable precision; the Giroux method is essentially the same using the method of IDW; the Qiu method and the Ender method have a higher precision than the Giroux method; and the Qiu and Ender methods with phase error compensation have even better precision than the methods without phase error compensation.

Then, comparing the ω-k algorithm based on the hyperbolic approximation with the Ender method or the Qiu method with phase error compensation, which can get a better image for the same scenario? To answer this question, a larger scenario is simulated while the resolution is also enhanced compared to simulation parameters in Table 5.1 and Figure 5.13. The new simulation parameters are listed in Table 5.7 and the simulated targets are shown in Figure 5.34.

The simulation results for target N are shown in Figure 5.35, from which we can see that the imaging result by the ω-k algorithm based on the hyperbolic approximation is totally dispersed, while the result of the Ender method with phase error compensation still compares fairly well. From Figure 5.35(a) it can be seen that the RCM of target N is not well corrected using the ω-k algorithm based on the hyperbolic approximation.

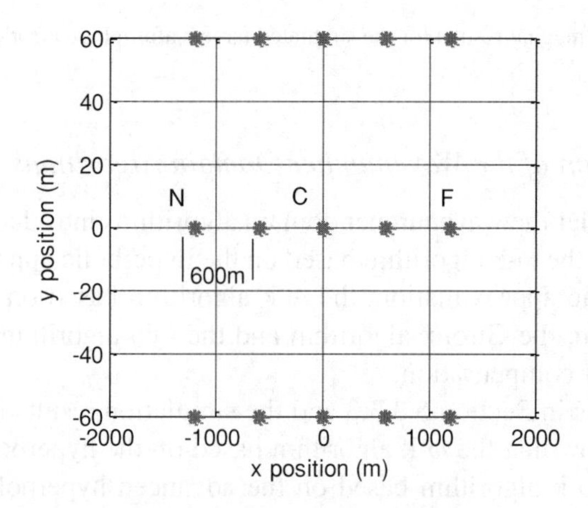

Figure 5.34 Simulated targets

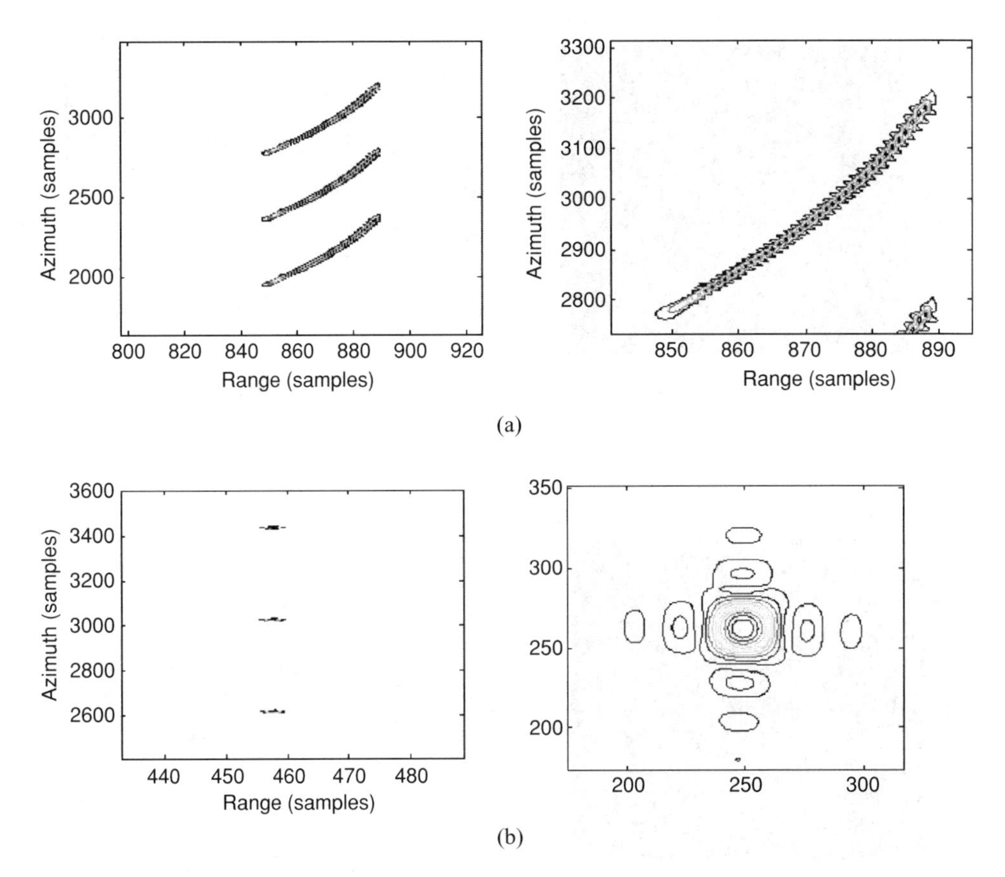

Figure 5.35 Imaging results of target N by different wavenumber domain algorithms: (a) ω-k algorithm based on the hyperbolic approximation, (b) Ender algorithm with phase error compensation

This can be explained according to Equation (5.113). The range variance of V_m is neglected while the Stolt transform (i.e., decomposition of the phase spectrum) causes a quite large residual RCM for the near-range target, so this algorithm cannot handle a wide range swath. On the other hand, the Ender or Qiu decomposition has a smaller residual phase error and the residual RCM caused by this phase error is smaller, so it can handle a wider swath than the ω-k algorithm based on the hyperbolic approximation.

5.6.2 Comparison of the Wavenumber Domain Algorithms and the Range-Doppler Domain Algorithms

According to Section 5.6.1, among the wavenumber domain algorithms introduced in this chapter, the Ender algorithm and the Qiu algorithm with phase error compensation have the best precision. According to Section 5.3, we found that the NLCS algorithm

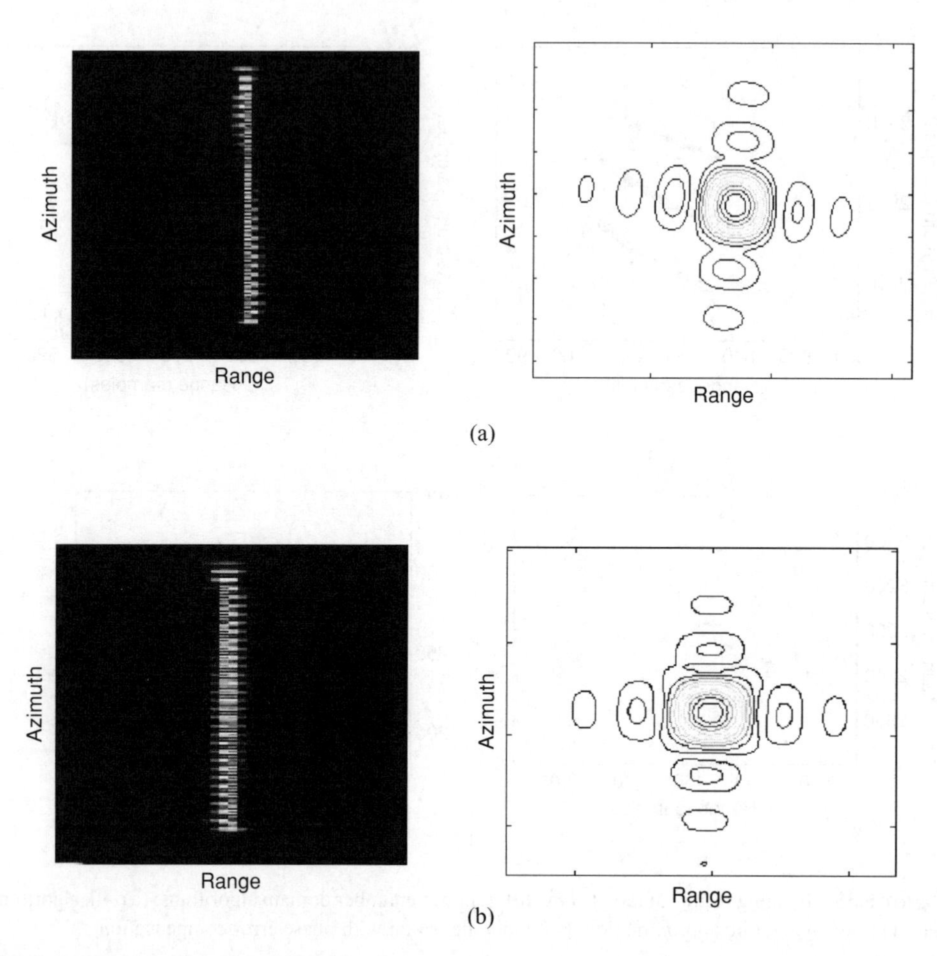

Figure 5.36 Imaging algorithms of target N using different algorithms: (a) NLCS algorithm based on AHRE, (b) Ender algorithm with phase error compensation

based on the advanced hyperbolic approximation has the best precision among all the range-Doppler domain algorithms based on range history simplification introduced in this chapter. Which has the better performance, the Ender (or Qiu) algorithm with phase error compensation or the NLCS algorithm based on the advanced hyperbolic approximation? To answer this question, the simulated echo data under the same condition as in Section 5.6.1 is processed by these two algorithms, respectively, and the imaging results are shown in the following.

It can be seen from Figure 5.36 that, in this simulation, the two algorithms both do an acceptable job, so they have comparable precision in this situation. However, what can be inferred from the principles of the two algorithms is that the NLCS algorithm is better at handling the range variance of the imaging parameters, while the wavenumber domain algorithm is better at handling the high-resolution, large squint angle cases. Readers can choose the correct algorithm based on the above characteristics.

5.7 Summary

In this chapter, the algorithms for bistatic SAR of the translational invariant configuration are categorized into six categories and the main four categories that are specifically developed (or mainly developed) for this configuration are introduced. The algorithms are also compared to show their similarities and differences. When this book was written, there were still some researchers working on the bistatic SAR algorithms for this configuration. Interested readers can continue to give further attention to this field.

References

[1] D'Aria, D., Guarnieri, A. M. and Rocca, F. (2004) Focusing bistatic synthetic aperture radar using dip move out. *IEEE Transactions on Geoscience and Remote Sensing*, **42** (7), 1362–1376.

[2] Ding, J., Zhang, Z., Xing, M., *et al.* (2008) A new look at the bistatic-to-monostatic conversion for tandem SAR image formation. *IEEE Geoscience and Remote Sensing Letters*, **5** (3), 392–395.

[3] Loffeld, O., Nies, H., Peters, V., *et al.* (2004) Models and useful relations for bistatic SAR processing. *IEEE Transactions on Geoscience and Remote Sensing*, **42** (10), 2031–2038.

[4] Loffeld, O., Nies, H., Gebhardt, U., *et al.* (2004) Bistatic SAR: some reflections on Rocca's smile. EUSAR Conference.

[5] Yan, H.H., Wang, Y.F., Yu, H.F., *et al.* (2005) An imaging method of distributed small satellites bistatic SAR based on range distance compensation. *Journal of Electronics and Information Technology*, **27** (5), 771–774.

[6] Yang, J. and Wang, J.G. (2005) Chirp scaling algorithm for airborne bistatic SAR. *Signal Processing*, **21** (4A), 522–525.

[7] Li, J.P., Bai, X. and Mao, S.Y. (2007) An algorithm of extended chirp scaling imaging of bistatic SAR. *Journal of Air Force Radar Academy*, **21** (2), 112–114.

[8] Sun, J.P., Bai, X. and Mao, S.Y. (2007) A new extended ETF imaging algorithm for bistatic SAR. *Acta Electronica Sinica*, **35** (12), 2394–2399.

[9] Wang, F., Tang, Z.Y., Zhu, Z.B., *et al.* (2007) The nonlinear chirp scaling algorithm for bistatic SAR imaging. *Journal of Electronics and Information Technology*, **29** (9), 2098–2100.

[10] Zhu, Z.B., Tang, Z.Y. and Jiang, X.Z. (2007) The imaging algorithm of bistatic SAR with parallel track. *Journal of Electronics and Information Technology*, **29** (11), 2702–2705.

[11] Chen, X.L., Ding, C.B., Liang, X.D., *et al.* (2008) An improved CS imaging algorithm of bistatic SAR with a translational invariant configuration. *Science of Surveying and Mapping*, **33** (3), 31–34.

[12] Zhang, S.K. and Yang, R.L., A squint mode bistatic synthetic aperture radar image formation algorithm based on second range compression. *Journal of Electronics and Information Technology*, **30** (7), 1717–1721.

[13] Bamler, R. and Boerner, E. (2005) On the use of numerically computed transfer functions for processing of data from bistatic SARs and high squint orbital SARs. In: *Proceedings of the IEEE International Geoscience and Remote Sensing Symposium (IGARSS'05)*, vol. 2. pp. 1051–1055.

[14] Geng, X., Yan, H. and Wang, Y. (2008) A two-dimensional spectrum model for general bistatic SAR. *IEEE Transactions on Geoscience and Remote Sensing*, **46** (8), 2216–2223.

[15] Huang, L.J., Qiu, X.,L. and Hu, D.H. (2011) Focusing of medium-earth-orbit SAR with advanced nonlinear chirp scaling algorithm. *IEEE Transactions on Geoscience and Remote Sensing*, **49** (1), 500–509.

[16] Loffeld, O., Nies, H., Peters, V., *et al.* (2003) Models and useful relations for bistatic SAR processing. In: Proceedings of the *IEEE International Geoscience and Remote Sensing Symposium (IGARSS'03)*, vol. 3), pp. 1442–1445.

[17] Natroshvili, K., Loffeld, O., Nies, H., *et al.* (2006) Focusing of general bistatic SAR configuration data with 2-D inverse scaled FFT. *IEEE Transactions on Geoscience and Remote Sensing*, **44** (10), 2718–2727.

[18] Neo, Y.L., Wong, F. and Cumming, I.G. (2007) A two-dimensional spectrum for bistatic SAR processing using series reversion. *IEEE Geoscience and Remote Sensing Letters*, **4** (1), 93–96.

[19] Neo, Y.L., Wong, F. and Cumming, I.G. (2008) Processing of azimuth-invariant bistatic SAR data using the range Doppler algorithm. *IEEE Transactions on Geoscience and Remote Sensing*, **46** (1), 14–21.

[20] Neo, Y.L., Wong, F. and Cumming, I.G. (2008) A comparison of point target spectra derived for bistatic SAR processing. *IEEE Transactions on Geoscience and Remote Sensing*, **46** (9), 2481–2492.

[21] Li, Y., Zhang, Z., Xing, M., *et al.* (2007) Focusing general bistatic SAR using analytically computed point target spectrum. 1st Asian and Pacific Conference on Synthetic Aperture Radar (APSAR), pp. 638–641.

[22] Zhang, Z., Xing, M., Ding, J., *et al.* (2007) Focusing parallel bistatic SAR data using the analytic transfer function in the wavenumber domain. *IEEE Transactions on Geoscience and Remote Sensing*, **45** (11), 3633–3645.

[23] Li, Y.P., Xing, M., Jing, W., *et al.* A Sr-Ecs imaging algorithm for bistatic SAR. *Progress in Natural Science*, **18** (3), 323–333.

[24] He, F., Liang, D.N. and Dong, Z. (2005) A wavenumber domain algorithm for spaceborne bistatic SAR imaging with large squint angle. *Acta Electronica Sinica*, **33** (6), 1011–1014.

[25] Liu, Z., Yang, J.Y., Pi, Y.M., *et al.* (2007) A wavenumber domain imaging algorithm for bistatic SAR system based on phase approximation. *Journal of Electronics and Information Technology*, **29** (9), 2094–2097.

[26] Soumekh, M. (1991) Bistatic synthetic aperture radar inversion with application in dynamic object imaging. *IEEE Transactions on Signal Processing*, **39** (9), 2044–2055.

[27] Soumekh, M. (1998) Wide-bandwidth continuous-wave monostatic/bistatic synthetic aperture radar imaging. In: *Proceedings of the International Conference on Image Processing (ICIP 98)*, pp. 361–365.

[28] Ender, J.H.G. (2003) Signal theoretical aspects of bistatic SAR. In: *Proceedings of the IEEE International Geoscience and Remote Sensing Symposium (IGARSS'03)*, vol. 3, pp. 1438–1441.

[29] Ender, J.H.G., Walterscheid, I. and Brenner, A.R. (2004) New aspects of bistatic SAR, processing and experiments. In: *Proceedings of the IEEE International Geoscience and Remote Sensing Symposium (IGARSS'04)*, vol. 3, pp. 1758–1762.

[30] Walterscheid, I., Ender, J.H.G., Brenner, A.R., *et al.* (2005) Bistatic SAR processing using an omega-K type algorithm. In: *Proceedings of the IEEE International Geoscience and Remote Sensing Symposium (IGARSS'05)*, vol. 2, pp. 1064–1067.

[31] Ender, J. (2004) A step to bistatic SAR processing. EUSAR Conference, pp. 356–359.

[32] Giroux, V., Cantalloube, H. and Daout, F. (2005) An omega-K algorithm for SAR bistatic systems. In: *Proceedings of the IEEE International Geoscience and Remote Sensing Symposium (IGARSS'05)*, vol. 2, pp. 1060–1063.

[33] Giroux, V., Cantalloube, H., and Daout, F. (2006) Frequency domain algorithm for bistatic SAR. EUSAR Conference, pp. 1–4.

[34] Qiu, X.L., Hu, D.H. and Ding, C.B. (2008) An omega-K algorithm with phase error compensation for bistatic SAR of a translational invariant case. *IEEE Transactions on Geoscience and Remote Sensing*, **46** (8), 2224–2232.

[35] Bamler, R., Meyer, F. and Liebhart, W. (2006) No math: bistatic SAR processing using numerically computed transfer functions. IEEE International Conference on Geoscience and Remote Sensing Symposium, pp. 1844–1847.
[36] Zeng, D., Zeng, T., Hu, C., *et al.* (2006) Back-projection algorithm characteristic analysis in forward-looking bistatic SAR. International Conference on Radar, pp. 1–4.
[37] Li, R., Pi, Y.M. and Zhang, X.L. (2006) An improved back projection imaging algorithm of airborne bistatic SAR. *Radar Science and Technology*, **4** (6), 348–352.
[38] Yu, D. and Munson, D.C. (2002) A fast back-projection algorithm for bistatic SAR imaging. *International Conference on Image Processing*, **442** (2), II-449–II-452.
[39] Chen, J., Xiong, J., Huang, Y., *et al.* (2004) Research on a novel fast backprojection algorithm for stripmap bistatic SAR imaging. 1st Asian and Pacific Conference on Synthetic Aperture Radar, pp. 622–625.
[40] Roderiguez-Cassola, M., Prats, P. and Moreira, A. (2011) Efficient time-domain image formation with precise topography accommodation for general bistatic SAR configures. *IEEE Transactions on Aerospace and Electronic System*, **47** (4), 2949–2966.
[41] Monti Guarnieri, A. and Rocca, F. (2006) Reduction to monostatic focusing of bistatic or motion uncompensated SAR surveys. *IEE Proceedings of Radar, Sonar and Navigation*, **153** (3), 199–207.
[42] Tang, Z.Y. and Zhang, S.R. (2003), *Theory of Bistatic Synthetic Aperture Radar System*, Science Publisher, Beijing.
[43] Jiao, C.P. (2008) Along track mode bistatic synthetic aperture radar imaging algorithm using chirp scaling. *Journal of System Simulation*, **20** (12), 3223–3228.
[44] Sun, Z.Y., Liang, D.N. and Zhang, Y.S. (2007) Phase-preserving imaging for spaceborne bistatic SAR by range Doppler algorithm. *Journal of National University of Defense Technology*, **29** (6), 81–85.
[45] Bamler, R., Meyer, F. and Liebhart, F. (2007) Processing of bistatic SAR data from quasi-stationary configurations. *IEEE Transactions on Geoscience and Remote Sensing*, **45** (11), 3350–3358.
[46] Qiu, X.L, Hu, D.H. and Ding, C.B. (2006) Focusing bistatic images use RDA based on hyperbolic approximating. International Conference on Radar, pp. 1–4.
[47] Cumming, I.G. and Wong, F.H. (2007) *Digital Processing Synthetic Aperture Radar*, Publishing House of Electronics Industry, Beijing.
[48] Runge, H. and Bamler, R. (1992) A novel high precision SAR focusing algorithm based on chirp scaling. International Geoscience and Remote Sensing Symposium, pp. 372–375.
[49] Cumming, I.G., Wong, F.H. and Raney, R.K. (1992) A SAR processing algorithm with no interpolation. International Geoscience and Remote Sensing Symposium, pp. 376–379.
[50] Jin, M.Y., Cheng, F. and Ming, C. (1993) Chirp scaling algorithms for SAR processing. *International Geoscience and Remote Sensing Symposium, Better Understanding of Earth Environment*, **3**, 1169–1172.
[51] Davidson, G.W., Cumming, I.G. and Ito, M.R. (1993) An approach for improved processing in squint mode SAR. *International Geoscience and Remote Sensing Symposium, Better Understanding of Earth Environment*, **3**, 1173–1179.
[52] Raney, R. K., Runge, H., Bamler, R., *et al.* Precision SAR processing using chirp scaling. *IEEE Transaction on Geoscience and Remote Sensing*, **32** (4), 786–799.
[53] Davidson, G.W. (1996) A chirp scaling approach for processing squint mode SAR data. *IEEE Transactions on Aerospace and Electronic Systems*, **32** (1), 121–133.
[54] Prati, C., Rocca, F., Guarnieri, A.M., *et al.* (1990) Seismic migration for SAR focusing: interferometrical applications. *IEEE Transactions on Geoscience and Remote Sensing*, **28** (4), 627–640.
[55] Cafforio, C., Prati, C. and Rocca, F. (1991) SAR data focusing using seismic migration techniques. *IEEE Transactions on Aerospace and Electronic Systems*, **27** (2), 194–207.
[56] Hu, Y.X., Ding, C.B. and Wu, Y.R. (2005) The wide swath spaceborne SAR imaging based on ω-k algorithm. *Acta Electronica Sinica*, **33** (6), 1044–1047.

[57] Dubois-Fernandez, H., Cantalloube, H., Vaizan, B., *et al.* (2006) ONERA-DLR bistatic SAR campaign: planning, data acquisition, and first analysis of bistatic scattering behaviour of natural and urban targets. *IEE Proceedings of Radar, Sonar and Navigation*, **153** (3), 214–223.

[58] Eldhuset, K. (1998) A new fourth-order processing algorithm for spaceborne SAR. *IEEE Transactions on Aerospace and Electronic Systems*, **34** (3), 824–835.

[59] Ul-Ann, Q., Loffeld, O., Nies, H., *et al.* (2008) A point target reference spectrum based on Loffeld's bistatic formula (LBF) for hybrid configurations. 4th International Conference on Emerging Technologies, pp. 74–77.

[60] Ul-Ann, Q., Loffeld, O., Nies, H., *et al.* (2008) Optimizing the individual azimuth contribution of transmitter and receiver phase terms in Loffeld's bistatic formula (LBF) for bistatic SAR processing. *IEEE International Geoscience and Remote Sensing Symposium*, **3**, III-455–III-458.

[61] Wang, R., Loffeld, O., Ul-Ann, Q., *et al.* (2008). A bistatic point target reference spectrum for general bistatic SAR processing. *IEEE Geoscience and Remote Sensing Letters*, **5** (3), 517–521.

[62] Wang, R., Loffeld, O., Nies, H., *et al.* (2009) Chirp-scaling algorithm for bistatic SAR data in the constant-offset configuration. *IEEE Transactions on Geoscience and Remote Sensing*, **47** (3), 952–964.

[63] Ding, J.S., Loffeld, O., Knedlik, S., *et al.* (2009) Focusing bistatic SAR data in the wavenumber domain using linearized weighted LBF. IEEE Radar Conference, pp. 1–6.

[64] Wang, R., Loffeld, O., Neo, Y.L., *et al.* (2010) Extending Loffeld's bistatic formula for the general bistatic SAR configuration. *IET Radar, Sonar and Navigation*, **4** (1), 74–84.

[65] Wang, R., Deng, Y.K., Loffeld, O., *et al.* (2011) Processing the azimuth-variant bistatic SAR data by using monostatic imaging algorithms based on two-dimensional principle of stationary phase. *IEEE Transactions on Geoscience and Remote Sensing*, **49** (10), 3504–3520.

[66] Wong, F.H., Cumming, I.G. and Neo, Y.L. (2009) Focusing bistatic SAR data using the non-linear chirp scaling algorithm. *IEEE Transactions on Geoscience and Remote Sensing*, **46** (9), 2493–2505.

[67] Zhang, H. and Liu, X.Z. (2006) A fourth-order imaging algorithm for spaceborne bistatic SAR. IEEE International Conference on Geoscience and Remote Sensing Symposium, pp. 1196–1199.

[68] Walterscheid, I., Brenner, A.R. and Ender, H.G. (2004) Results on bistatic synthetic aperture radar. *Electronics Letters*, **40** (19), 1224–1225.

[69] Ender, J.H.G., Walterscheid, I. and Brenner, A.R. (2006) Bistatic SAR – translational invariant processing and experimental results. *IEE Proceedings of Radar, Sonar and Navigation*, **153** (3), 177–183.

[70] Walterscheid, I., Ender, J.H.G., Brenner, A.R., *et al.* (2006) Bistatic SAR processing and experiments. *IEEE Transactions on Geoscience and Remote Sensing*, **44** (10), 2710–2717.

[71] Giroux, V., Cantalloube, H. and Daout, F. (2005) An omega-K algorithm for SAR bistatic systems. In: *Proceedings of the IEEE International Geoscience and Remote Sensing Symposium (IGARSS'05)*, vol. 2, pp. 1060–1063.

6

Imaging Algorithm for Translational Variant Bistatic SAR

6.1 Introduction

In Chapter 5, algorithms for translational invariant bistatic SAR are described. Here in this chapter, the imaging algorithms for the more complex configuration, which is the translational variant configuration, will be focused on. In the translational variant configuration, the relative position between the transmitter antenna and the receiver antenna varies with time during the data acquisition. According to the classification of bistatic SAR configurations in Chapter 3, the translational variant configuration includes the one-stationary case and the constant velocity case.

In addition to that the double-square-root range history and the topography sensitive imaging geometry, which exist in the translational invariant configuration, also exist in the translational variant case and there is another characteristic that makes its imaging more difficult: the azimuth variance of the echo data. In the translational invariant case, the transmitter and the receiver are relatively stationary and so the imaging geometry will not change with time. Thus the targets with the same bistatic range but different azimuth positions will have the same bistatic range history, so the bistatic echo is azimuth invariant. Based on this characteristic, the same azimuth matched filter can be used to focus the targets with the same bistatic range in batches, and so the range-Doppler domain algorithms and the wavenumber domain algorithms introduced in the previous chapter are derived. However, in the translational variant configuration, the imaging geometry is changing with time, so the targets at different positions have different range histories. This should make the azimuth matched filter change with the azimuth position, and so the imaging turns out to be a two-dimensional variant matched filtering process, which makes it difficult to be processed efficiently.

In this chapter, algorithms for the translational variant configuration will be introduced. Section 6.2 describes the algorithms for the one-stationary case and Section 6.3

Bistatic SAR Data Processing Algorithms, First Edition. Xiaolan Qiu, Chibiao Ding and Donghui Hu.

describes the algorithms for the constant velocity case. Finally, Section 6.4 concludes this chapter.

6.2 Imaging Algorithms for One-Stationary Bistatic SAR

The one-stationary case is a relatively simple case of the translational variant configuration, and it is not very difficult to realize, so it is the preferred configuration to start with. The SARBRINA (synthetic aperture radar bistatic receiver for interferometrical applications) system, which was built at the UPC (Universitat Politècnica de Catalunya), Spain [1], has carried out several one-stationary bistatic SAR experiments since 2005 using ESAs ENVISAT and ERS-2 C-band SAR satellites as opportunity transmitters [1–6]. The HITCHHIKER system, which is developed at the University of Siegen, has also carried out several one-stationary bistatic SAR experiments since 2009 [7, 8]. In August 2011, the IECAS (Institute of Electronics, Chinese Academy of Sciences) carried out its first field bistatic SAR experiment, which is also a one-stationary configuration [9].

In the one-stationary case, the bistatic range history is in the form of a single square root and thus the range from the target to the stationary antenna is constant. Then, the main difficulties to focusing the data of this configuration are the two-dimensional variant (mainly the azimuth variant) characteristic and the tomography sensitive characteristic of the echo data. Several algorithms for processing these experimental datasets are also published, such as the time-domain back-projection algorithm [10] and the subaperture algorithm [11] to deal with the azimuth variance. The common disadvantage of these algorithms is their inefficiency. In this section, efficient algorithms for one-stationary bistatic SAR will be introduced.

6.2.1 *Imaging Geometry and Signal Model*

The one-stationary bistatic SAR geometry can be sketched as shown in Figure 6.1; in some earth fixed coordinates, the vector of the stationary antenna location is $\mathbf{R}_1 = (X_1, Y_1, Z_1)^{\mathrm{T}}$. The vector of the moving antenna is $\mathbf{R}_2(\eta) = [X_2(\eta), Y_2(\eta), Z_2(\eta)]^{\mathrm{T}}$, when the azimuth time (slow time) is η. An arbitrary point scatter on the earth is denoted as $\mathbf{r} = (x, y, z)^{\mathrm{T}}$. Because it does not matter whether the moving station is the transmitter or the receiver, here we use subscript "1" and "2" to denote the different stations. To solve the problem of synchronization between the transmitter and the receiver, an additional channel receiving the direct pulses from the transmitter, which is called the direct channel, is usually built. The range history of the direct pulses and that of the target $\mathbf{r}$ can be respectively represented as follows:

$$R_D(\eta) = |\mathbf{R}_1 - \mathbf{R}_2(\eta)| \tag{6.1}$$

$$R_{bi}(\eta; \mathbf{r}) = |\mathbf{r} - \mathbf{R}_2(\eta)| + |\mathbf{r} - \mathbf{R}_1| \tag{6.2}$$

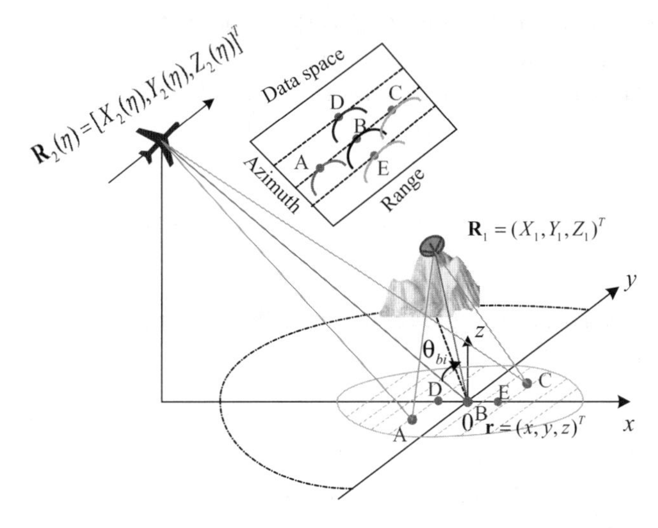

Figure 6.1 Imaging geometry of one-stationary bistatic SAR

If the nominal frequency of the transmitted signal is f_{0T} and the transmitted signal has a time duration T and a wide bandwidth, the received signal in the direct channel and the echo channel before down-conversion at time $\eta + \tau$ can be written, respectively, as follows:

$$\begin{aligned} g_D(\eta, \tau) \approx\ & w_D(\eta, \tau) p\left[\tau - R_D(\eta)/c\right] \\ & \times \exp\{j2\pi f_{0T}[\eta + \tau - R_D(\eta)/c] + j\phi_T[\eta + \tau - R_D(\eta)/c]\} \end{aligned} \tag{6.3}$$

$$g(\eta, \tau) \approx \iint\limits_{x,y} \left\{ \begin{array}{l} w(\eta, \tau; \mathbf{r})\sigma(\mathbf{r}) p\left[\tau - R_{bi}(\eta; \mathbf{r})/c\right] \\ \times \exp\{j2\pi f_{0T}[\eta + \tau - R_{bi}(\eta; \mathbf{r})/c]\} \\ \times \exp\{j\phi_T[\eta + \tau - R_{bi}(\eta; \mathbf{r})/c]\} \end{array} \right\} dxdy \tag{6.4}$$

Where

$p(\tau)$ represents the transmitted waveform;

τ represents the fast time (range time), whose origin is when the pulse is transmitted;

ϕ_T is the phase error of the transmitter oscillator;

$\sigma(\mathbf{r})$ is the scattering coefficient of target $\mathbf{r}$;

$w_D(\eta, \tau)$ and $w(\eta, \tau; \mathbf{r})$ are the weight due to the transmitter and receiver antenna patterns, propagation loss, and so on.

The approximate equal marks in Equations (6.3) and (6.4) come from neglecting the variation of R_D and R_{bi} with τ, respectively.

If the nominal frequency of the receiver is f_{0R}, the down conversion signal at $\eta + \tau$ is then

$$g_R(\eta + \tau) = \exp\{-j2\pi f_{0R}(\eta + \tau) - j\phi_R(\eta + \tau)\} \tag{6.5}$$

where ϕ_R is the phase error of the receiver oscillator. Hence, after down-conversion, the direct channel signal is

$$\begin{aligned} s_D(\eta, \tau) = {} & w_D(\eta, \tau) p\left[\tau - R_D(\eta)/c\right] \\ & \times \exp\{-j2\pi f_{0T} R_D(\eta)/c\} \\ & \times \exp\{j2\pi (f_{0T} - f_{0R})(\eta + \tau)\} \\ & \times \exp\{j\phi_T[\eta + \tau - R_D(\eta)/c] - j\phi_R(\eta + \tau)\} \end{aligned} \tag{6.6}$$

and the echo channel signal is

$$s(\eta, \tau) = \iint_{x,y} \left\{ \begin{aligned} & \sigma(\mathbf{r}) w(\eta, \tau; \mathbf{r}) p\left[\tau - \frac{R_{bi}(\eta; \mathbf{r})}{c}\right] \\ & \times \exp\{-j2\pi f_{0T} R_{bi}(\eta; \mathbf{r})/c\} \\ & \times \exp\{j2\pi (f_{0T} - f_{0R})(\eta + \tau)\} \\ & \times \exp\{j\phi_T[\eta + \tau - R_{bi}(\eta; \mathbf{r})/c] - j\phi_R(\eta + \tau)\} \end{aligned} \right\} dxdy \tag{6.7}$$

6.2.2 *NLCS Algorithm Based on Azimuth Perturbation for the Strip Mode*

If the stationary antenna is far away from the scenario (e.g., it is placed on a geo-stationary satellite or it is placed on a stationary balloon in the near-space), while the moving antenna is relatively close to the scenario and acquisition data in the strip mode, the scenario can be quite large, so this configuration can be of practicality. Wong proposed an azimuth perturbation method to cope with the azimuth variation of this configuration [12] and the authors improved the perturbation and proposed an NLCS algorithm with azimuth perturbation and perturbation error compensation [13, 14]. In the following text, the NLCS algorithm for the strip mode will be introduced.

We suppose that the transmitted signal is a chirp $ss(\tau) = \mathrm{rect}(\tau/T)$ $\exp\left\{-j\pi b\tau^2 + j2\pi f_{0T}\tau\right\}$, where b is the frequency modulation rate. We also suppose that the synchronization errors, that is, $\exp\{j2\pi(f_{0T} - f_{0R})(\eta + \tau)\}$ and

$\exp\{j\phi_T[\eta + \tau - R_{bi}(\eta)/c] - j\phi_R(\eta + \tau)\}$ in Equation (6.7) have been well compensated. Then, the echo from target $\sigma(\mathbf{r})$ can be written as

$$d(\tau, \eta; \mathbf{r}) = \sigma(\mathbf{r})w(\eta, \tau; \mathbf{r})\text{rect}\left[\frac{\tau - R_{bi}(\eta; \mathbf{r})/c}{T}\right]\exp\left\{-j\pi b\left[\tau - \frac{R_{bi}(\eta; \mathbf{r})}{c}\right]^2\right\}$$
$$\times \exp\left\{-j2\pi f_{0T}\frac{R_{bi}(\eta; \mathbf{r})}{c}\right\} \tag{6.8}$$

Let

$$R_{bi}(\eta; \mathbf{r}) = R_1(\mathbf{r}) + R_2(\eta; \mathbf{r}) \tag{6.9}$$

$$R_1(\mathbf{r}) = \sqrt{(x - X_1)^2 + (y - Y_1)^2 + (z - Z_1)^2} \tag{6.10}$$

$$R_2(\eta; \mathbf{r}) = \sqrt{(x - X_2)^2 + (y + V\eta - Y_2)^2 + (z - Z_2)^2} \tag{6.11}$$

By not losing generality, $w(\eta, \tau; \mathbf{r})$ is approximated to be constant and is omitted in the following analysis for the sack of simplicity. By denoting the minimal range from target $\mathbf{r}$ to the moving antenna as R_{20} and the minimal bistatic range as R_{bi0}, then

$$R_{20}(x, y) = \sqrt{(x - X_2)^2 + (z - Z_2)^2} \tag{6.12}$$

$$R_{bi0}(R_{20}, y) = R_{20} + \sqrt{\left[\sqrt{R_{20}^2 - (z - Z_2)^2} + X_2 - X_1\right]^2 + (y - Y_1)^2 + (z - Z_1)^2} \tag{6.13}$$

It can be seen that R_{20} is a function of both R_{bi0} and y. As the Doppler frequency modulation rate and the RCM depends on R_{20}, they both vary with R_{bi0} and y. In other words, for targets having the same R_{bi0}, as long as their azimuth positions (y) are different, their RCMs and the Doppler frequency modulation rates are different, as shown in Figure 6.1.

By using the principle of stationary phase, after a two-dimensional Fourier transform and a range inverse Fourier transform, the signal in the range-Doppler can be expressed as

$$d(\tau, f_\eta; R_{bi0}, y) = \sigma(x, y)\text{rect}\left[\tau - \frac{R_{bi0} + R_{20}C_s(f_\eta)}{c}\right]$$
$$\times \exp\left\{-j\pi K_e\left[\tau - \frac{R_{bi0} + R_{20}C_s(f_\eta)}{c}\right]^2\right\} \tag{6.14}$$
$$\times \exp\left\{-j\frac{2\pi}{\lambda}[R_{bi0} + R_{20}(D(f_\eta) - 1)]\right\}$$

where $\lambda = c/f_{0T}$ and

$$K_e = \frac{1}{\frac{1}{b} + \frac{\lambda R_{20}}{c^2} \frac{(\lambda f_\eta / V)^2}{D^3(f_\eta)}} \tag{6.15}$$

$$D(f_\eta) = \sqrt{1 - (\lambda f_\eta / V)^2} \tag{6.16}$$

$$C_s(f_\eta) = 1/D(f_\eta) - 1 \tag{6.17}$$

Here, the constant introduced by the Fourier transform, which is not important to the following analysis, has been omitted for simplicity.

6.2.2.1 Range Nonlinear Chirp Scaling Function

As can be seen from Equation (6.14), the RCM in the range-Doppler domain is $R_{20}(R_{bi0}, y)C_s(f_\eta)$. If the RCMs of targets at the same R_{bi0} can be considered as the same, only the different RCMs in different ranges need to be identified. Usually, when differential RCM is within half a range resolution cell, these RCMs can be considered as identical, so the restriction is as follows:

$$\begin{aligned} \left|[R_{20}(y_{\max}; R_{bi0}) - R_{20}(y_{ref}; R_{bi0})]C_s(f_\eta)\right| < 0.5\rho_r \\ \left|[R_{20}(y_{\min}; R_{bi0}) - R_{20}(y_{ref}; R_{bi0})]C_s(f_\eta)\right| < 0.5\rho_r \end{aligned} \tag{6.18}$$

Here, ρ_r means the range resolution and $y_{\max}$ and $y_{\min}$ are the azimuth edges of the imaging area. In the following text, we assume that the above equations can be satisfied in the whole imaging area. This is a precondition of this algorithm.

Based on Equation (6.13), it is not a linear relationship between $R_{20}(y_{ref})$ and $R_{bi0}(y_{ref})$(or τ). Thus, the chirp scaling function for this kind of bistatic SAR is different from the one for monostatic SAR [18, 19] and should be renewed. Usually, a quadratic polynomial is accurate enough to express the relationship between $R_{20}(y_{ref})$ and $R_{bi0}(y_{ref})$, and to guarantee a good imaging quality. Thus, the range chirp scaling function can be written as

$$\begin{aligned} \Phi_1(\tau, f_\eta; R_{bi0\_ref}, y_{ref}) = \exp\left\{-j\pi K_e C_s p \left[\tau - \frac{R_{bi0\_ref} + R_{20\_ref} C_s}{c}\right]^2\right\} \\ \times \exp\left\{-j\pi K_e C_s q \left[\tau - \frac{R_{bi0\_ref} + R_{20\_ref} C_s}{c}\right]^3\right\} \end{aligned} \tag{6.19}$$

where R_{bi0_ref} and y_{ref} are usually chosen as the middle bistatic range and the middle azimuth position of the dataset, respectively. R_{20_ref} is determined by R_{bi0_ref} and y_{ref} through Equation (6.13), while p and q are coefficients waiting to be resolved. The intention of introducing these two coefficients will be explained later.

After multiplying Equation (6.14) by (6.19), the signal can be written as

$$\begin{aligned} d_1(\tau, f_\eta; R_{bi0}, y) = \sigma(x, y)\mathrm{rect}\left[\tau - \frac{R_{bi0} + R_{20}C_s(f_\eta)}{c}\right] \\ \times \exp\{j\Theta(\tau, f_\eta; R_{bi0}, y)\} \\ \times \exp\left\{-j\frac{2\pi}{\lambda}[R_{bi0} + R_{20}(D(f_\eta) - 1)]\right\} \end{aligned} \tag{6.20}$$

where

$$\begin{aligned} \Theta(\tau, f_\eta; R_{bi0}, y) = -\pi K_e\left[\tau - \frac{R_{bi0} + R_{20}C_s(f_\eta)}{c}\right]^2 \\ -\pi K_e C_s p\left[\tau - \frac{R_{bi0\_ref} + R_{20\_ref}C_s}{c}\right]^2 \\ -\pi K_e C_s q\left[\tau - \frac{R_{bi0\_ref} + R_{20\_ref}C_s}{c}\right]^3 \end{aligned} \tag{6.21}$$

The purpose of scaling in the CS algorithm is to correct the differential RCM between targets in different range gates, so that all RCM curves will be the same as the RCM of the reference target (i.e., $R_{20\_ref}C_s$), and meanwhile, the range position corresponding to the reference frequency (which is zeros here) remains R_{bi0}. Thus, we expand Θ at

$$\tau = \frac{R_{bi0} + R_{20\_ref}C_s}{c}$$

to its Taylor series and reserve the quadratic term, and then we have

$$\Theta(\tau, f_\eta; R_{bi0}, y) = \Theta_0(f_\eta) + \Theta_1(\tau, f_\eta) + \Theta_2(\tau, f_\eta) \tag{6.22}$$

where

$$\begin{aligned} \Theta_0(f_\eta) = -\pi K_e C_s^2\left[\frac{R_{20\_ref} - R_{20}}{c}\right]^2 - \pi K_e C_s p\left[\frac{R_{bi0} - R_{bi0\_ref}}{c}\right]^2 \\ -\pi K_e C_s q\left[\frac{R_{bi0} - R_{bi0\_ref}}{c}\right]^3 \end{aligned} \tag{6.23}$$

$$\begin{aligned} \Theta_1(\tau, f_\eta) = -2\pi K_e C_s\left\{p\left[\frac{R_{bi0} - R_{bi0\_ref}}{c}\right] + \frac{3}{2}q\left[\frac{R_{bi0} - R_{bi0\_ref}}{c}\right]^2 \right. \\ \left. - \frac{R_{20} - R_{20\_ref}}{c}\right\}\left[\tau - \frac{R_{bi0} + R_{20\_ref}C_s}{c}\right] \end{aligned} \tag{6.24}$$

$$\Theta_2(\tau, f_\eta) = -\pi K_e\left(1 + C_s p + 3qC_s\frac{R_{bi0} - R_{bi0\_ref}}{c}\right)\left[\tau - \frac{R_{bi0} + R_{20\_ref}C_s}{c}\right]^2 \tag{6.25}$$

Since both q and C_s are very small, the third term between the parenthesis in Equation (6.25) can be ignored as long as the equation

$$\pi K_e 3qC_s\frac{R_{bi0} - R_{bi0\_ref}}{c}\frac{T^2}{4} < \frac{\pi}{4} \tag{6.26}$$

(where T is the pulse duration of the transmitted signal) is satisfied within the imaging swath. Under the simulating parameters later, the imaging swath meeting Equation (6.26) can reach 10^{13} m. Thus, Equation (6.25) can be simplified as follows:

$$\tilde{\Theta}_2(\tau, f_\eta) = -\pi K_e(1 + C_s p)\left[\tau - \frac{R_{bi0} + R_{20\_ref}C_s}{c}\right]^2 \tag{6.27}$$

It can be seen that if p and q could make $\Theta_1(\tau, f_\eta) \equiv 0$, then all RCM curves will be identical. Recall that R_{20} is approximately a quadratic equation of R_{bi0}; thus, if the curve ($R_{bi0} - R_{bi0\_ref}$ versus $R_{20} - R_{20\_ref}$) has been calculated through Equation (6.13), the linear and quadratic coefficients can be found through the fit method, and p and q can be settled.

After range chirp scaling, a range Fourier transform is performed and then a range process factor

$$\Phi_2(f_\tau, f_\eta; R_{bi0\_ref}, y_{ref}) = \exp\left\{-j\frac{\pi f_\tau^2}{K_e(1 + pC_s)}\right\}\exp\left\{j\frac{2\pi}{\lambda}R_{20\_ref}C_s\right\} \tag{6.28}$$

is multiplied by the signal in a two-dimensional frequency domain, which completes the range compression and the bulk RCM correction.

6.2.2.2 Azimuth Frequency Perturbation

After range processing, the signal in the range-Doppler domain can be expressed as

$$d_2(\tau, f_\eta; R_{bi0}, y) = \sigma(x, y)\text{sinc}\left(\tau - \frac{R_{bi0}}{c}\right)\exp\left\{-j\frac{2\pi}{\lambda}[R_{bi0} + R_{20}(D(f_\eta) - 1)]\right\} \tag{6.29}$$

As previously explained, R_{20} changes with y under the same R_{bi0}, which makes $f_r(R_{bi0}, y) \neq f_r(R_{bi0}, y_{ref})$. Thus, targets having the same R_{bi0} cannot be compressed with the same filter unless their Doppler frequency modulation rates are preprocessed to be the same.

Wong proposed a way to identify $f_r(R_{bi0}, y)$ to be $f_r(R_{bi0}, y_{ref})$. This method is said to be azimuth frequency perturbation. The perturbation function proposed by Wong in [12] can be described as follows:

$$h_{pert}(t) = \exp\{j\pi\alpha_{bi}(\eta - y_{ref}/V)^4\}, \quad \text{where} \quad \alpha_{bi} = \frac{f_r(R_{bi0}, y_{ref}) - f_r(R_{bi0}, y_b)}{6(y_b - y_{ref})^2/V^2} \tag{6.30}$$

Here y_b represents the azimuth edge of the imaging scenario. This perturbation function is symmetrical about the azimuth time corresponding to y_{ref}, so it implies that the changing of $f_r(R_{bi0}, y)$ is symmetrical about y_{ref}. It is known from one-stationary bistatic SAR geometry that $f_r(R_{bi0}, y)$ is symmetrical about Y_1, which means that Y_1 should be chosen as y_{ref}. When Y_1 is near to the azimuth scene center, the perturbation function above works pretty well, but when the bistatic angle is large and Y_1 is far from the azimuth scene center (as Figure 6.1), it becomes invalid. This is because the relationship between $f_r(R_{bi0}, y)$ and y is rather complicated and a quadratic fit is not accurate enough for the large azimuth scale. Therefore, in this section, a perturbation function generated through the local fit method will solve this problem and make the NLCS algorithm suitable for a large azimuth swath or a large bistatic angle case. The perturbation function can be generated as follows:

1. For each R_{bi0}, by choosing a series of azimuth positions y_i, the differential Doppler frequency modulation rates can be calculated as follows:

$$\Delta f_r(R_{bi0}, y_i) = f_r(R_{bi0}, y_i) - f_r(R_{bi0}, y_{ref}) \tag{6.31}$$

   As the differential azimuth time $\eta_i = (y_i - y_{ref})/V$, the curve ($\Delta f_r$ versus η) can be obtained.
2. $\Delta f_r(R_{bi0}, \eta_i)$ can be expressed as a polynomial function of η through the polynomial fit method

$$\Delta f_r(R_{bi0}, \eta) \approx \sum_{n=1}^{N} P_n \eta^n \tag{6.32}$$

   Since $\Delta f_r = 0$ when $\eta = 0$, this equation does not have a constant term.
3. The phase of a perturbation function is then generated by applying an integral with η to Equation (6.32) twice, which is

$$\varphi(\eta; R_{bi0}) = -2\pi \left[\sum_{n=1}^{N} \frac{1}{(n+1)(n+2)} P_n \eta^{n+2} \right] \tag{6.33}$$

Then the perturbation function is

$$\Phi_3(\eta; R_{bi0}) = \exp\{j\varphi(\eta; R_{bi0})\} \tag{6.34}$$

After multiplying the signal by Equation (6.34) in the azimuth time domain, all the $f_r(R_{bi0}, y)$ are equal to $f_r(R_{bi0}, y_{ref})$. Thus, all of the targets in the range gate R_{bi0} can be compressed with the same matched filter (i.e., Equation 6.35) in the range-Doppler domain:

$$\Phi_4(f_\eta; R_{bi0}, y_{ref}) = \exp\left\{j\frac{2\pi}{\lambda}R_{20}(R_{bi0}, y_{ref})[D(f_\eta) - 1]\right\} \tag{6.35}$$

One thing that needs to be paid attention to is that, as the azimuth signal is not a neat chirp and the perturbation can only identify the quadratic phase term of different targets, higher-order phase errors still exist.

6.2.2.3 Negative Effects of Azimuth Perturbation

It can be seen from Equation (6.34) that φ is an $N(N \geq 3)$-order function of η. Thus, while identifying the Doppler frequency modulation rate (i.e., the quadratic phase term of η), this perturbation function also introduces the constant, the linear and the higher-order phase errors. As φ changes with η(or y), this perturbation function affects targets differently at different azimuth positions. For a target located at y_d (corresponding to azimuth time η_d), the phase introduced by $\Phi_3(\eta; R_{bi0})$ within its synthetic aperture time T_a can be expressed by the Taylor series around η_d:

$$\varphi_d(\eta; R_{bi0}) = \frac{1}{n!}\sum_{n=0}^{N+2}\varphi^{(n)}(\eta_d; R_{bi0})(\eta - \eta_d)^n, \quad -T_a/2 \leq \eta - \eta_d \leq T_a/2 \tag{6.36}$$

Here the superscript (n) means the Nth-order derivative. Thus, the effects of $\varphi_d(\eta; R_{bi0})$ can be listed as follows:

1. The constant phase at η_d, which is $\varphi(\eta_d; R_{bi0})$, destroys the phase term that stores the information of the bistatic range (i.e., R_{bi0}), and it needs to be eliminated for interferometrical applications.
2. The linear phase term in Equation (6.36) causes a frequency shift, whose amount is

$$\Delta f(\eta_d; R_{bi0}) = \varphi^{(1)}(\eta_d; R_{bi0})/2\pi \tag{6.37}$$

This frequency shift will finally cause a shift in the azimuth position of the compressed target, which is

$$\Delta y(\eta_d; R_{bi0}) = V\Delta f(\eta_d; R_{bi0})/f_r(\eta_{ref}; R_{bi0}) \tag{6.38}$$

To avoid aliasing, the remnant of PRF subtracted by the Doppler bandwidth must be larger than the maximal shifting frequency.

3. The quadratic term of Equation (6.36) identifies the Doppler frequency modulation rate and, on the other hand, it causes the change in Doppler bandwidth. Before perturbation, the Doppler bandwidths of all targets are coincident, although the Doppler frequency modulation rates and the synthetic times are both different. The perturbation identifies the Doppler frequency modulation rates while the synthetic times remain the same; thus, the Doppler bandwidths become different. The Doppler bandwidth of the target at η_d after perturbation is

$$B_a'(\eta_d; R_{bi0}) = B_a f_r(\eta_{ref}; R_{bi0})/f_r(\eta_d; R_{bi0}) \tag{6.39}$$

Again, to avoid aliasing, PRF must be larger than the biggest Doppler bandwidth after perturbation. Another thing that needs to be paid attention to is that, although the Doppler bandwidth (thereby, the azimuth time resolution) is modified, the azimuth resolution in distance remains the same due to the azimuth shift previously discussed. As an example, a target appears at η_d before perturbation will appear at η_e after perturbation:

$$\begin{aligned}\eta_e &= \eta_d + \Delta y(\eta_d; R_{bi0})/V \\ &= \eta_d f_r(\eta_d; R_{bi0})/f_r(\eta_{ref}; R_{bi0})\end{aligned} \tag{6.40}$$

Thus, the equivalent velocity can be written as

$$V_e = \eta_d V/\eta_e = V f_r(\eta_{ref}; R_{bi0})/f_r(\eta_d; R_{bi0}) \tag{6.41}$$

Thus the azimuth resolution in distance is

$$V_e(\eta_d; R_{bi0})/B_a'(\eta_d; R_{bi0}) = V/B_a \tag{6.42}$$

which remains the same.

4. The higher-order phase terms introduced by the perturbation function deteriorate the imaging quality. Simulation indicates that the cubic phase error is the main component, when Y_1 is far away from the azimuth scene center. The cubic phase error will cause the dissymmetry of the sidelobes and raise the peak sidelobe ratio (PSLR) and the integral sidelobe ratio (ISLR). The higher-order phase errors also cause a phase error at the peak of the mainlobe, which makes the image not perfect for interferometrical applications.

These effects can be shown in Figure 6.2. Three different targets, namely, 1, 2 and 3, at the same range gate are denoted by the black dots. The thick solid line is the frequency–time curve before perturbation; they have different slopes but the same bandwidth. The dash–dot curve is the frequency–time relationship of the perturbation function (taking target 2 as the referential target). After perturbation,

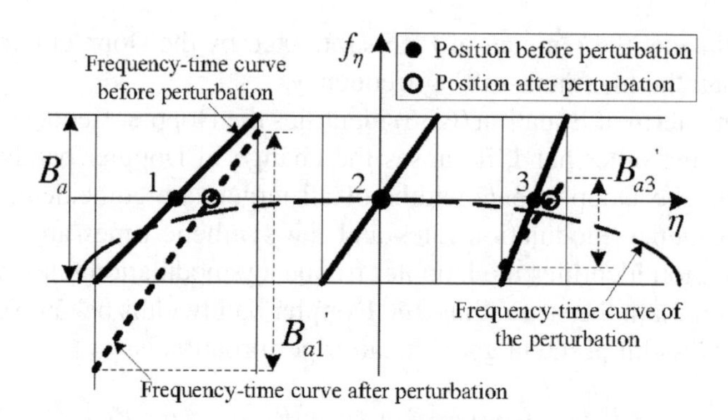

Figure 6.2 Illustration of azimuth perturbation (© 2008 IEEE. Reprinted with permission from X. Qiu, D. Hu, C. Ding, "An Improved NLCS Algorithm With Capability Analysis for One-Stationary BiSAR" *IEEE Transaction on Geoscience and Remote Sensing*, **46**, 10, pp. 3179–3186, 2008.)

the three frequency–time curves become the dashed lines in the figure. It is clear that the slope of the dashed lines becomes the same and their bandwidths become different (the bandwidth of target 1 increases, whereas the bandwidth of target 3 decreases). After compressing, the targets will appear at their zero Doppler times, which are marked by the circles in Figure 6.2. It can be seen that the reference target remains in its original position, whereas targets 1 and 3 shift to other places. In addition, there are higher-order terms in the perturbation curve; thus, it will inevitably introduce higher-order phase errors.

6.2.2.4 Compensating the Perturbation Errors

To minimize the phase error and to improve the imaging quality, methods should be found to cope with those negative effects previously discussed. Here, some simple methods will be proposed, which will improve the imaging to some degree:

1. Each azimuth line (which corresponds to each range gate) in the image should be multiplied by $\Phi_3^*(\eta_d; R_{bi0})$ to eliminate the constant phase introduced by the perturbation at each η_d. Here, the "$*$" means conjugation.
2. The PRF in this configuration should be high enough to endure the frequency shift and the change in Doppler bandwidth.
3. A geometric correction can be done along the azimuth direction to cope with the azimuth shift caused by the perturbation. As the azimuth shift for each target can be calculated according to Equation (6.38), the correction can be done by a resample for each azimuth line of the image.
4. A simple compensation for the higher-order phase errors can be made as follows to improve the imaging quality:
   a. For each R_{bi0}, calculate the coefficients of the higher-order phase terms at each azimuth time η_d, which are $(1/n!)\varphi^{(n)}(\eta_{di}; R_{bi0})$, $n \geq 3$, and calculate the mean value of all these coefficients, that is, $P_n = \overline{(1/N!)\varphi^{(n)}(\eta_{di}; R_{bi0})}$.

b. According to the frequency–time relationship, the compensation factor in the azimuth frequency domain can be

$$\Phi_5(f_\eta; R_{bi0}) = \exp\left\{-j\sum_{n=3}^{N+2} P_n[f_\eta/f_r(y_{ref}; R_{bi0})]^n\right\} \tag{6.43}$$

c. If the imaging quality still is not good enough for the whole scene, the image can be cut into subpatches along the azimuth direction, and then the average higher-order phase for each subpatch can be compensated separately to make the residual phase error even smaller.

6.2.2.5 Block Diagram of the Algorithm

A block diagram of the NLCS algorithm with azimuth perturbation and perturbation error compensation is shown in Figure 6.3.

This algorithm is suitable for the strip mode, but may not be valid in the spotlight mode. In the strip mode, the signals of different targets with different azimuth positions span different azimuth time regions. Then, the azimuth perturbation in the time domain can have different effects on different targets, so as to identify their Doppler frequency modulation rate. While in the spotlight mode, the signals of different targets with different azimuth positions have the same azimuth time region, so the azimuth perturbation method turns out to be invalid.

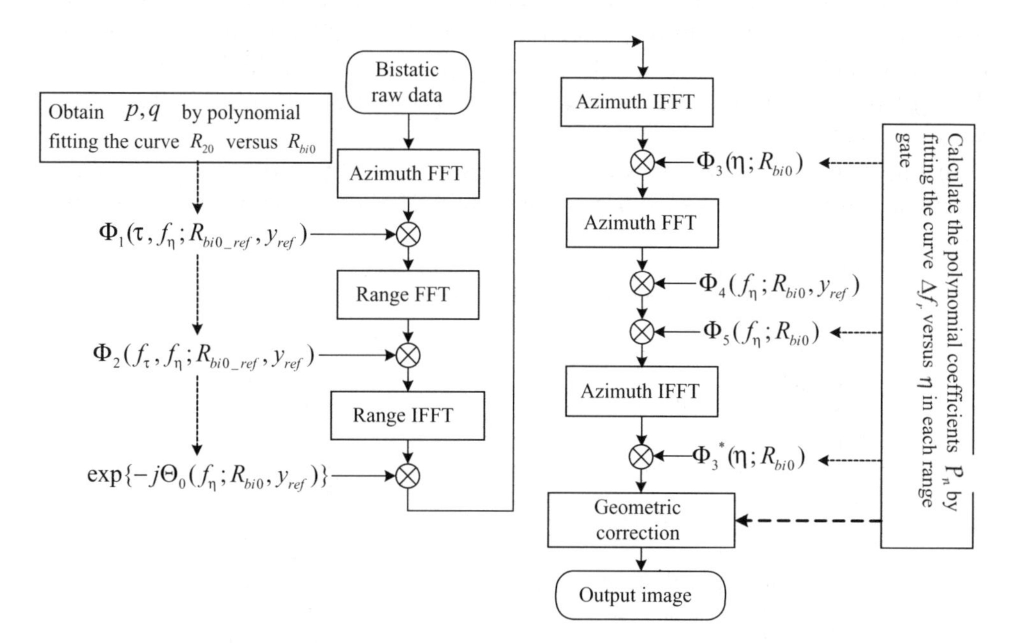

Figure 6.3 Block diagram of the improved NLCS algorithm

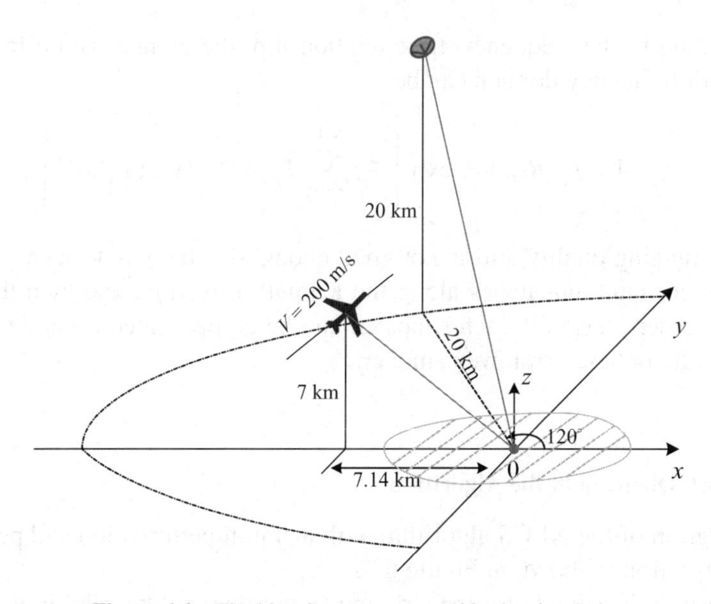

Figure 6.4 The bistatic geometry used in the simulation

6.2.2.6 Simulations

To verify the improved NLCS algorithm proposed here and to testify its capability in a large bistatic configuration, a peculiar bistatic SAR geometry is designed, in which the transmitter and the receiver are placed at the adjacent side of the imaging scenario to make the azimuth position of the fixed antenna far away from the scene center. The fixed antenna is assumed to be placed on a stationary balloon in the near-space and the moving antenna is assumed to be placed on an airplane (as shown in Figure 6.4). The system parameters are listed in Table 6.1 and the simulated points in the scenario are shown in Figure 6.5.

In Figure 6.6, the curve in which $R_{20} - R_{20\_ref}$ changes with $R_{bi0} - R_{bi0\_ref}$ is given, and three different methods are analyzed. The left column shows the results

Table 6.1 Parameters for the simulation

Parameters	Value
Wavelength	0.031 25 m
Transmitted bandwidth	300 MHz
Sample rate	330 MHz
Azimuth length of the moving antenna	0.8 m
PRF	400 Hz
Sample number for each pulse	32 768
Total pulse number	8192
Position of moving antenna when $\eta = 0$	[−7141.4, 0, 7000] m
Position of stationary antenna	[10, 1.732, 20] km
Sample delay	1.148 44e-004 s

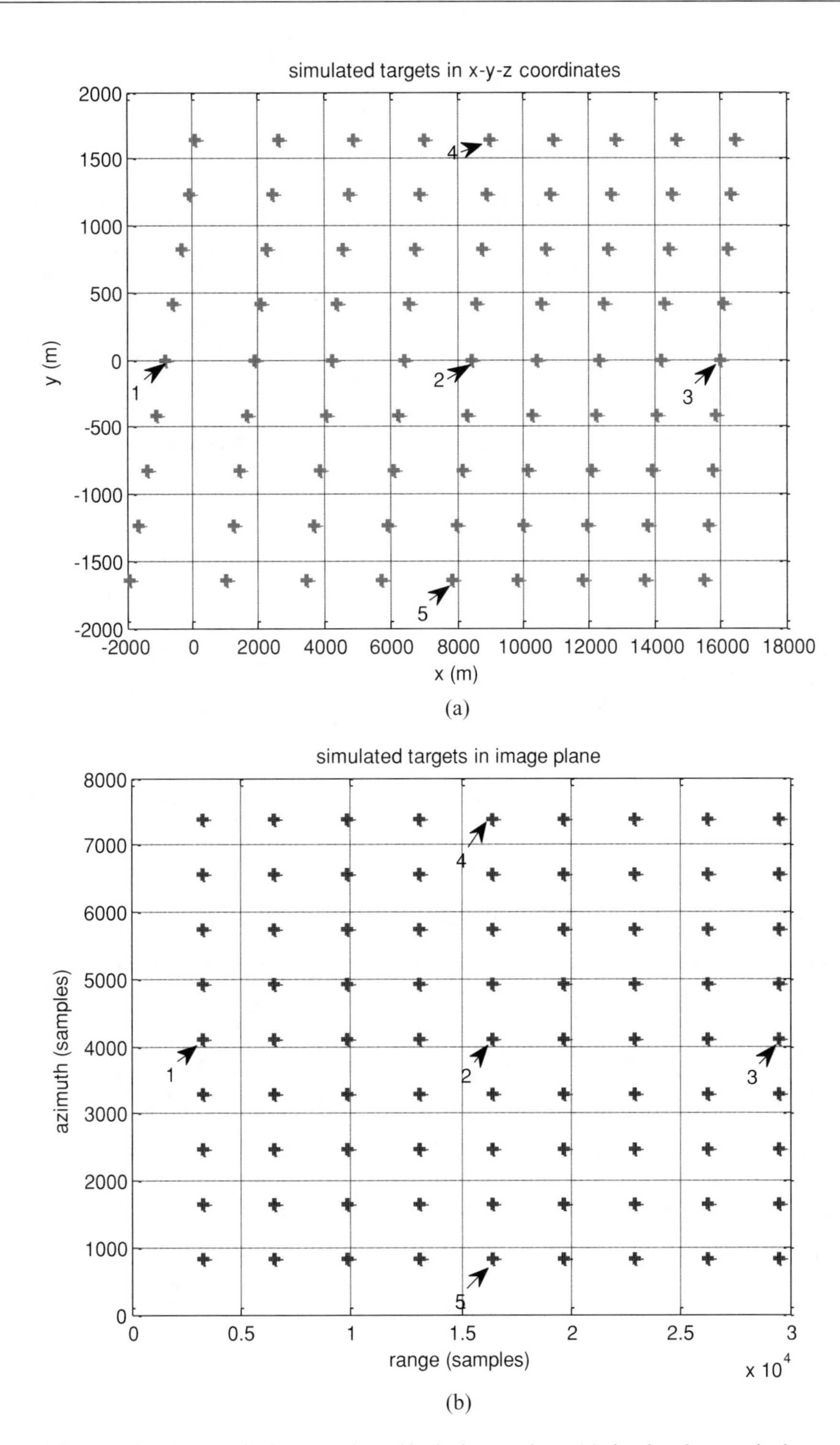

Figure 6.5 Simulated targets in the scenario and in the image plane: (a) simulated targets in the scenario, (b) simulated targets in the image plane

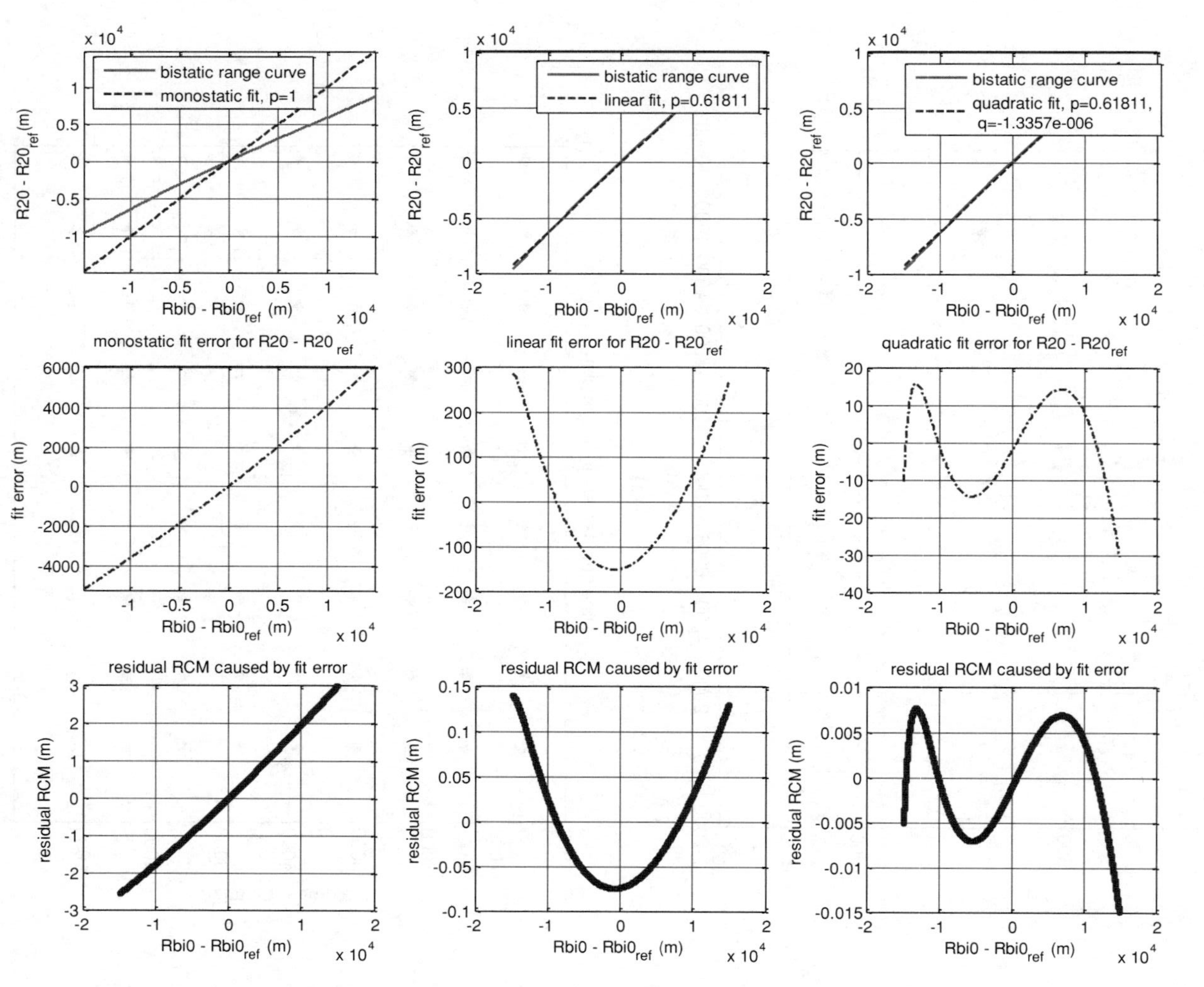

Figure 6.6 Illustrating the relationship between $R_{20} - R_{20\_ref}$ and $R_{bi0} - R_{bi0\_ref}$

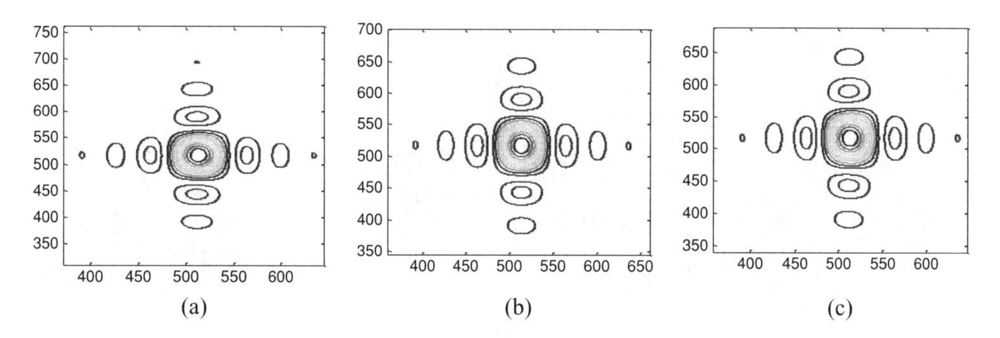

Figure 6.7 Imaging results of targets 1, 2 and 3: (a) target 1, (b) target 2, (c) target 3

if we consider this bistatic SAR signal as the traditional monostatic SAR signal. In monostatic SAR, the RCM depends on the sum of the transmitting range and the receiving range (i.e., R_{bi0}), so p equals one. The residual RCM will be about 3 m at the edge of the range swath. Therefore, it is clear that this bistatic SAR data cannot be processed using the monostatic chirp scaling method in the range direction. The linear fit to the curve ($R_{20} - R_{20\_ref}$ versus $R_{bi0} - R_{bi0\_ref}$) causes a fit error that is about 300 m at the edge of the range swath, and thus causes a residual RCM that is about 0.15 m at the edge of the swath. The bistatic slant range resolution in this simulation is about 0.443 m, so this residual RCM is acceptable. Further, a quadratic fit to the curve leaves a fit error that is only about 30 m at the edge of the swath, which causes the residual RCM to be 0.015 m, which can surely be ignored. Hence, the NLCS algorithm based on a quadratic polynomial fit to the bistatic range can do a good job in handling the range variance of the bistatic data. This can also be validated by the focusing results shown in Figure 6.7 and Table 6.2, where the two-dimensional responses of targets 1, 2 and 3 are all in very good shape and the quality indexes are also nearly perfect.

To clarify the ability of the azimuth perturbation and the perturbation phase error compensation, the different imaging results of targets with different azimuth positions under three different processing flows, which are NLCS processing without

Table 6.2 Imaging qualities of targets 1, 2 and 3

	Range				Azimuth			
Target	Resolution (m)	Widen ratio	PSLR (dB)	ISLR (dB)	Resolution (m)	Widen ratio	PSLR (dB)	ISLR (dB)
1	0.886	0.886	−13.19	−9.97	0.708	0.886	−13.23	−10.18
2	0.887	0.887	−13.23	−9.98	0.708	0.886	−13.25	−10.19
3	0.886	0.886	−13.23	−9.98	0.708	0.886	−13.24	−10.18

Note: The range resolution here is the bistatic slant range resolution, and no weighted window is used in imaging.

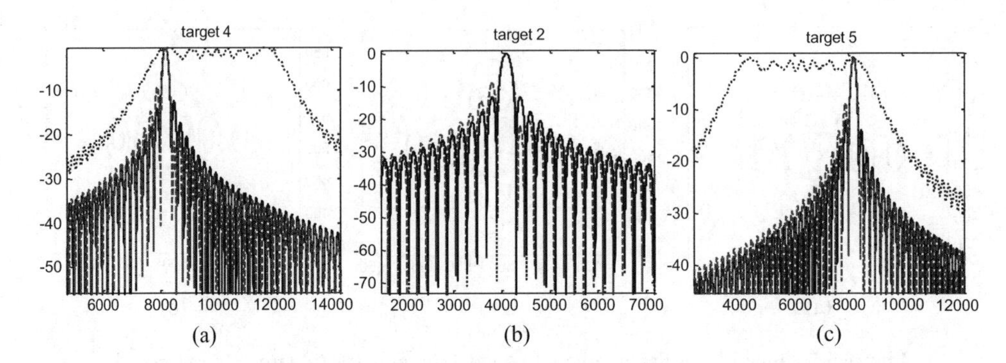

Figure 6.8 Azimuth response of targets 4, 2 and 5 (the dotted line is the azimuth response using method I, the dashed line is the azimuth response using method II and the solid line is the azimuth response using method III): (a)target 4, (b) target 2, (c) target 5 (© 2008 IEEE. Reprinted with permission from X. Qiu, D. Hu, C. Ding, "An Improved NLCS Algorithm With Capability Analysis for One-Stationary BiSAR" *IEEE Transaction on Geoscience and Remote Sensing*, **46**, 10, pp. 3179–3186, 2008.)

azimuth perturbation (method I), NLCS processing with azimuth perturbation but without perturbation phase error compensation (method II) and NLCS processing with azimuth perturbation and perturbation error compensation (method III), are shown in Figure 6.8 and Table 6.3. As can be seen from Figure 6.8 and Table 6.3, the NLCS algorithm without azimuth perturbation will only focus the targets at the referential azimuth position, the targets at the azimuth edge of the scene being totally unfocused. When using the NLCS without phase compensation, targets can be focused better, but they suffer from the severe dissymmetry of sidelobes and the high PSLR and ISLR. Once adding the phase compensation step into the NLCS, the phase error of the referential targets is almost eliminated completely, and its imaging quality is close to the ideal value. Meanwhile, the imaging qualities of the other two targets are improved to some degree.

Moreover, Figure 6.9 shows the azimuth spectra of different targets before and after perturbation, where the frequency shift and the Doppler bandwidth modification can

Table 6.3 Azimuth qualities of targets 2, 4 and 5 under different imaging conditions

	Target	Resolution (m)	PSLR (dB)	ISLR (dB)
Method I	4	16.89	−0.00	11.35
	2	0.709	−13.24	−10.19
	5	18.86	−0.00	11.16
Method II	4	0.695	−9.28	−8.68
	2	0.715	−9.06	−8.49
	5	0.742	−8.81	−8.31
Method III	4	0.711	−12.35	−9.99
	2	0.708	−13.25	−10.19
	5	0.710	−12.22	−9.95

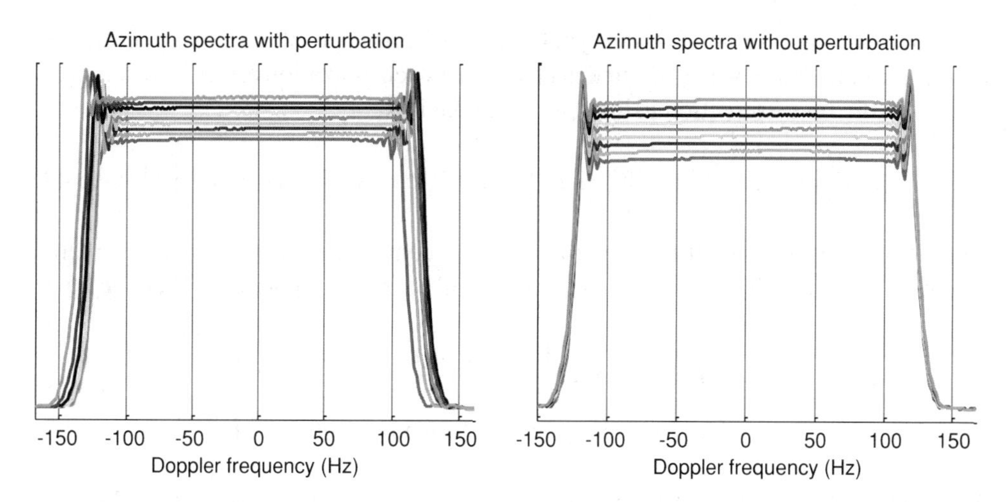

Figure 6.9 Illustration of the azimuth frequency shift caused by azimuth perturbation

be seen. This figure validates the correctness of the analysis about the perturbation effects. Figure 6.10 shows the azimuth position shifts of the targets at the referential range. It can be seen that all the targets except the target at the referential azimuth position have an azimuth shift. The maximal shift happens at the azimuth edge where $y = -1638$ m and is 54 pixels along the azimuth. After geometry correction according to Equation (6.38), the targets go back to their ideal positions, which validates the correctness of the analysis about the effects of the perturbation.

6.2.3 Algorithm Based on the Keystone Transform for the Spotlight Mode

The NLCS algorithm introduced in the previous section is more suitable for the strip mode. Moreover, as mentioned in Section 6.2.2, the NLCS algorithm introduced above

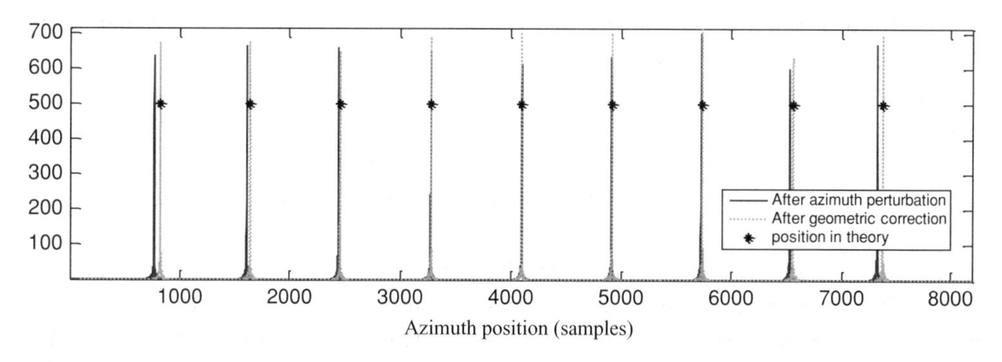

Figure 6.10 Illustration of the azimuth position shift caused by azimuth perturbation. Azimuth position (samples)

is based on the assumption that the synchronization errors are all well compensated. In practice, a convenient way to compensate for the synchronization errors is to compress $s(\eta, \tau)$ in the range direction using $s_D(\eta, \tau)$, which is

$$s_1(\eta, \tau) = IFFT^{\tau}\left[\text{shift}\left\{FFT^{\tau}\left[s(\eta, \tau)\right] \cdot \text{conj}\left\{FFT^{\tau}\left[s_D(\eta, \tau)\right]\right\}\right\}\right] \quad (6.44)$$

Here, "shift" means to shift the spectrum which is centered at $f_{0T} - f_{0R}$ to be centered at zero in the range frequency domain and "conj" means conjugation. Then we have

$$s_1(\eta, \tau) \approx \iint_{x,y} \left\{ \begin{array}{l} \sigma(\mathbf{r}) \text{ sinc}\left[\tau - \dfrac{R_{bi}(\eta; \mathbf{r}) - R_D(\eta)}{c}\right] \\ \times \exp\{-j2\pi f_{0T}[R_{bi}(\eta; \mathbf{r}) - R_D(\eta)]/c\} \end{array} \right\} dxdy \quad (6.45)$$

The approximation in the above equation comes from omitting the phase difference between

$$\text{rect}\left[\frac{\tau - R_D(\eta)/c}{T}\right] \exp\{-j\phi_R(\eta + \tau)\} \text{ and rect}\left[\frac{\tau - R_{bi}(\eta; \mathbf{r})/c}{T}\right] \exp\{-j\phi_R(\eta + \tau)\}$$

which can usually be ignored.

It can be seen from Equation (6.45) that, after range compression using the reference signal from the direct channel, the time synchronization problem is solved because all the received echo pulses are aligned to the known $R_D(\eta)$. The frequency synchronization problem is solved because the direct pulses and the echo pulses are generated by the same transmitter. The phase synchronization is also solved as long as no additional phase difference is introduced by the two receive channels. This is because the phase errors caused by the transmitter are the same in a direct pulse and the corresponding echo pulse and hence can be counteracted; besides, as the direct pulse and the echo pulse are received with a small time difference, the phase errors caused by the receiver oscillator are nearly the same and can also be well counteracted. Therefore, as stated previously, this is a convenient and commonly used synchronization method in practice. Therefore, it is reasonable to find an algorithm that starts bistatic SAR focusing from $s_1(\eta, \tau)$.

After range compression using direct pulses, the residual range history of target $\mathbf{r}$ is as follows:

$$R_{bic}(\eta; \mathbf{r}) = R_{bi}(\eta; \mathbf{r}) - R_D(\eta) \quad (6.46)$$

It can be seen that, if the receiver is relatively near the scenario (e.g., in the spaceborne–ground bistatic configuration in which the transmitter is a spaceborne SAR and the receiver is located on some hill or high building near the scenario), the higher-order terms in $R_{bic}(\eta; \mathbf{r})$ will be much smaller than those in the original range history $R_{bi}(\eta; \mathbf{r})$, so the Doppler bandwidth of the targets after range compression by the reference pulses is much smaller than the original Doppler bandwidth. In the spotlight mode, where the PRF is smaller than the original Doppler bandwidth of each target, this range compression step also can fulfill the job of deramping the aliasing Doppler frequency.

In the following text, an algorithm based on the Keystone transform that is more suitable for the spotlight mode will be introduced.

6.2.3.1 Keystone Transform for Linear RCM Correction

In one-stationary spotlight bistatic SAR cases, such as spaceborne–ground cases and straight path airborne–ground cases, the residual range history can be approximated as follows:

$$R_{bic}(\eta;\mathbf{r}) \approx R_{bic}(0;\mathbf{r}) + A\eta + B\eta^2 + C\eta^3 + D\eta^4 \tag{6.47}$$

where A, B, C and D are functions of the target position (i.e., $\mathbf{r}$). In the case of simple trajectories such as uniform linear motion, the formula of these parameters A, B, C and D can be obtained as derivatives of $R_{bic}(\eta;\mathbf{r})$. For example, if the moving antenna flies along the y axis with a constant velocity V, then the parameters A and B are as follows:

$$A(\mathbf{r}) = V\left[\frac{Y_2(0) - y}{|\mathbf{r} - \mathbf{R}_2(0)|} - \frac{Y_2(0) - Y_1}{|\mathbf{R}_1 - \mathbf{R}_2(0)|}\right] \tag{6.48}$$

$$B(\mathbf{r}) = V_2^2\left[\frac{\cos^2\theta_2(\mathbf{r})}{|\mathbf{r} - \mathbf{R}_2(0)|} - \frac{\cos^2\theta_2(\mathbf{R}_1)}{|\mathbf{R}_1 - \mathbf{R}_2(0)|}\right] \tag{6.49}$$

where

$$\theta_2(\mathbf{r}) = \cos^{-1}\left\{\frac{\sqrt{[X_2(0) - x]^2 + [Z_2(0) - z]^2}}{|\mathbf{r} - \mathbf{R}_2(0)|}\right\} \tag{6.50}$$

$$\theta_2(\mathbf{R}_1) = \cos^{-1}\left\{\frac{\sqrt{[X_2(0) - X_1]^2 + [Z_2(0) - Z_1]^2}}{|\mathbf{R}_1 - \mathbf{R}_2(0)|}\right\} \tag{6.51}$$

It can be seen that $A(\mathbf{r})$ nearly linearly depends on y, which means it is strongly azimuth variant, while $B(\mathbf{r})$ depends on $|\mathbf{r} - \mathbf{R}_\mathrm{T}(0)|$ and $\theta_T(\mathbf{r})$, which is more slightly azimuth variant than $A(\mathbf{r})$. The higher-order terms are much smaller than the first two-order terms, so here we pay more attention to $A(\mathbf{r})$ and $B(\mathbf{r})$.

In the case of a complex trajectory, these parameters can be numerically obtained by polynomial fit to the residual range history.

After performing a range Fourier transform on $s_1(\eta, \tau)$,

$$
\begin{aligned}
S_1(\eta, f) &= \iint_{x,y} \sigma(\mathbf{r}) \exp\left\{-j2\pi \frac{f_{0T}+f}{c} R_{bic}(\eta; \mathbf{r})\right\} dxdy \\
&\approx \iint_{x,y} \left\{ \begin{array}{l} \sigma(\mathbf{r}) \exp\left\{-j2\pi \frac{f_{0T}+f}{c} R_{bic}(0; \mathbf{r})\right\} \\ \times \exp\left\{-j2\pi \frac{f_{0T}+f}{c} \left(A\eta + B\eta^2 + C\eta^3 + D\eta^4\right)\right\} \end{array} \right\} dxdy
\end{aligned} \tag{6.52}
$$

The Keystone transform for linear decoupling is as follows. Let

$$
\xi = \frac{f_{0T}+f}{f_{0T}} \eta \tag{6.53}
$$

then

$$
S_1(\xi, f) \approx \iint_{x,y} \left\{ \begin{array}{l} \sigma(\mathbf{r}) \exp\left\{-j2\pi \frac{f_{0T}+f}{c} R_{bic}(0; \mathbf{r})\right\} \\ \times \exp\left\{-j\frac{2\pi}{\lambda} A\xi\right\} \exp\left\{-j\frac{2\pi}{\lambda}\left(\frac{f_{0T}}{f_{0T}+f}\right) B\xi^2\right\} \\ \times \exp\left\{-j\frac{2\pi}{\lambda}\left(\frac{f_{0T}}{f_{0T}+f}\right)^2 C\xi^3\right\} \\ \times \exp\left\{-j\frac{2\pi}{\lambda}\left(\frac{f_{0T}}{f_{0T}+f}\right)^3 D\xi^4\right\} \end{array} \right\} dxdy \tag{6.54}
$$

In the general SAR case, $|f| \ll f_{0T}$, so the following approximation can be applied:

$$
\frac{f_{0T}}{f_{0T}+f} \approx 1 - \frac{f}{f_{0T}} + \left(\frac{f}{f_{0T}}\right)^2 - \left(\frac{f}{f_{0T}}\right)^3 \tag{6.55}
$$

$$
\left(\frac{f_{0T}}{f_{0T}+f}\right)^2 \approx 1 - 2\frac{f}{f_{0T}} + 3\left(\frac{f}{f_{0T}}\right)^2 - 4\left(\frac{f}{f_{0T}}\right)^3 \tag{6.56}
$$

$$
\left(\frac{f_{0T}}{f_{0T}+f}\right)^3 \approx 1 - 3\frac{f}{f_{0T}} + 6\left(\frac{f}{f_{0T}}\right)^2 - 10\left(\frac{f}{f_{0T}}\right)^3 \tag{6.57}
$$

Substituting Equations (6.55) to (6.57) into Equation (6.54) and reorganizing yields

$$S_1(\xi, f) \approx \iint_{x,y} \left\{ \begin{array}{l} \sigma(\mathbf{r}) \exp\left\{-j\frac{2\pi}{\lambda}\frac{f_{0T}+f}{f_{0T}}R_{bic}(0;\mathbf{r})\right\} \\ \times \exp\{-j\psi_A(\xi;\mathbf{r})\} \\ \times \exp\left\{-j\psi_{coup}(\xi, f;\mathbf{r})\right\} \end{array} \right\} dxdy \tag{6.58}$$

where

$$\psi_A(\xi;\mathbf{r}) = \frac{2\pi}{\lambda}\left(A\xi + B\xi^2 + C\xi^3 + D\xi^4\right) \tag{6.59}$$

is the azimuth phase, which only depends on ξ. Then

$$\begin{aligned} \psi_{coup}(\xi, f;\mathbf{r}) = {} & \frac{2\pi}{\lambda}\frac{f}{f_{0T}}\left(-B\xi^2 - 2C\xi^3 - 3D\xi^4\right) \\ & + \frac{2\pi}{\lambda}\left(\frac{f}{f_{0T}}\right)^2\left(B\xi^2 + 3C\xi^3 + 6D\xi^4\right) \\ & + \frac{2\pi}{\lambda}\left(\frac{f}{f_{0T}}\right)^3\left(-B\xi^2 - 4C\xi^3 - 10D\xi^4\right) \end{aligned} \tag{6.60}$$

is the phase containing the azimuth-range coupling. It can be seen from this phase that no linear coupling between f and ξ exists, which means that the linear RCMs for all targets have been corrected by the Keystone transform. Then bulk phase compensation according to the reference target can be done using the following function:

$$\begin{aligned} H_{comp}(\xi, f) = {} & \exp\left\{-j\frac{2\pi}{\lambda}\left(\frac{f}{f_{0T}}\right)\left(B_{ref}\xi^2 + 2C_{ref}\xi^3 + 3D_{ref}\xi^4\right)\right\} \\ & \times \exp\left\{j\frac{2\pi}{\lambda}\left(\frac{f}{f_{0T}}\right)^2\left(B_{ref}\xi^2 + 3C_{ref}\xi^3 + 6D_{ref}\xi^4\right)\right\} \\ & \times \exp\left\{-j\frac{2\pi}{\lambda}\left(\frac{f}{f_{0T}}\right)^3\left(B_{ref}\xi^2 + 4C_{ref}\xi^3 + 10D_{ref}\xi^4\right)\right\} \end{aligned} \tag{6.61}$$

where the subscript "*ref*" means the value calculated at the reference position (usually the scene center). Here we denote $S_2(\xi, f) = S_1(\xi, f)H_{comp}(\xi, f)$.

Because B, C and D are all changing with $\mathbf{r}$, there is a coupled phase remaining in $S_2(\xi, f)$. The second- and higher-order terms of f/f_{0T} in the remaining coupled phase will decrease the range focusing quality. Usually if the quadratic phase error is smaller than $\pi/4$, the imaging quality is acceptable, so the following condition is used to restrict the processing area:

$$\max\left\{\frac{2\pi}{\lambda}\left(\frac{f_B}{2f_{0T}}\right)^2\left|\left[B(\mathbf{r})-B_{ref}\right]\xi^2+3\left[C(\mathbf{r})-C_{ref}\right]\xi^2+6\left[D(\mathbf{r})-D_{ref}\right]\xi^2\right|\right\}\leq\frac{\pi}{4} \tag{6.62}$$

This ensures that in each processing area, all targets $\mathbf{r}$ at each ξ have a quadratic phase error that is smaller than $\pi/4$, so the imaging quality is guaranteed.

6.2.3.2 Residual RCM Correction

If the higher-order term phases of $S_2(\xi, f)$ are restricted and their effects can be neglected, then, applying an inverse Fourier transform to $S_2(\xi, f)$, the signal in the time domain is

$$s_2(\xi,\tau)\approx\iint_{x,y}\left\{\begin{array}{l}\sigma(\mathbf{r})\exp\left\{-j\dfrac{2\pi}{\lambda}R_{bic}(0;\mathbf{r})\right\}\\ \times\text{sinc}\left(\tau-\dfrac{R_{bic}(0;\mathbf{r})-R^r_{RCM}(\xi;\mathbf{r})}{c}\right)\\ \times\exp\left\{-j\psi_A(\xi;\mathbf{r})\right\}\end{array}\right\}dxdy \tag{6.63}$$

where

$$R^r_{RCM}(\xi;\mathbf{r})=(B-B_{ref})\xi^2+2(C-C_{ref})\xi^3+3(D-D_{ref})\xi^4 \tag{6.64}$$

It can be seen that the position of a target in the image plane along the fast time axis is determined by $R_{bic}(0;\mathbf{r})$, and the residual RCM is a function of the target location, which is range-azimuth variant. As has been said in the previous section, usually the RCM, which is within half a slant range resolution cell ρ_r, can be neglected in the SAR imaging field. Therefore, if the following restriction

$$\max\left\{\left|R^r_{RCM}(\xi;\mathbf{r})\right|\right\}<\frac{\rho_r}{2} \tag{6.65}$$

is satisfied for any $\mathbf{r}$, then the residual RCM correction step can be omitted. If this cannot be satisfied for the whole scene, then the subswath strategy can be used to deal with the azimuth variance and interpolation in the two-dimensional time domain can correct the range variance of $R^r_{RCM}(\xi;\mathbf{r}_i)$.

The subswath dealing with the azimuth variance is as follows. Firstly, a reference position $\mathbf{r}_i$ is chosen for each range gate i (usually chosen to be in the middle of the azimuth swath). Then the azimuth processing subswath is restricted according to the following condition:

$$\max\left\{\left|R_{RCM}^{r}(\xi;\mathbf{r}) - R_{RCM}^{r}(\xi;\mathbf{r}_i)\right|\right\} < \rho_r/2 \tag{6.66}$$

Consider the main component (i.e., the second-order term) in $|R_{RCM}^{r}(\xi;\mathbf{r}) - R_{RCM}^{r}(\xi;\mathbf{r}_i)|$ and calculate the above restriction; we have

$$\begin{aligned}
&\max\left\{\left|R_{RCM}^{r}(\xi;\mathbf{r}) - R_{RCM}^{r}(\xi;\mathbf{r}_i)\right|\right\} \\
&\approx \max\left\{V^2\xi^2\left|\frac{\cos^2\theta_2(\mathbf{r})}{|\mathbf{r}-\mathbf{R}_2(0)|} - \frac{\cos^2\theta_2(\mathbf{r}_i)}{|\mathbf{r}_i-\mathbf{R}_2(0)|}\right|\right\} \\
&\approx \max\left\{\frac{V^2\xi^2}{|\mathbf{r}_i-\mathbf{R}_2(0)|}\left|\cos^2\theta_2(\mathbf{r}) - \cos^2\theta_2(\mathbf{r}_i)\right|\right\} < \frac{\rho_r}{2}
\end{aligned} \tag{6.67}$$

It can be seen that the azimuth subswath can be restricted through $\theta_2(\mathbf{r})$ and the limitation of $\theta_2(\mathbf{r})$ is variant with range.

This azimuth restriction is the same as that in the NLCS algorithm introduced in the previous section when used in the nonsquint strip mode where no linear RCM exists.

6.2.3.3 Azimuth Compression

After the above steps, the final azimuth compression step can be done. The higher-order term of $\psi_A(\xi;\mathbf{r})$, which is

$$\psi_A^h(\xi;\mathbf{r}) = \frac{2\pi}{\lambda}\left(B\xi^2 + C\xi^3 + D\xi^4\right) \tag{6.68}$$

is desired to be removed for fine focusing. If the azimuth variation of $\psi_A^h(\xi;\mathbf{r})$ is smaller than $\pi/4$, which means the following restriction is satisfied,

$$\left|[B(\mathbf{r}) - B(\mathbf{r}_i)]\,\xi^2 + [C(\mathbf{r}) - C(\mathbf{r}_i)]\,\xi^3 + [D(\mathbf{r}) - D(\mathbf{r}_i)]\,\xi^4\right| \le \lambda/8 \tag{6.69}$$

then for each range gate the following phase compensation filter is applied:

$$H_{Ac}(\xi;\mathbf{r}_i) = \exp\left\{j\frac{2\pi}{\lambda}\left[B(\mathbf{r}_i)\xi^2 + C(\mathbf{r}_i)\xi^3 + D(\mathbf{r}_i)\xi^4\right]\right\} \tag{6.70}$$

Then the final focused image can be obtained through the azimuth Fourier transform. The final image can be written as

$$s_3(f_\xi, \tau) = \iint_{x,y} \left\{ \begin{array}{c} \sigma(\mathbf{r}) \exp\left\{-j\dfrac{2\pi}{\lambda} R_{bic}(0;\mathbf{r})\right\} \\ \times \operatorname{sinc}\left(\tau - \dfrac{R_{bic}(0;\mathbf{r})}{c}\right) \operatorname{sinc}\left(f_\xi + \dfrac{A(\mathbf{r})}{\lambda}\right) \end{array} \right\} dxdy \tag{6.71}$$

If the azimuth variation of $\psi_A^h(\xi;\mathbf{r})$ is larger than $\pi/4$, an additional step of subsegment azimuth processing is needed. This means that the above image is cut into several small segments along the azimuth direction (according to the azimuth variation of $\psi_A^h(\xi;\mathbf{r})$), and then an inverse Fourier transform is performed in order to get the azimuth time-domain signal for each segment. The differential phase between $\psi_A^h(\xi;\mathbf{r})$ of this segment and the phase of $H_{Ac}(\xi;\mathbf{r}_i)$ for each range gate can be calculated through the range history, and be compensated for this segment. Finally, all the segments are transformed to the frequency domain and combined together to obtain the final fine focused image.

Let us now consider the restrictions of the imaging area in this algorithm, which are Equations (6.66) and (6.69). Because ρ_r should be much larger than λ, the restriction Equation (6.69) is more rigorous than the restriction Equation (6.66). This means that the azimuth swath is mainly limited by the restriction Equation (6.69).

6.2.3.4 Geocoding

As can be seen from Equation (6.71), the axes of the focused image are τ and f_ξ, whose limitations are $[\tau_1 : 1/f_s : \tau_1 + (N_r - 1)/f_s]$ and $[-f_{prf}/2 : f_{prf}/N_a : f_{prf}/2)$, respectively. Here N_r is the image size in the range direction and N_a is the size in the azimuth direction, f_s is the sample rate and f_{prf} is the pulse repetition frequency, and τ_1 corresponds to the sample delay of the first range gate after range compression. When the trigger time of the echo channel is the same as that of the direct pulse channel, then $\tau_1 = R_D(0)/c$.

The position of target $\mathbf{r}$ in the image plane is as follows:

$$\begin{cases} \tau_{\mathbf{r}} = \dfrac{R_{bic}(0;\mathbf{r})}{c} \\ f_{\xi_{\mathbf{r}}} = -\dfrac{A(\mathbf{r})}{\lambda} \end{cases} \tag{6.72}$$

where

$$A(\mathbf{r}) = \left.\frac{dR_{bic}(\eta;\mathbf{r})}{dt}\right|_{\eta=0} = \frac{-\mathbf{V}(0)\cdot[\mathbf{r} - \mathbf{R}_2(0)]}{|\mathbf{r} - \mathbf{R}_2(0)|} - \frac{-\mathbf{V}(0)\cdot[\mathbf{R}_1 - \mathbf{R}_2(0)]}{|\mathbf{R}_1 - \mathbf{R}_2(0)|} \tag{6.73}$$

Here $\mathbf{V}$ is the transmitter velocity vector in the chosen coordinates. If a DEM of the scenario is provided, then geocoding can be done according to Equations (6.72), (6.73) and (6.46), (6.1) and (6.2).

6.2.3.5 Block Diagram of the Algorithm

The block diagram of the algorithm for one-stationary bistatic SAR of the spotlight mode is shown in Figure 6.11. It can be seen that the algorithm is quite efficient as only a pair of range Fourier transforms and a single azimuth Fourier transform are needed for the nonsevere azimuth variant case.

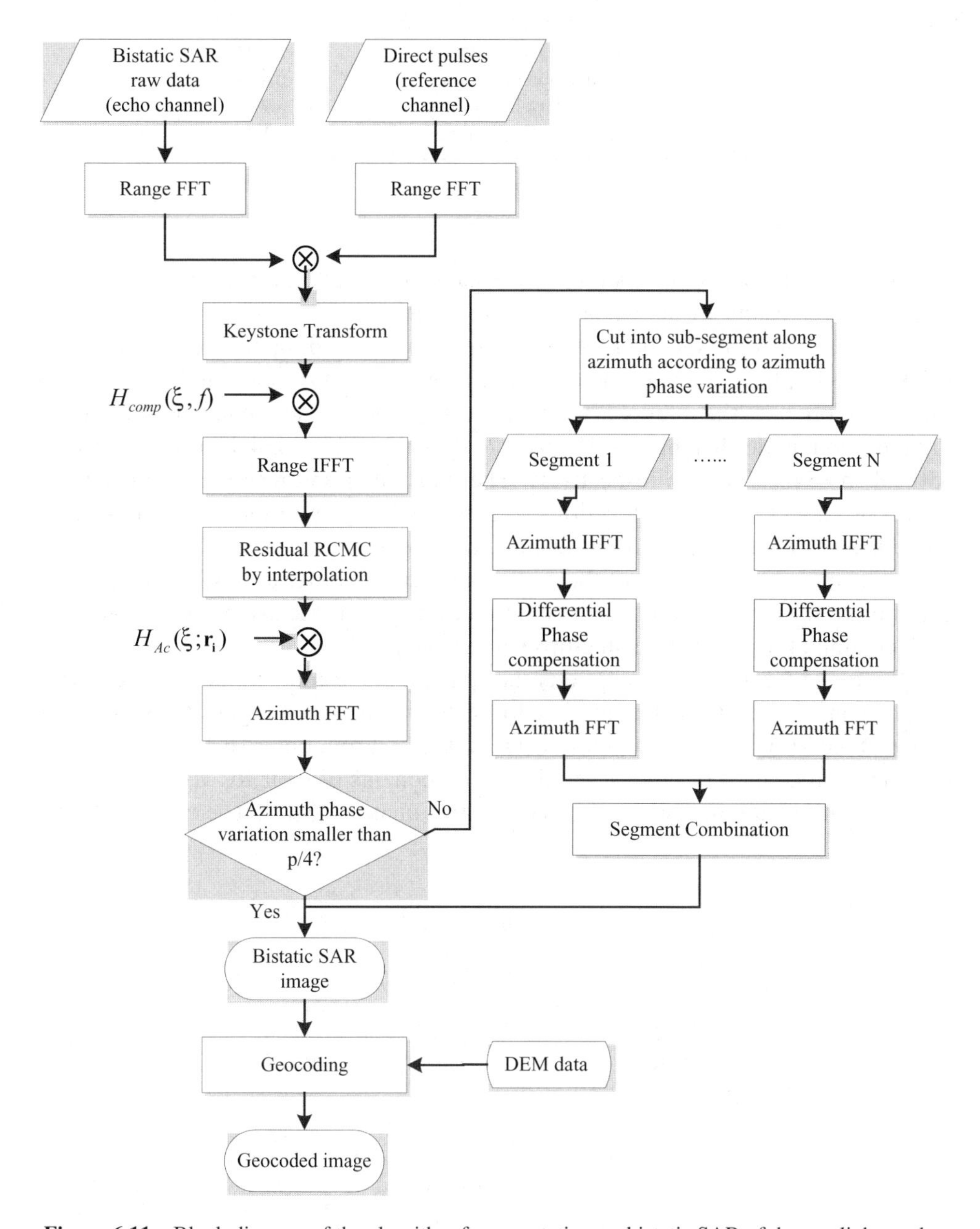

Figure 6.11 Block diagram of the algorithm for one-stationary bistatic SAR of the spotlight mode

Table 6.4 Transmitter and receiver parameters for simulation

Symbol	Meaning	Quantity
f_{0T}	Transmitter frequency	9.65 GHz
f_B	Signal bandwidth	300 MHz
f_s	Sample rate	330 MHz
PRF	Pulse repetition frequency	3224.35 Hz
f_{0R}	Receiver frequency	9.70 GHz
T_a	Data acquisition time	2.48 s
N_a	Azimuth pulse number	8000
N_r	Range sample number	30 000

6.2.3.6 Simulations

To testify for this algorithm, simulated bistatic SAR data are processed using this algorithm and results are shown and analyzed here.

The parameters of the point target simulation according to the TerraSAR-X/HITCHHIKER bistatic experiment [7] in Siegen, Germany, are listed in Table 6.4. The locations of the stationary receiver and the scene center described by [latitude, longitude, height] are [50.910 787°, 8.027 111°, 434.91 m] and [50.913 650°, 8.059 843°, 292 m], respectively. The transmitter and receiver parameters are shown in Table 6.4 and the locations of simulated targets in the east–north–height coordinates with the receiver as the origin are shown in Figure 6.12. These targets represent the main scenario area of the real experiment. In order to observe the azimuth variation of the range history, each column of targets has the same bistatic range, so each column of targets are on an ellipse arc when $\eta = 0$.

After range compression using reference pulses, the residual range histories (without the constant term) of the simulated targets are shown in Figure 6.13. It can be seen that the RCMs are different for different targets. From Figure 6.13(a), we can see that, for each column of targets, the linear components of their RCMs changes with their azimuth location. This is consistent with Equation (6.48). The residual RCMs

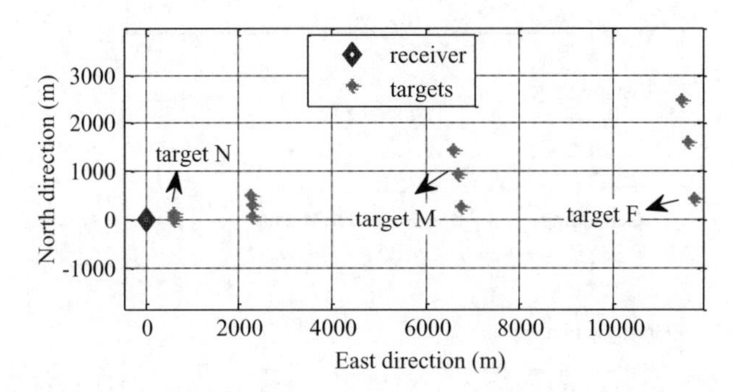

Figure 6.12 Locations of the targets and the receiver

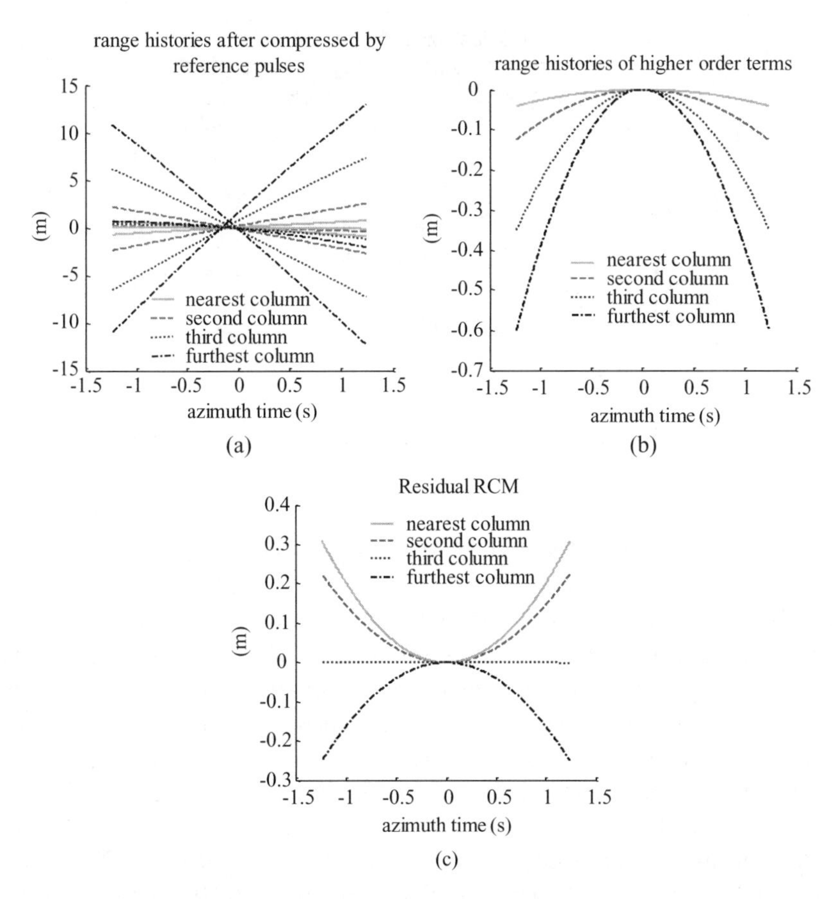

Figure 6.13 Range history on different processing stages: (a) range histories after compression by reference pulses (without constant terms), (b) range histories after the linear term is removed by the Keystone transform, (c) residual RCMs after bulk correction by H_{comp}

after the linear terms have been removed by the Keystone transform are shown in Figure 6.13(b). It can be seen that the residual RCMs are mainly range variant. The targets within the same column (which have the same bistatic range when $\eta = 0$) have almost the same residual RCMs. After bulk residual RCM correction by H_{comp}, the residual RCMs for the simulated targets are all smaller than 0.35 m (see Figure 6.13 (c)). This is smaller than half a range resolution cell (a range resolution cell is about 0.886 m here), so the range-variant residual RCM correction can be omitted in this case. The residual azimuth phases after azimuth phase compensation for each range gate using H_{Ac} are shown in Figure 6.14. It can be seen that the azimuth phase errors for all the targets in the scene are smaller than 25°, so the subsegment processing step is not needed in this simulation. The imaging results are shown in Figure 6.15. It can be seen that targets at different locations are all well focused.

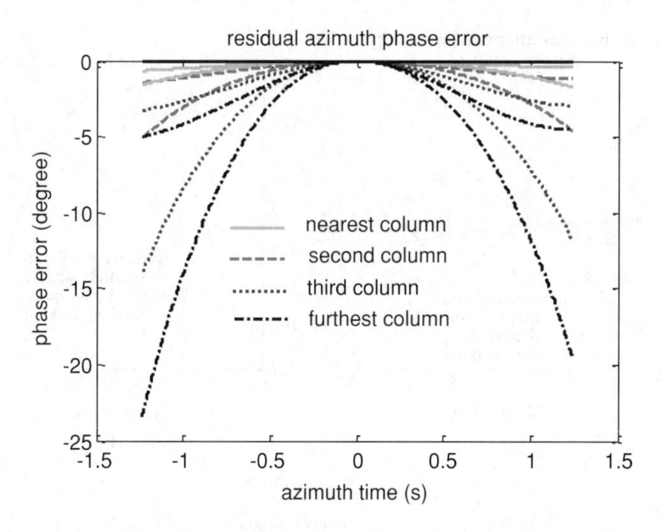

Figure 6.14 Residual azimuth phase errors

To testify the phase preserving ability of this algorithm and to answer the question of whether the image processed by this algorithm is able to do further interferometry processing, another set of raw data is simulated. All the parameters are the same as the first raw data, except that the receiver's location described by [latitude, longitude, height] is [50.910 787°, 8.027 111°, 435.91 m]. The differential phases of targets N, m and F are shown in Table 6.5, from which we can see that the differential phases are very close to the ideal differential phases calculated by $-2\pi/\lambda\left[R_{bic}(0;\mathbf{r}) - R_{bic}^{new}(0;\mathbf{r})\right]$, where R_{bic}^{new} means the range corresponding to the second receiver. The variance of the differential phase errors of all the simulated targets is 0.0919°. This means that the algorithm introduced here preserves the phase well and can support interferometry processing.

On August 1, 2011, IECAS carried out its first field bistatic SAR experiment. In this experiment, the transmitter was placed on an airplane that flew at about 3600 m above sea level and the receiver was stationary and placed on a mountain top that was about

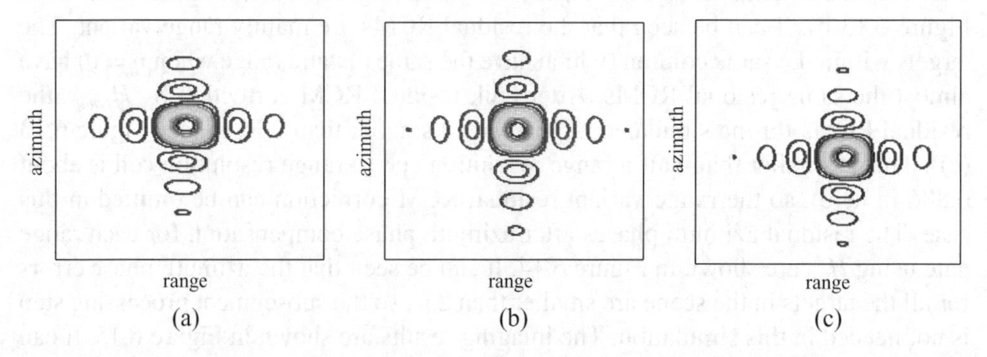

Figure 6.15 Imaging results of the simulated targets: (a) target N, (b) target m, (c) target F

Table 6.5 Interferometry phases of the simulated targets

Target	Differential phase (degree)		Phase error (degree)
	Measured from images	Ideal value	
N	156.9894	157.1157	−0.1263
M	−55.1917	−55.0924	−0.0993
F	76.2087	76.1398	0.0689

Table 6.6 Interferometry phases of the simulated targets

Parameters	Value
Carrier frequency	9.6 GHz
Velocity of the transmitter	114 m/s
Transmitted bandwidth	100 MHz
Height of the transmitter	3600 m
Height of the receiver	530 m
PRF	2000 Hz
Sample rate	115 MHz
Ground range from the transmitter to the scene center	2400 m

530 m above sea level. The parameters of this experiment are listed in Table 6.6, while the imaging geometry and the receiver system are shown in Figures 6.16 and 6.17, respectively. A patch of the imaging result of the IECAS bistatic SAR experimental data processed by the algorithm introduced in this section is exhibited in Figure 6.18, where an image with quite a good focusing quality is obtained.

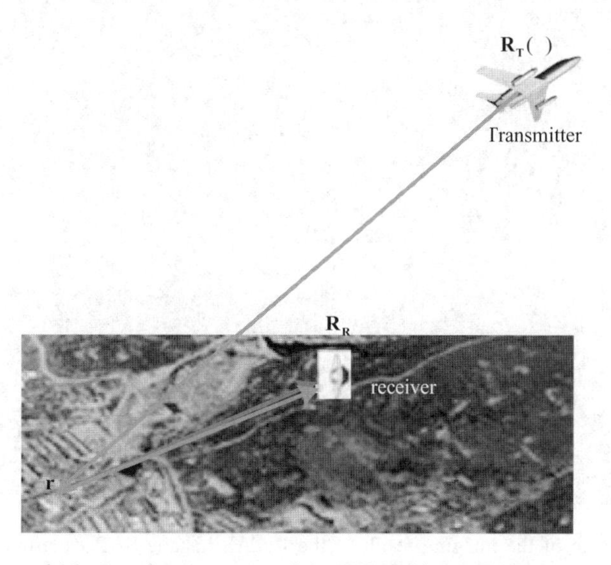

Figure 6.16 Imaging geometry of the IECAS bistatic SAR experiment

Figure 6.17 Picture of the receiver system

6.2.4 *Algorithm for the One-Stationary Forward-Looking Configuration*

Forward-looking imaging has many potential applications. For example, it can be used in airplane navigation and landing, independently of weather and daytime. In the usual monostatic SAR configuration, it is impossible to form a two-dimensional image in forward-looking imaging because the directions of the Doppler resolution and the range resolution are the same. However, in bistatic SAR, the transmitter and the

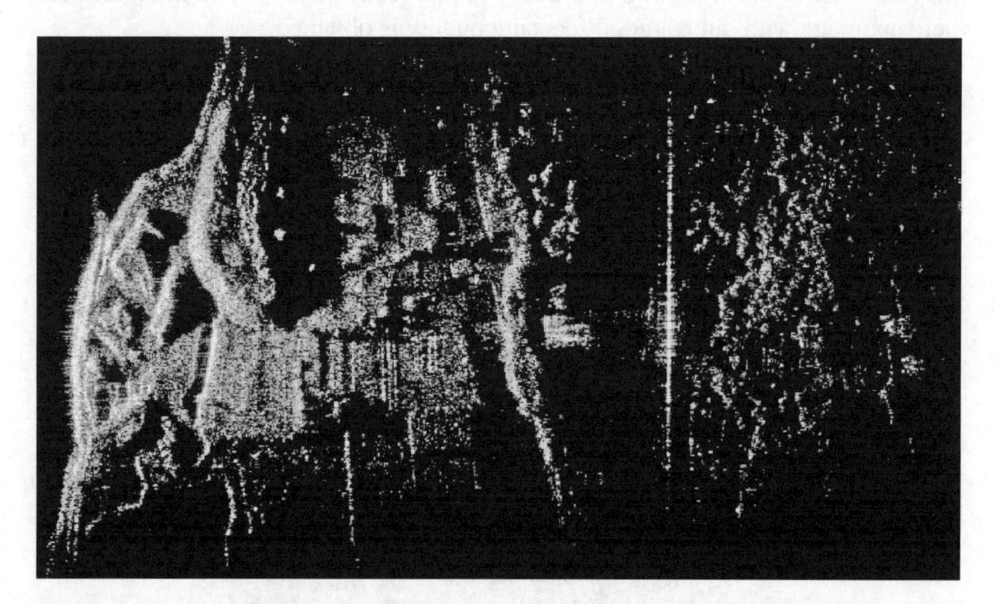

Figure 6.18 A patch of the imaging result of the IECAS bistatic SAR experimental data (vertical, azimuth; horizontal, range)

receiver are placed on different platforms, so the directions of the two resolutions are determined both by the transmitter and the receiver. When one of the platforms works in the forward-looking mode, as long as the other platform can ensure the difference between the two resolution directions, the two-dimensional image can be generated.

The advantages of the bistatic forward-looking configuration have been mentioned in [15] and [16]. Experiments have already been carried out to test the bistatic forward-looking imaging ability [17–19]. In this section, the characteristics of one-stationary forward-looking bistatic SAR are firstly analyzed and then an RD algorithm for this configuration is given.

6.2.4.1 Signal Characteristic of the Forward-Looking Configuration

The bistatic configuration with a stationary transmitter (or receiver) and a forward-looking moving receiver (or transmitter) is shown in Figure 6.19. The moving antenna is assumed to be moving along the y axis with velocity V.

Because this configuration is quite special, let us firstly take a look at its resolution ability. When $\eta = 0$, the isogram of the bistatic range is as follows:

$$\sqrt{(x - X_1)^2 + (y - Y_1)^2 + (z - Z_1)^2} + \sqrt{[x - X_2(0)]^2 + [y - Y_2(0)]^2 + [z - Z_2(0)]^2} = R_{bi} \tag{6.74}$$

and the isogram of Doppler frequency is as follows:

$$-\frac{1}{\lambda}\frac{\mathbf{V}\cdot(\mathbf{R}_2 - \mathbf{r})}{|\mathbf{R}_2 - \mathbf{r}|} = -\frac{1}{\lambda}\frac{V[Y_2(0) - y]}{\sqrt{[x - X_2(0)]^2 + [y - Y_2(0)]^2 + [z - Z_2(0)]^2}} = \frac{V}{\lambda}\sin\phi_2\cos\theta_2 = f_\eta \tag{6.75}$$

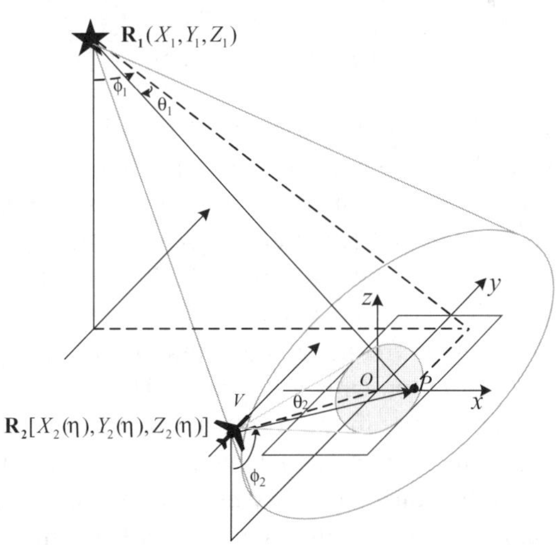

Figure 6.19 One-stationary forward-looking bistatic SAR geometry (© 2008 IEEE. Reprinted with permission from X. Qiu, D. Hu, C. Ding, "Some Reflections on Bistatic SAR of Forward-looking Configuration" *IEEE Geoscience and Remote Sensing Letters*, **5**, 4, pp. 735–739, 2008.)

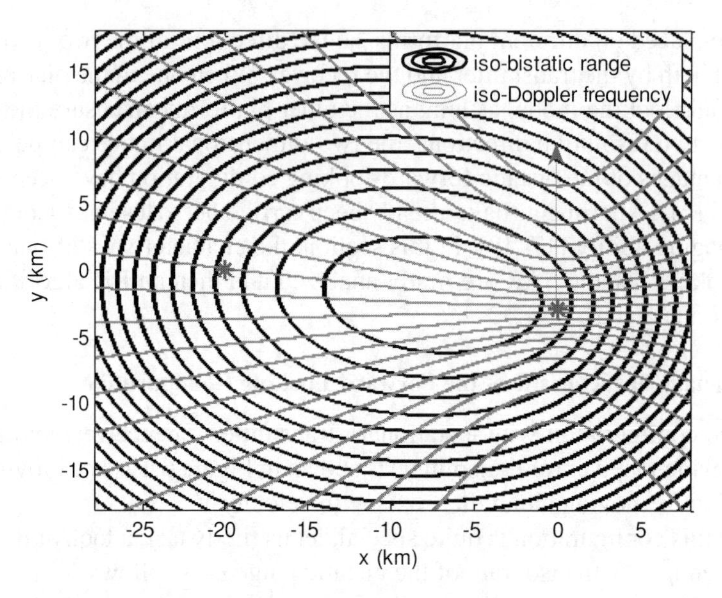

Figure 6.20 Isograms of the bistatic range and Doppler frequency for the forward-looking configuration (the "*" are the stationary and moving antennas projected on to the ground)

where

$$\sin \phi_2 = \frac{y - Y_2(0)}{\sqrt{[y - Y_2(0)]^2 + [z - Z_2(0)]^2}} \tag{6.76}$$

$$\cos \theta_2 = \frac{\sqrt{[y - Y_2(0)]^2 + [z - Z_2(0)]^2}}{\sqrt{[x - X_2(0)]^2 + [y - Y_2(0)]^2 + [z - Z_2(0)]^2}} \tag{6.77}$$

If the scenario is flat and $z \cong 0$, the isogram of the bistatic range on the ground is a group of ellipses and the isogram of the Doppler frequency on the ground is a group of hyperbolas that are symmetrical with the y axis. Figure 6.20 shows the isograms of the bistatic range and Doppler frequency on the ground where $z = 0$, while the stationary antenna is at [−20 km, 0, 10 km] and the moving antenna is at [0, −3 km, 5 km] with a constant velocity [0, 200 m/s, 0]. It can be seen that when ϕ_2 is small, the Doppler frequency contours do not bend very easily, which ensures that acceptable resolution cells are formed. However, when ϕ_2 is large, the isolines of the Doppler frequency in the forward-looking area bend severely, which makes the resolution cell in some areas become very big, so only poor resolution can be obtained. Thus, in this one-stationary bistatic forward-looking configuration, the forward-looking angle of the moving antenna is a key parameter that dominates the two-dimensional imaging ability. To realize the two-dimensional imaging on both sides of the flying path (which makes a real forward-looking mode), ϕ_2 cannot be very large, which means the receiver cannot "see" targets too far away.

Suppose that the forward-looking moving antenna works in the strip mode. The echo signal after synchronization error compensation is rewritten here as follows:

$$d(\tau,\eta;\mathbf{r}) = \sigma(\mathbf{r})w(\eta,\tau;\mathbf{r})\text{rect}\left[\frac{\tau - R_{bi}(\eta;\mathbf{r})/c}{T}\right]\exp\left\{-j\pi b\left[\tau - \frac{R_{bi}(\eta;\mathbf{r})}{c}\right]^2\right\} \tag{6.78}$$

$$\times\exp\left\{-j2\pi f_{0T}\frac{R_{bi}(\eta;\mathbf{r})}{c}\right\}$$

By making a reasonable approximation, which is

$$w(\eta,\tau;\mathbf{r}) \approx \text{rect}\left[\frac{\eta + Y_2(0)/V}{T_{syn}}\right]$$

where T_{syn} is the synthetic time, the Doppler centroid of target $\mathbf{r}$ is as follows:

$$f_{dc}(x,z) = -\frac{V}{\lambda}\frac{Y_2(0)}{\sqrt{[x - X_2(0)]^2 + Y_2(0)^2 + [z - Z_2(0)]^2}} \tag{6.79}$$

which is a function of x. In the forward-looking mode, the imaging scenario should be ahead of the moving antenna, so here f_{dc} may be quite large. Therefore the problems appearing in high-squint angle mode SAR processing [20, 21] will also appear here. Moreover, as can be seen from (6.79), the Doppler centroid changes with x and z, which should be taken into consideration while imaging.

6.2.4.2 RD Algorithm for the Forward-Looking Configuration

The echo signal after range compression can be written as follows:

$$s(\eta,t;\mathbf{r}) = \text{rect}\left[\frac{\eta + Y_2(0)/V}{T_{syn}}\right]\text{sinc}\left[t - \frac{R_{bi}(\eta;\mathbf{r}) - R_{near}}{c}\right]\exp\left\{-j\frac{2\pi}{\lambda}R_{bi}(\eta;\mathbf{r})\right\} \tag{6.80}$$

Adapting the principle of the stationary phase to the azimuth Fourier transform, the echo signal in the two-dimensional frequency domain is as follows:

$$\begin{aligned} s(f_\eta,f;\mathbf{r}) &= \exp\{j2\pi f R_{near}/c\}\exp\left\{-j2\pi f_\eta\frac{y - Y_2(0)}{V}\right\} \\ &\times\exp\left\{-j\frac{2\pi(f_{0T} + f)}{c}R_1(\mathbf{r})\right\} \\ &\times\exp\left\{-j\frac{2\pi}{c}R_{2min}(x,z)\sqrt{(f + f_{0T})^2 - (f_\eta c/V)^2}\right\} \end{aligned} \tag{6.81}$$

where

$$R_{2min}(x,z) = \sqrt{[x - X_2(0)]^2 + [z - Z_2(0)]^2} \tag{6.82}$$

$$f_\eta \in [f_{dc}(x) - B_a/2,\ f_{dc}(x) + B_a/2] \tag{6.83}$$

and B_a is the Doppler frequency bandwidth. From Equation (6.81), the phase term of the echo signal in the two-dimensional frequency domain has the linear terms of f, f_η and $\sqrt{(f+f_{0T})^2-(f_\eta c/V)^2}$ at the same time. Therefore, it is difficult to find two new variables that make the phase term have only two linear phase terms versus the two variables (which was done in the Stolt transform). This means that it is quite difficult to find wavenumber domain algorithms (such as the ω-k algorithm) for one-stationary bistatic SAR. As a result, here we expand the last phase term in Equation (6.81) at $f=0$ to its Taylor series, giving

$$\begin{aligned} S(f_\eta, f; \mathbf{r}) &= \exp\{j2\pi f R_{near}/c\} \exp\left\{-j2\pi f_\eta \frac{y-Y_2(0)}{V}\right\} \\ &\times \exp\left\{-j\frac{2\pi(f_{0T}+f)}{c}[R_1(\mathbf{r})+R_2(\eta_{dc};\mathbf{r})]\right\} \; (1) \\ &\times \exp\left\{-j\frac{2\pi}{\lambda}\left[R_{2min}(x,z)\sqrt{1-(f_\eta\lambda/V)^2}-R_2(\eta_{dc};\mathbf{r})\right]\right\} \; (2) \\ &\times \exp\left\{-j\frac{2\pi f}{c}\left[\frac{R_{2min}(x,z)}{\sqrt{1-(f_\eta\lambda/V)^2}}-R_2(\eta_{dc};\mathbf{r})\right]\right\} \; (3) \\ &\times \exp\left\{j\frac{\pi}{c}\frac{(\lambda f_\eta/V)^2}{[\sqrt{1-(f_\eta\lambda/V)^2}]^3}\frac{f^2}{f_{0T}}R_{2min}(x,z)\right\} \; (4) \\ &\times \exp\left\{-j\frac{\pi}{c}\frac{(\lambda f_\eta/V)^2}{[\sqrt{1-(f_\eta\lambda/V)^2}]^5}\frac{f^3}{f_{0T}^2}R_{2min}(x,z)\right\} \; (5) \ldots \end{aligned} \tag{6.84}$$

where η_{dc} is the azimuth time when the Doppler frequency of target $\mathbf{r}$ happens to be its Doppler centroid. Here, term (1) determines the location in range direction, phase term (2) is the azimuth phase history, phase term (3) determines the RCM, phase term (4) is the secondary coupling phase term, which determines the secondary range compression (SRC), and phase term (5) is the tertiary coupling phase term, which affects the symmetry of the sidelobes.

It can be seen from Equation (6.84) that:

- Similar to the ordinary one-stationary bistatic SAR, the echo signal is azimuth variant. This can be seen from the term (2), where the phase is variant with $\mathbf{r}$. Therefore the azimuth perturbation should be applied if needed.
- As stated previously, in the forward-looking configuration the forward-looking angle ϕ_R may not be very small, so the problems appearing in the high-squint case will also appear here. The range–azimuth coupling in this configuration may be

severe, so the imaging algorithm should have a high precision. The imaging algorithms proposed for the large-squint-angle mode [20,21] can be used for reference.

- Another thing that needsattention to be paid to it is that the Doppler centroid f_{dc} nonlinearly depends on x and z, and it changes acutely when ϕ_R is large, so RCMs are quite different at different ranges. If bulk RCM correction is done in the range frequency domain, the targets will be displaced in the range direction, which will bring further difficulties to azimuth compression and geocoding. This phenomenon can be interpreted by formulas. The filter that corrects the bulk RCM in the range frequency domain is as follows:

$$H_{RCMref} = \exp\left\{ j\frac{2\pi f}{c}\left[\frac{R_{2min}(x_{ref}, y_{ref})}{\sqrt{1-(f_\eta\lambda/V)^2}} - R_2[f_{dc}(\mathbf{r}_{ref}); x_{ref}, z_{ref}]\right]\right\} \tag{6.85}$$

The range shift for target $\mathbf{r}(x, y, z)$ at its Doppler centroid is then

$$\Delta R(\mathbf{r}) = \frac{R_{2min}(x, z)}{\sqrt{1-[f_{dc}(\mathbf{r})\lambda/V]^2}} - R_2[f_{dc}(\mathbf{r}_{ref}); x, z] \neq 0 \tag{6.86}$$

This is not equal to zero when $x \neq x_{ref}$. Moreover, $\Delta R(\mathbf{r})$ is a nonlinear function of x, which makes the following processing steps more difficult. Therefore, using the chirp scaling method to correct the RCM is also not suitable here.

Based on the above analysis, we find that the chirp scaling imaging algorithm and the wavenumber domain imaging algorithm are both not very suitable for this special configuration. Therefore, here we choose the range-Doppler (RD) algorithm, whose main property is doing RCM correction in the range-Doppler domain by interpolation. This enables the RD algorithm to handle the variant Doppler centroid. To make the algorithm have higher preciseness, the secondary range compression (SRC) is accomplished in the range-Doppler domain as well, which can take the dependence of SRC on the range and on the azimuth frequency into consideration.

The RD algorithm can be shown in Figure 6.21, where

$$k_{src} = \frac{\left[\sqrt{1-\left(f_\eta\lambda/V\right)^2}\right]^3}{\left(f_\eta\lambda/V\right)^2}\frac{cf_{0T}}{R_{2min}(x, z)} \tag{6.87}$$

The algorithm above ignores the forth and higher coupling phase term and also the variation of the third term. However, it takes the range variation of the second order coupling term into consideration, so it is able to achieve a higher preciseness at such a high-squint mode as in the forward-looking configuration.

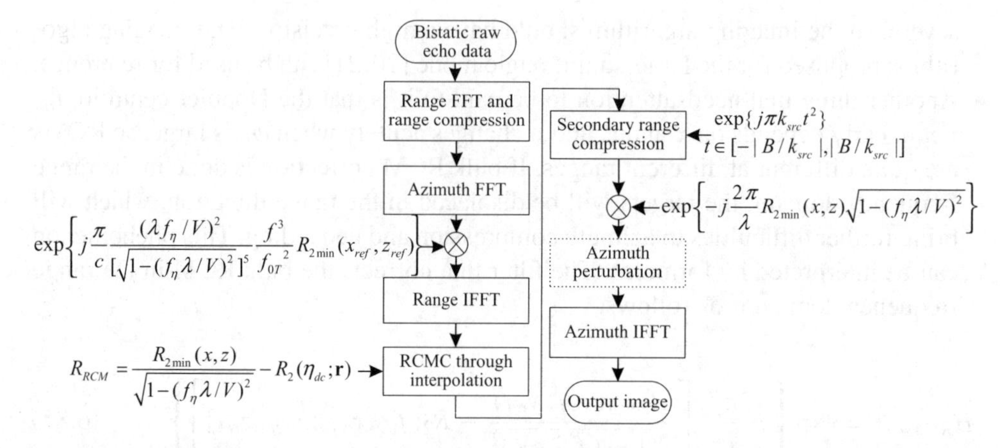

Figure 6.21 Block diagram of the RD algorithm for the bistatic forward-looking SAR (© 2008 IEEE. Reprinted with permission from X. Qiu, D. Hu, C. Ding, "Some Reflections on Bistatic SAR of Forward-looking Configuration" *IEEE Geoscience and Remote Sensing Letters*, **5**, 4, pp. 735–739, 2008.)

6.2.4.3 Simulations

To test the validity of this RD algorithm and to verify the two-dimensional imaging ability of the bistatic forward-looking SAR, the echo signals of the targets distributing as shown in Figure 6.22 in configuration I and configuration II (parameters of these two configurations are listed in Table 6.7) are simulated and then the images are obtained using this RD algorithm. To obtain an imaging area that is relatively large, the stationary antenna is placed far away from the scenario.

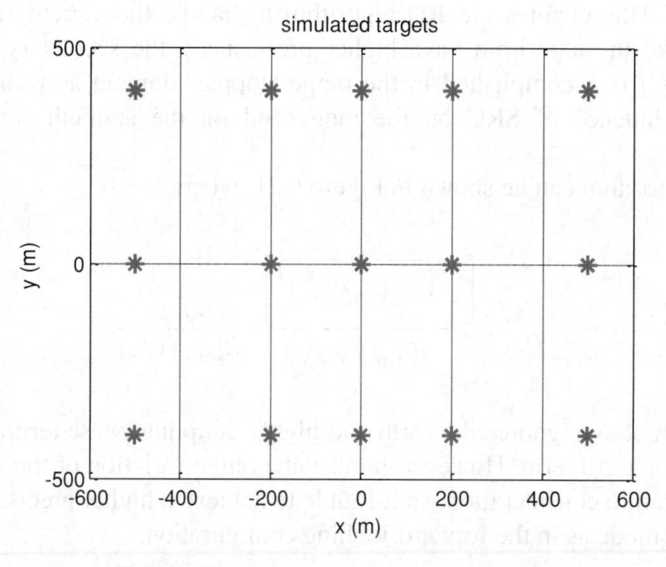

Figure 6.22 Positions of the simulated targets

Table 6.7 Simulation parameters of the two different configurations (© 2008 IEEE. Reprinted with permission from X. Qiu, D. Hu, C. Ding, "Some Reflections on Bistatic SAR of Forward-looking Configuration" *IEEE Geoscience and Remote Sensing Letters*, **5**, 4, pp. 735–739, 2008.)

Parameter	Configuration I	Configuration II
$[X_2(0), Y_2(0), Z_2(0)]$	(−200, 0, 20) km	(−200, 0, 20) km
(X_1, Y_1, Z_1)	(0, −3, 5) km	(0, −3, 1) km
Forward-looking angle ϕ_2	30.96°	71.56°
Transmitted bandwidth	100 MHz	100 MHz
Wavelength	3.1 cm	3.1 cm
Azimuth length of the moving antenna	2 m	2 m

The bistatic range contours and the Doppler frequency contours when $\eta = 0$ are shown in Figure 6.23. It can be seen that when ϕ_2 is small, the resolution cells in the scene are all very good, but when ϕ_2 is large, the direction of the Doppler resolution varies severely and the resolution cells in the $x < 0$ area are very large. Therefore, the two-dimensional imaging ability in this area is rather poor.

The RD imaging results of configuration I are shown in Figure 6.24. It can be seen that, in this configuration where resolution cells are small, all the targets are well focused and the two-dimensional image is obtained. Therefore, the bistatic forward-looking SAR does have the two-dimensional imaging ability. However, as can be seen from Figure 6.25(a), in configuration II, where the resolution cells for $x < 0$ are large, only the targets at $x > 0$ are normally well focused, while others are not. It is not that the RD algorithm causes the defocusing, but the configuration itself determines that those targets at $x < 0$ have not got the two-dimensional resolution ability. This can be verified by the imaging result shown in Figure 6.25(b) using the precise BP algorithm, which is quite similar to the result of the RD algorithm.

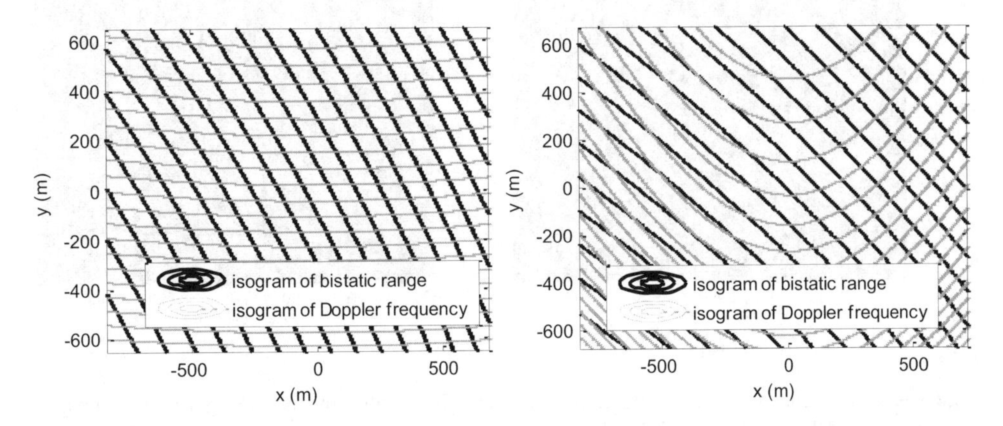

Figure 6.23 Bistatic range contours and Doppler frequency contours when $\eta = 0$. Left figure shows the contours of configuration I. Right figure shows the contours of configuration II (© 2008 IEEE. Reprinted with permission from X. Qiu, D. Hu, C. Ding, "Some Reflections on Bistatic SAR of Forward-looking Configuration" *IEEE Geoscience and Remote Sensing Letters*, **5**, 4, pp. 735–739, 2008.)

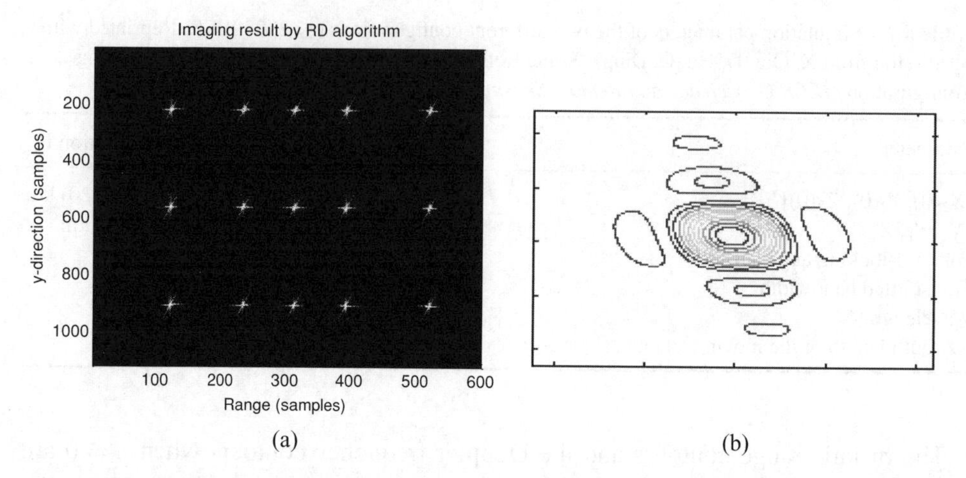

Figure 6.24 Imaging results of configuration I: (a) imaging result using the RD algorithm, (b) zoom into the target at the top left corner (© 2008 IEEE. Reprinted with permission from X. Qiu, D. Hu, C. Ding, "Some Reflections on Bistatic SAR of Forward-looking Configuration" *IEEE Geoscience and Remote Sensing Letters*, **5**, 4, pp. 735–739, 2008.)

Figure 6.26 shows the signals in the range-Doppler domain after range compression in the two configurations. The range variance of the Doppler frequency centroid can be clearly seen, especially in configuration II, where ϕ_2 is large. The results before and after bulk RCM correction using filter Equation (6.85) in the range frequency domain is shown in Figure 6.27. The targets here are displaced because of this bulk RCM correction step. The targets with $x = -500$ m appear close to the targets with $x = -200$ m in the image. The targets with $x = 500$ m are shifted out of the image and

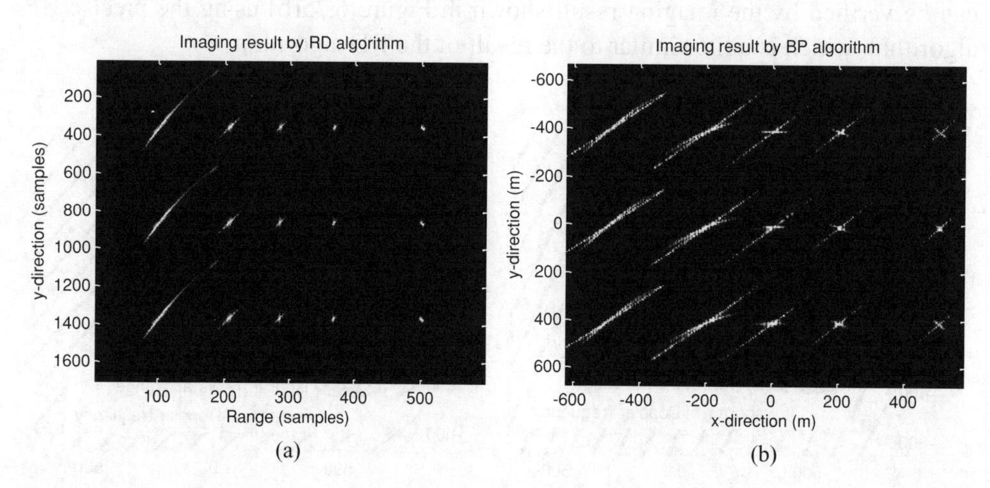

Figure 6.25 Imaging results of configuration II: (a) imaging result using the RD algorithm, (b) imaging result using the BP algorithm (© 2008 IEEE. Reprinted with permission from X. Qiu, D. Hu, C. Ding, "Some Reflections on Bistatic SAR of Forward-looking Configuration" *IEEE Geoscience and Remote Sensing Letters*, **5**, 4, pp. 735–739, 2008.)

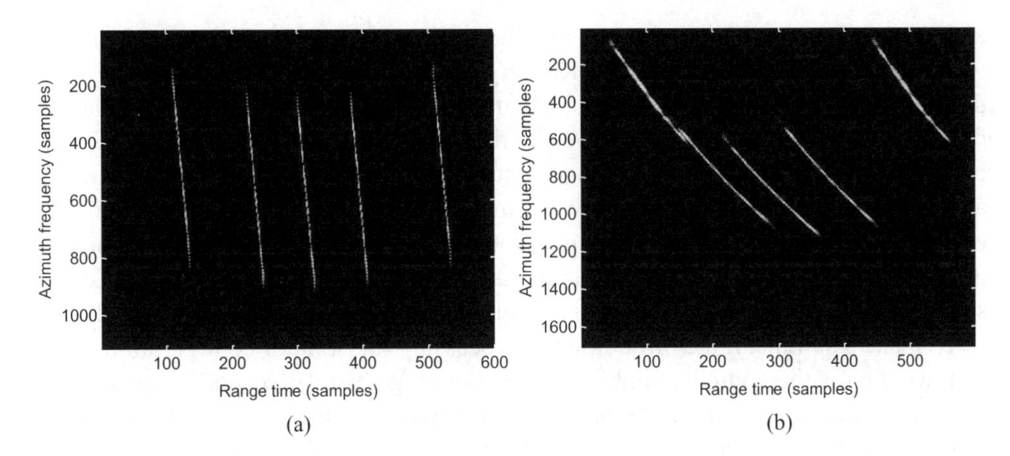

Figure 6.26 Signal in the range-Doppler domain after range compression: (a) configuration I, (b) configuration II

then wrapped into the near range of the image due to digital processing. This result verifies that the bulk RCM in the range frequency domain is not suitable here and validates the correctness of the analysis in the previous sections.

6.3 Imaging Algorithms for Translational Variant Bistatic SAR with Constant Velocities

In the previous section, the translational variant bistatic SAR with a one-stationary antenna is discussed. The one-stationary configuration is a relatively simple case of translational variant configurations, because the bistatic range history in this case is

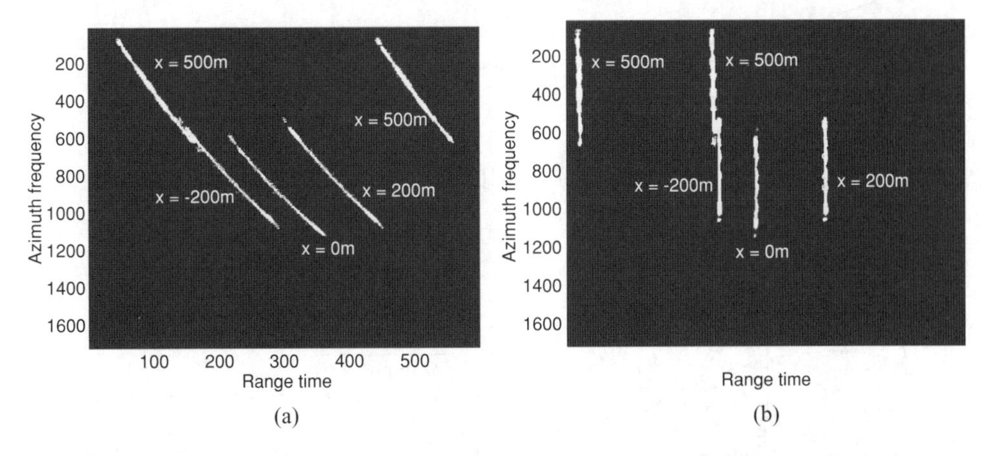

Figure 6.27 Signal in the range-Doppler domain before and after bulk RCM correction in the range frequency domain for configuration II: (a) before bulk RCM correction, (b) after bulk RCM correction (© 2008 IEEE. Reprinted with permission from X. Qiu, D. Hu, C. Ding, "Some Reflections on Bistatic SAR of Forward-looking Configuration" *IEEE Geoscience and Remote Sensing Letters*, **5**, 4, pp. 735–739, 2008.)

a single-square-root equation thanks to the stationary antenna. This makes it easy to obtain the two-dimensional BPTRS according to the POSP. Therefore the main difficulty of one-stationary bistatic SAR imaging is the range and azimuth variant property of the bistatic echo. However, in the translational variant bistatic SAR with constant velocities, the bistatic range history in a double-square-root equation is applied, as discussed in Chapter 5. What make the double-square-root range history even more complex than the range history in the translational invariant configuration is that the velocities of the transmitter and the receiver are not the same as each other, so the range histories caused by the transmitter and the receiver have different shapes.

To make a summery, the difficulties in processing this kind of configuration are as follows: first is the complex double-square-root range history, second is the azimuth and range variant property of the bistatic echo and third is the tomography sensitivity of the imaging parameters. As to the third problem, the DEM of the scenario should be provided and taken into consideration while imaging if necessary. If a flat scene is talked about, the former two difficulties should be resolved. In the following text, the assumption of a flat scenario is made.

6.3.1 *Imaging Geometry and the Signal Model*

The imaging geometry of the translational variant configuration with constant velocities is shown in Figure 6.28. The locations of the transmitter and the receiver are $\mathbf{R_T} : [X_T(\eta), Y_T(\eta), Z_T(\eta)]^{\mathrm{T}}$ and $\mathbf{R_R} : [X_R(\eta), Y_R(\eta), Z_R(\eta)]^{\mathrm{T}}$, respectively. The

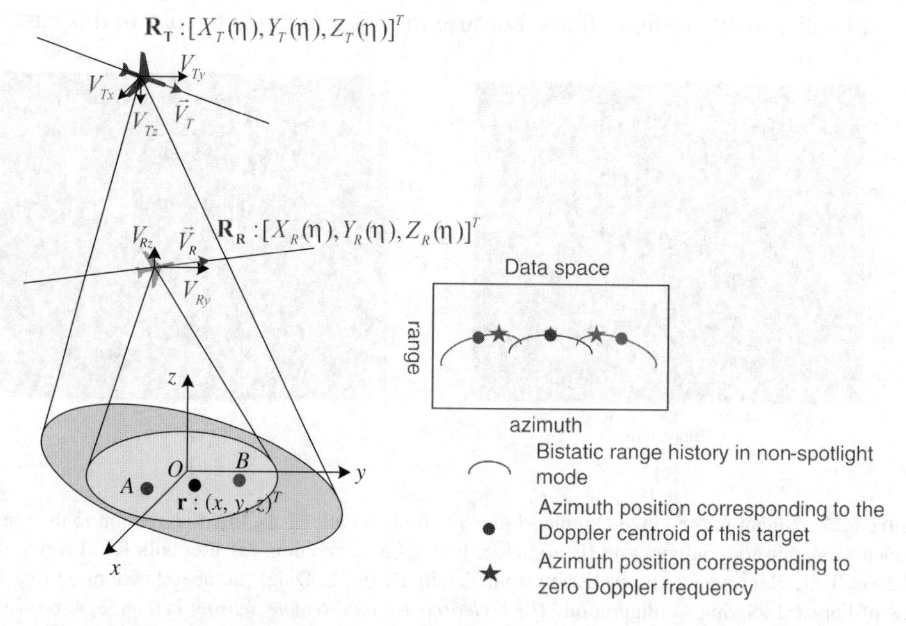

Figure 6.28 Imaging geometry of the translational variant case with constant velocities

velocities of the transmitter and the receiver are $\mathbf{V_T} : [V_{Tx}, V_{Ty}, V_{Tz}]^T$ and $\mathbf{V_R} : [V_{Rx}, V_{Ry}, V_{Rz}]^T$, respectively. Then the bistatic echo can be written as follows:

$$s(\eta, \tau) = \iint_{x,y} \left\{ \sigma(\mathbf{r}) w(\eta, \tau; \mathbf{r}) p\left[\tau - \frac{R_{bi}(\eta; \mathbf{r})}{c}\right] \exp\{-j2\pi f_0 R_{bi}(\eta; \mathbf{r})/c\} \right\} dxdy \tag{6.88}$$

where

$$R_{bi}(\eta; \mathbf{r}) = R_T(\eta; \mathbf{r}) + R_R(\eta; \mathbf{r}) \tag{6.89}$$

$$R_T(\eta; \mathbf{r}) = \sqrt{(X_T + V_{Tx}\eta - x)^2 + (Y_T + V_{Ty}\eta - y)^2 + (Z_T + V_{Tz}\eta - z)^2} \tag{6.90}$$

$$R_R(\eta; \mathbf{r}) = \sqrt{(X_R + V_{Rx}\eta - x)^2 + (Y_R + V_{Ry}\eta - y)^2 + (Z_R + V_{Rz}\eta - z)^2} \tag{6.91}$$

In Equation (6.88), $w(\eta, \tau; \mathbf{r})$ has the same meaning as that of Equation (6.7), which is the weight due to the transmitter and receiver antenna pattern, propagation loss and so on. In the spotlight mode, $w(\eta, \tau; \mathbf{r})$ can be approximately considered as not changing with η, while in the strip mode or sliding spotlight mode, $w(\eta, \tau; \mathbf{r})$ is changing with η and determines the azimuth time span and the Doppler frequency bandwidth of each target. The Doppler frequency for target $\mathbf{r}$ at the moment η is as follows:

$$f_\eta(\eta; \mathbf{r}) = -\frac{1}{\lambda} \frac{dR_{bi}(\eta; \mathbf{r})}{d\eta} \tag{6.92}$$

Because of the time variation of the bistatic geometry, the RCMs and the Doppler FM rates of different targets in the scenario are all different. Besides, positions of the targets in the image will be distorted in this configuration. To clarify this phenomenon, let us firstly recall the translational invariant configuration and the one-stationary configuration. Without losing universality, we take the nonsquint case as an example. As shown in Figure 6.29, in the translational invariant configuration and the one-stationary configuration, the targets A, P and B with the same azimuth

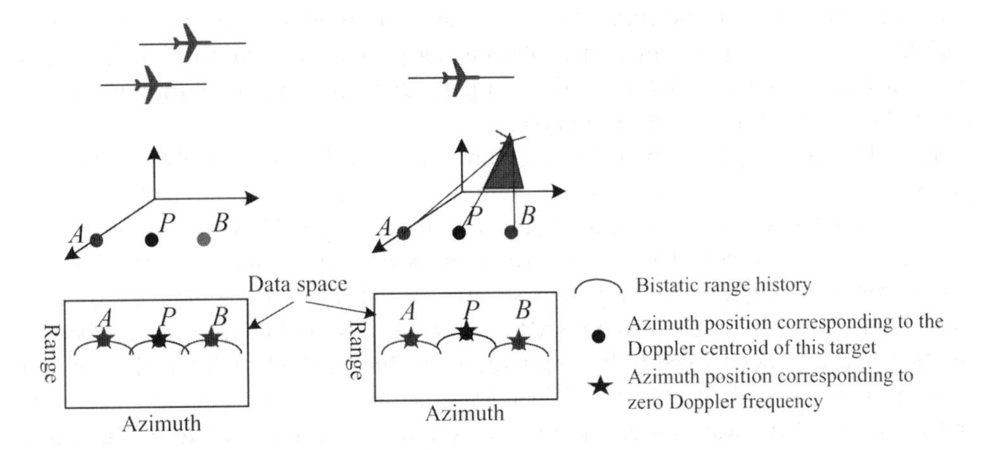

Figure 6.29 Illustration of the target positions in the images of the translational invariant and one-stationary case.

interval in the scenario will get echoes in the data space as shown in Figure 6.29. Because the moving velocities of the transmitter and the receiver are the same in the translational invariant configuration, the azimuth time interval between the data acquisitions of the targets are linearly dependent on the azimuth position interval between these targets, and the relationship between the time interval and the azimuth position interval will not change with time. In the one-stationary configuration, only one antenna is moving with a constant velocity, so the azimuth time interval between the data acquisitions of the targets are also linearly and constantly depending on the azimuth position interval. To get a clearer idea, one can look at Figure 6.29. It can be seen that all these targets have the same Doppler center, which is zero in the nonsquint case. If the targets are focused on the azimuth moment corresponding to zero Doppler frequency, the targets will appear at their centroid synthetic time. This means that the azimuth time axis has a quite simple relationship with the azimuth location axis on the ground, which is $\Delta y = \Delta\eta V_g$, where V_g is the velocity of the combined antenna beam center (translational invariant case) or the moving antenna beam center (one-stationary case) moving on the ground, and V_g is a constant. However, in the translational variant configuration with constant velocities, if the targets are still focused on their zero Doppler frequency, the targets will not all appear at their centroid synthetic time and the time interval of the targets in the image will change differently, as can be seen from Figure 6.28. Therefore, in the translational variant configuration with constant velocities, the image will be distorted nonlinearly. The geo-mapping of the image is a little more complex than the former two configurations and should be carefully done by considering the imaging algorithms used in processing.

6.3.2 *Imaging Processing Based on BPTRS*

As the bistatic echo of the translational variant configuration is a two-dimensional variant, the wavenumber domain algorithm is not very suitable for this configuration if the azimuth variation is large, while the range-Doppler domain algorithms such as RD and CS algorithms are more suitable.

To deduce the range-Doppler domain algorithms, it is best to first obtain a bistatic point target reference spectrum of this configuration. Because of the double-square-root range history of this configuration, this is not an easy job. However, as introduced in Chapter 5, a number of methods have been developed to deal with the double-square-root range history. Among them, several can also be used to obtain the BPTRS of the translational variant bistatic SAR, such as the method of hyperbolic approximation and advanced hyperbolic approximation, the method of IDW, the method of LBF and the method of MSR. When introducing these methods in Chapter 5, the velocities of the transmitter and the receiver were not restricted to be the same. So the derivation of BPTRS described in Chapter 5 can be directly adapted to this translational variant configuration. Hence, here the BPTRS by these methods will not be specified again. To find out more about BPTRS and the algorithms based on the method of LBF, interested readers can refer to [22] to [24] while information on BPTRS and the NLCS algorithm based on the method of MSR they can refer to [25]. Regarding BPTRS based on the method of hyperbolic approximation and advanced hyperbolic approximation, readers can look back to Chapter 5. An azimuth perturbation step combined with azimuth subswath processing as introduced in Section 6.2.2 should be added to the algorithms based on the hyperbolic approximation and the advanced hyperbolic approximation, which are described in Chapter 5, so as to handle the azimuth variation of the bistatic signal. As these are the combination of methods introduced in Chapter 5 and Section 6.2, here we will not go into the details. In the following text, a more universal method, which is called "NuSAR", will be introduced.

6.3.3 Imaging Processing Based on NuSAR

The concept of the "numerical SAR (NuSAR)" processor has been mentioned in the introduction to Chapter 5. It was first proposed by Bamler and colleagues [26,27] and is based on the idea that all the transfer functions used in imaging are obtained by numerical calculation. The authors studied and slightly modified the NuSAR processor in [28] to handle the azimuth variation of the signal. Here, the NuSAR processor will be introduced in this section.

6.3.3.1 The Processing Flow of NuSAR

The NuSAR processor proposed by Bamler has three key steps, each corresponding to a key transfer function. The key transfer functions are the bulk range processing transfer function, the differential RCM map and the azimuth transfer functions. To deal with the azimuth variation of the signal, the fourth key function, the azimuth perturbation functions, are introduced into the NuSAR processor. The flow of the NuSAR processor is shown in Figure 6.30 and the key transfer functions are interpreted as follows:

1. The range processing function $H_R(f, f_\eta; P_{ref})$ does the range compression, RCMC, SRC and all the higher-order range phase compensation for the reference target P_{ref}.
2. The differential range cell migration correction (DRCMC) function $\delta R(\tau, f_\eta; P_i, P_{ref})$ completes the RCMC for targets away from the reference target.
3. The azimuth perturbation function $H_{pert}(\eta; \tau)$ and the perturbation compensation function $H_{pert}^{cmp}(f_\eta; t)$ are used to identify the Doppler FM rate within the same range gate and to minimize the perturbation phase error, respectively.
4. The azimuth compression function $H_A(f_\eta; t)$ fulfills the azimuth compression for targets at each range gate.

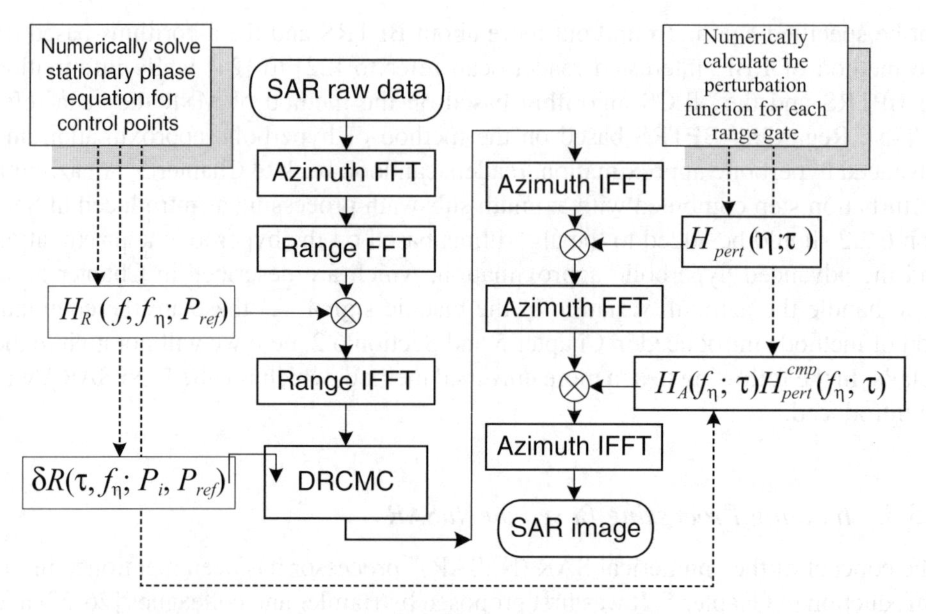

Figure 6.30 Block diagram of NuSAR for the translational variant bistatic SAR

6.3.3.2 The Calculation Methods of the Transfer Functions

For the transfer functions $H_R(f, f_\eta; P_{ref})$, $H_A(f_\eta; \tau)$ and $\delta R(\tau, f_\eta; P_i, P_{ref})$, literature [27] provides a set of numerical calculation methods. It begins by calculating the echo signal for a reference target in the time domain (denoted as $u(\tau, \eta; P_{ref})$). Then the range processing function $H_R(f, f_\eta; P_{ref})$ is obtained as follows:

$$H_R(f, f_\eta; P_{ref}) = \{FFT_\tau\{FFT_\eta\{u(\tau, \eta; P_{ref})\}\} \cdot H_A[f_\eta; \tau(P_{ref})]\}^* \quad (6.93)$$

where

$$H_A[f_\eta; \tau(P_{ref})] = FFT_\eta\left\{\exp\left\{j\frac{4\pi}{\lambda}\Delta R(\eta; P_{ref})\right\}\right\} \quad (6.94)$$

and

$$\Delta R(\eta; P_{ref}) = R_{bi}(\eta; P_{ref}) - R_{bi\_\min}(P_{ref}) \quad (6.95)$$

$R_{bi\_\min}(P_{ref})$ is the minimal bistatic range of the reference target. To store the reference echo $u(\tau, \eta; P_{ref})$, an additional memory equal to the SAR raw echo data has to be allocated. To save memory, a different calculating method is introduced here:

- The translational variant bistatic SAR raw echo can be written as follows:

$$s(\tau,\eta)=\iint \sigma(x,y)w(\eta,\tau;x,y)p\left[\tau-\frac{R_{bi}(\eta;x,y)}{c}\right]\exp\left\{-j\frac{2\pi R_{bi}(\eta;x,y)}{\lambda}\right\}dxdy \tag{6.96}$$

- Firstly, $H_R(f,f_\eta;P_{ref})$ can be written by the following equation:

$$H_R(f,f_\eta;P_{ref})=\left(H_A[f_\eta;\tau(P_{ref})]\cdot\int\exp\{-j\Phi(f,f_\eta,\eta)\}d\eta\right)^* \tag{6.97}$$

where

$$\Phi(f,f_\eta,\eta)=\frac{2\pi(f_0+f)}{c}R_{bi}(\eta;x,y)+2\pi f_\eta\eta \tag{6.98}$$

According to the principle of the stationary phase, the range processing function can be obtained by numerically solving the following phase stationary equation:

$$\begin{aligned}&\frac{V_{Tx}[X_T(0)+V_{Tx}\eta-x_{ref}]+V_{Ty}[Y_T(0)+V_{Ty}\eta-y_{ref}]+V_{Tz}[Z_T(0)+V_{Tz}\eta-z]}{R_T(\eta;x_{ref},y_{ref})}\Big|_{\eta=\eta_c}\\&+\frac{V_{Rx}[X_R(0)+V_{Rx}\eta-x_{ref}]+V_{Ry}[Y_R(0)+V_{Ry}\eta-y_{ref}]+V_{Rz}[Z_R(0)+V_{Rz}\eta-z]}{R_R(\eta;x_{ref},y_{ref})}\Big|_{\eta=\eta_c}=-\xi\end{aligned} \tag{6.99}$$

where

$$\xi=\frac{cf_\eta}{f_0+f}$$

From the equation, it can be seen that, for a selected target $P(x,y)$, the stationary time η_c only depends on ξ. So once the curve of $\eta_c(\xi;x_{ref},y_{ref})$ is obtained for the reference target, $H_R(f,f_\eta;P_{ref})$ at each (f,f_η), which corresponds to a unique ξ, can be found by substituting $\eta_c(\xi;x_{ref},y_{ref})$ for the phase term $\Phi(f,f_\eta,\eta_c)$. This means that only a curve of $\eta_c(\xi;x_{ref},y_{ref})$ needs to be stored in the memory, so it is memory saving. Hence, the detailed calculation method of $H_R(f,f_\eta;P_{ref})$ is as follows:

1. A reference target $P_{ref}(x_{ref},y_{ref})$ is chosen, which usually is chosen to be the scene center.
2. Calculate the Doppler centroid $f_{\eta dc}$ of this reference target and specify the azimuth frequency axis according to $f_{\eta dc}$.
3. Calculate the scope of ξ according to the scopes of f and f_η, and then calculate the stationary points for a set of ξ_i, $i=0,\ldots,N_\xi-1$, which are $\eta_c(\xi_i;x_{ref},y_{ref})$.

Because the relationship between η_c and ξ_i can easily be fitted by a polynomial function with 3 or 4 order terms, N_ξ does not have to be very large. In the following simulation, $N_\xi = 32$.

4. Calculate the corresponding ξ for each (f, f_η) in the two-dimensional frequency domain and obtain $\eta_c(\xi; x_{ref}, y_{ref})$ through interpolation. Then the transfer function can be obtained, which is

$$H_R(f, f_\eta; P_{ref}) = \exp\{j\varphi_R(f, f_\eta; P_{ref})\} \tag{6.100}$$

where

$$\begin{aligned}\varphi_R(f, f_\eta; P_{ref}) = & \frac{2\pi(f_0 + f)}{c} R_{bi}[\eta_c(\xi; x_{ref}, y_{ref}); x_{ref}, y_{ref}] + 2\pi f_\eta \eta_c(\xi; x_{ref}, y_{ref}) \\ & - \frac{2\pi f_0}{c} R_{bi}[\eta_c(cf_\eta/f_0; x_{ref}, y_{ref}); x_{ref}, y_{ref}] - 2\pi f_\eta \eta_c(cf_\eta/f_0; x_{ref}, y_{ref}) \\ & - \frac{2\pi f}{c} R_{bi}[\eta_c(cf_{\eta dc}/f_0; x_{ref}, y_{ref}); x_{ref}, y_{ref}]\end{aligned} \tag{6.101}$$

In Equation (6.101), the first row of terms are the phase terms of the matched filter of the reference target. The second row of phase terms keeps the azimuth phase. The third row of phase terms ensure that the target will appear at its bistatic range position corresponding to its Doppler centroid:

- Secondly, the differential RCMC function can be calculated as follows:
  1. The bistatic range history and the Doppler frequency curves of the reference target (i.e., $R_{bi,ref}(\eta; x_{ref}, y_{ref})$ and $f_{\eta;ref}(\eta; x_{ref}, y_{ref})$) can be calculated according to the target position and the transmitter and receiver trajectories and according to the following equation:

$$f_\eta = -\frac{1}{\lambda}\frac{dR_{bi}(\eta; x, y)}{d\eta} \tag{6.102}$$

  Then the RCM curve versus the Doppler frequency can be obtained as follows:

$$\Delta R_{bi;ref}(f_\eta; x_{ref}, y_{ref}) = R_{bi,ref}(f_\eta; x_{ref}, y_{ref}) - R_{bi,ref}(f_{\eta dc}; x_{ref}, y_{ref}) \tag{6.103}$$

  2. For the range gate i, the bistatic range $R_{bi,i}$ of this range gate can be calculated through the sample time delay and the sample rate. Then, together with the known y_{ref}, the location of the reference target of this range gate, that is, (x_i, y_{ref}), can be obtained. Hence, the RCM curve versus the Doppler frequency of this target can be obtained by calculating $R_{bi,i}(\eta; x_i, y_{ref})$ and $f_{\eta;ref}(\eta; x_{ref}, y_{ref})$, and then

$$\Delta R_{bi,i}(f_\eta; x_i, y_{ref}) = R_{bi,i}(f_\eta; x_i, y_{ref}) - R_{bi,i}(f_{\eta dc}; x_i, y_{ref}) \tag{6.104}$$

3. The differential RCM is then

$$\delta R(\tau_i, f_\eta; P_i, P_{ref}) = \Delta R_{bi;i}(f_\eta; x_i, y_{ref}) - \Delta R_{bi;ref}(f_\eta; x_{ref}, y_{ref}) \tag{6.105}$$

- Thirdly, the calculations of the azimuth perturbation and compensation function are described in Section 6.2.2.2. The only difference is that the target $(x_{i,j}, y_j)$, which has the same bistatic range as the target (x_i, y_{ref}) but appears at azimuth time η_j, should be calculated by solving the following equations:

$$\begin{cases} f_\eta(\eta_j; x_{i,j}, y_j) = f_{\eta dc} \\ R_{bi,i}(\eta_j; x_{i,j}, y_j) = R_{bi,i}(0; x_i, y_{ref}) \end{cases} \tag{6.106}$$

- Finally, the azimuth compression function for each range gate is $H_A(f_\eta; \tau_i) = \exp\{j\varphi_A(f_\eta; \tau_i)\}$, where

$$\begin{aligned} \varphi_A(f_\eta; \tau_i) = {} & \frac{2\pi f_0}{c} R_{bi}[\eta_c(cf_\eta/f_0; x_i, y_{ref}); x_i, y_{ref}] + 2\pi f_\eta \eta_c(cf_\eta/f_0; x_i, y_{ref}) \\ & - \frac{2\pi f_0}{c} R_{bi}[\eta_c(cf_{\eta dc}/f_0; x_i, y_{ref}); x_i, y_{ref}] \end{aligned} \tag{6.107}$$

6.3.3.3 Image Distortion and Geometrical Correction

As mentioned in Section 6.3.1, the image of the translational variant configuration will be distorted. The way of distortion is related to the imaging algorithm. For the NuSAR processer introduced here, the image distortion is as follows.

In this NuSAR processor, the RCM correction is done in the Doppler domain and the RCM correction is done while keeping the range corresponding to the Doppler centroid as immovable (see Equations (6.104) and (6.103)). The target will therefore appear at its $R_{bi}(f_{\eta dc}; x, y)$ in the range direction. In the azimuth, the target will appear at the azimuth time (noted as $\eta_{dc}(x, y)$) corresponding to its Doppler centroid $f_{\eta dc}$ if no azimuth perturbation is performed. When the azimuth perturbation is performed, the target appears at η'_{dc}, where

$$\eta'_{dc} = \eta_{dc} + \frac{\Delta f_\eta(\eta_{dc})}{f_r(x, y_{ref})} \tag{6.108}$$

and $\Delta f_\eta(\eta_{dc})$ is the Doppler frequency shift at η_{dc} caused by azimuth perturbation.

For a target with position $(x, y, 0)$ on the scenario, its location in the image processed by this NuSAR is (I_t, I_η), $0 \le I_t \le N_R - 1$, $0 \le I_\eta \le N_A - 1$(N_R and N_A are the image range size and image azimuth size, respectively), which can be calculated as follows:

1. Calculate $R_{bi}(\eta; x, y)$ and $f_\eta(\eta; x, y)$ so as to obtain $\eta_{dc}(x, y)$ corresponding to $f_{\eta dc}$.

2. The range pixel location of the target in the image is I_t, where $I_t = [R_{bi}(\eta; x, y)/c - \tau_{dly}]f_s$.
3. The azimuth pixel location of the target in the image is I_n, where

$$I_\eta = \left(\frac{\sum_{n=0}^{N_p} P_n \eta_{dc}^{n+1}/(n+1)}{f_r(I_t, y_{ref})} + \eta_{dc} \right) f_{prf} + N_A/2 \qquad (6.109)$$

Here, $P_n, n = 0, \ldots, N_p$ are the parameters used in the perturbation in the range gate I_t and $f_r(I_t, y_{ref})$ is the reference Doppler FM rate used while imaging in the range gate I_t.

According to the above analysis, the geo-mapping can be done by calculating the image position of each target or, for the sake of simplicity, by calculating the image positions $(I_{t,i}, I_{\eta,i})$ for some control points (x_i, y_i). A proper model (such as the quadratic polynomial model) is then chosen and the model parameters are resolved based on these control points, which are as follows:

$$(I_t, I_\eta) = \left(1, x, y, x^2, xy, y^2\right) \begin{pmatrix} a_{11} & a_{12} \\ a_{21} & a_{22} \\ a_{31} & a_{32} \\ a_{41} & a_{42} \\ a_{51} & a_{52} \\ a_{61} & a_{62} \end{pmatrix} \text{and } (x, y) = \left(1, I_t, I_\eta, I_t^2, I_t I_\eta, I_\eta^2\right) \begin{pmatrix} b_{11} & b_{12} \\ b_{21} & b_{22} \\ b_{31} & b_{32} \\ b_{41} & b_{42} \\ b_{51} & b_{52} \\ b_{61} & b_{62} \end{pmatrix} \qquad (6.110)$$

Once the model parameters are settled, the geo-mapping can be done according to this known model.

6.3.4 Simulations

In this section, two typical translational variant configurations are simulated. The parameters of the first configuration are listed in Table 6.8, The transmitting antenna

Table 6.8 Simulated parameters of the balloon–aircraft configuration

Parameters	Value	Parameters	Value
Balloon position when $\eta = 0$	(−250, 10, 20) km	Transmitted centroid frequency	9.6 GHz
Velocity of the balloon	(4, 10, 3) m/s	Transmitted bandwidth	500 MHz
Aircraft position when $\eta = 0$	(−10, 0, 10) km	Sample rate	555.5 MHz
Velocity of the aircraft	(0, 200, 0) m/s	PRF	325.057 Hz
Azimuth length of the aircraft's antenna	0.8 m	Sample time delay	0.880 37 ms

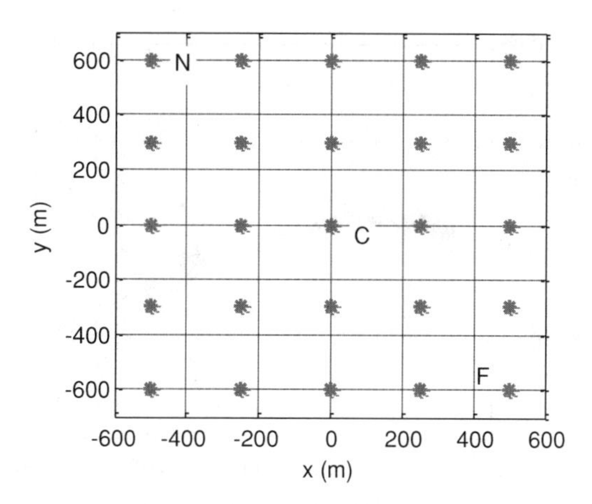

Figure 6.31 Simulated targets

is placed on a balloon that is slowing floating and far away from the scenario and the receiving antenna is placed on an aircraft that is relatively close to the scenario. The transmitter and the receiver are working in the strip mode. The simulated targets are shown in Figure 6.31.

In this simulation, $y_{ref} = 0$ and the curve $R_{bi}(x; y_{ref})$ versus x is shown in Figure 6.32, where it can be seen that $R_{bi}(x; y_{ref})$ is nearly linearly changing with x. Therefore, when calculating x for each $R_{bi}(y_{ref})$, the quadratic polynomial fit method is used. The differential RCMs of the targets in the reference range gate with different azimuth positions are shown in Figure 6.33. It can be seen that within this simulated scenario, the differential RCMs are smaller than 0.02 m, which is less than the half

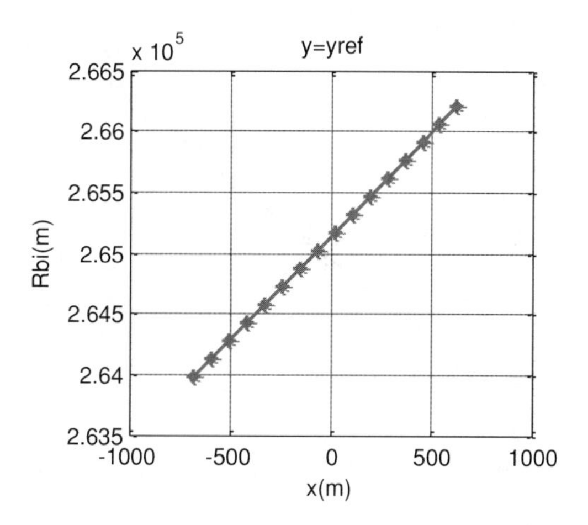

Figure 6.32 Relationship between the bistatic range $R_{bi}(x; y_{ref})$ and the azimuth position x

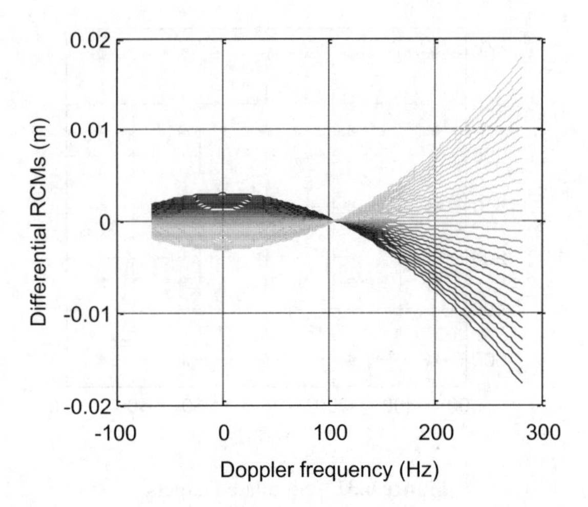

Figure 6.33 The differential RCMs of the targets within the referential range gate at different azimuth positions

bistatic range resolution cell (0.3 m), so there is no need to cut the image into a sub-swath in the azimuth. What can also be seen from Figure 6.30 is that the differential RCMs have different linear terms, which implies that the targets with the same range gate with different azimuth positions have different Doppler centroids in this translational variant configuration. The maximal phase error caused by the differential range histories is 50.692° (the linear phase term is not considered), so azimuth perturbation is done to obtain fine focusing results. Figure 6.34 shows the relationship between the ξ and the stationary time η_c for the referential target. It can be seen that though the range history is a double-square-root equation, which is quite complex, the curve

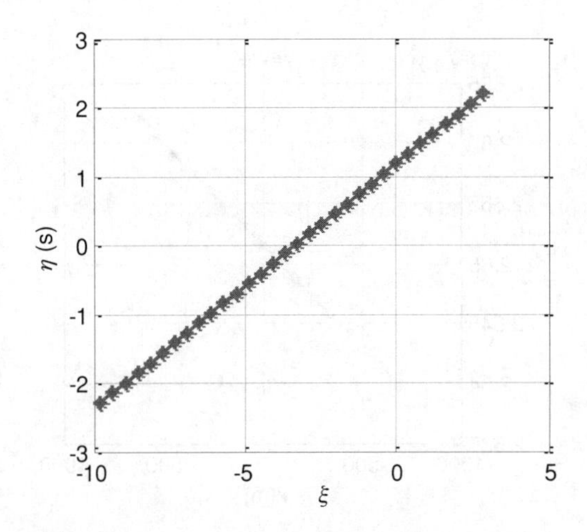

Figure 6.34 The relationship between ξ and the stationary time η_c for the referential target

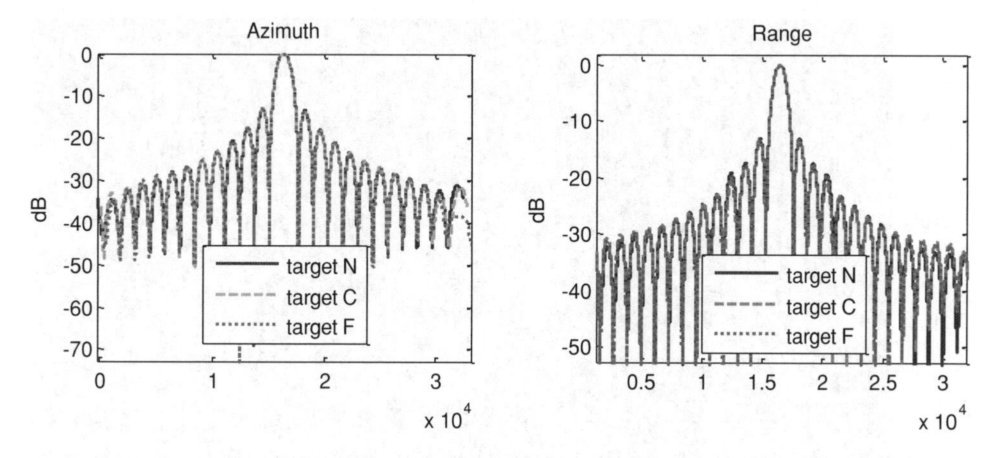

Figure 6.35 The azimuth and range responses of the simulated targets

ξ versus η_c is not a complicated curve where the linear term is dominating. In this simulation, the quadratic polynomial interpolation is used to get η_c for each ξ.

The imaging results for targets N, C and F are shown in Figure 6.35 and the quality indexes are listed in Table 6.9. It can be seen that all the targets are focused quite well. The focused images without and with geometrical correction are shown in Figure 6.36(a) and (b), respectively. If a linear model is used while geo-mapping, the mean position error of all the simulated targets is 0.870 m in the x direction and 0.0149 m in the y direction. If a quadratic model is used, the mean error is then 0.0948 m in the x direction and 0.0147 m in the y direction. It can be seen that in this translational variant configuration, the image is nonlinearly distorted, especially in the x direction.

The second configuration simulated here is the spaceborne–airborne case. In the spaceborne SAR in low orbit with not a very high resolution (e.g., larger than 1 m), the satellite usually can be considered as moving with a constant velocity within the synthetic time. So the spaceborne–airborne case can be considered as a typical translational variant configuration with constant velocities. Though the spaceborne–airborne configuration suffers from its short co-illuminating time, it is still a good experimental configuration by using the existing spaceborne SAR as the transmitter. The spaceborne–airborne bistatic SAR experiments have been carried out in Germany

Table 6.9 Imaging qualities of the simulated targets

	Range			Azimuth		
Target	Widen ratio	PSLR (dB)	ISLR (dB)	Widen ratio	PSLR (dB)	ISLR (dB)
N	0.884	−13.132	−9.762	0.884	−13.147	−9.629
C	0.884	−13.204	−10.031	0.886	−13.283	−10.241
F	0.882	−13.165	−9.877	0.887	−13.048	−9.554

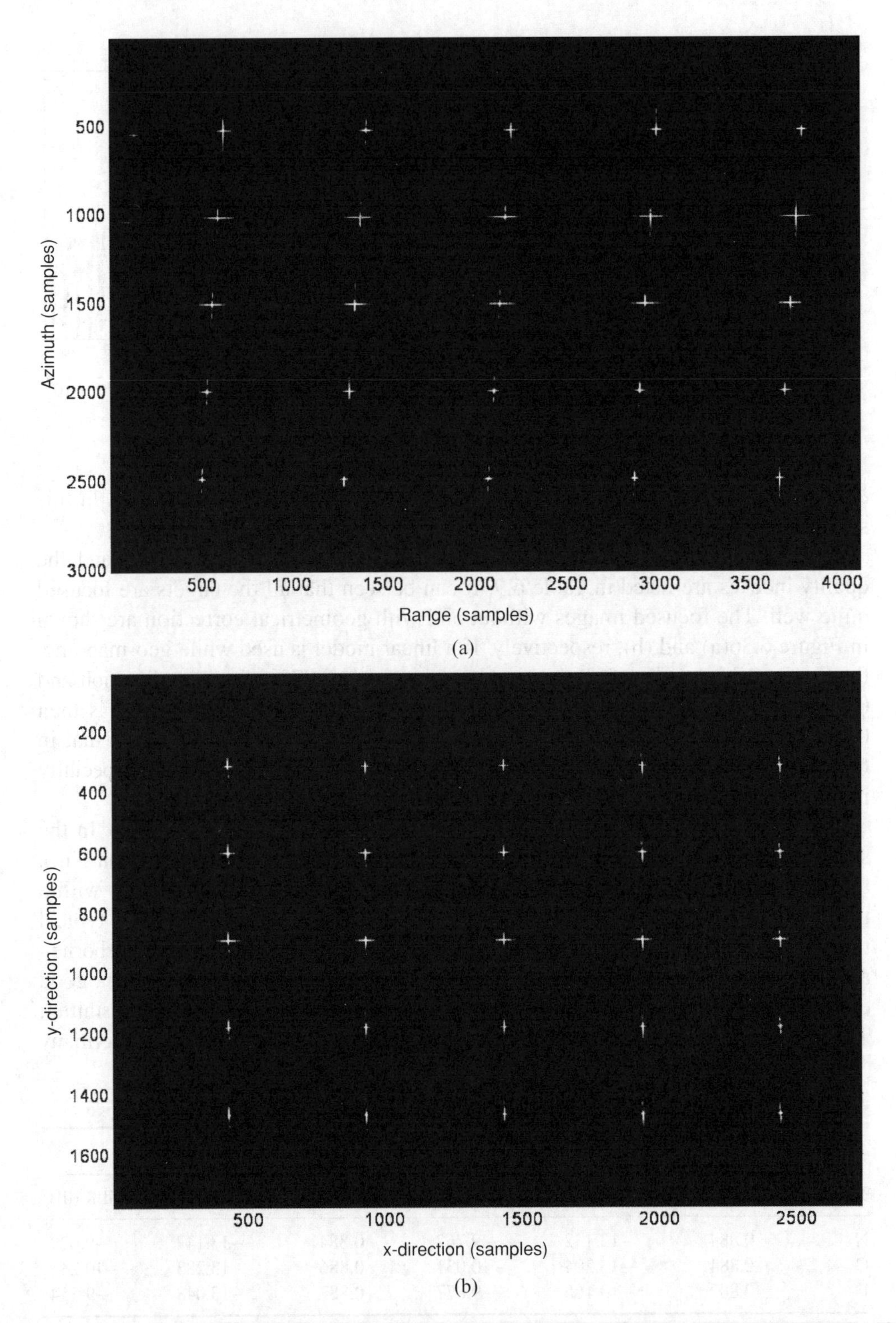

Figure 6.36 Focused image without and with geometrical correction (horizontal is the range in samples and vertical is the azimuth in pulses): (a) focused image without geometrical correction, (b) focused image with geometrical correction

Table 6.10 Simulated parameters of the spaceborne–airborne configuration

Spaceborne transmitter		Airborne receiver		System parameters	
Orbit height	514.8 km	Aircraft height	17 km	Carrier frequency	9.65 GHz
Antenna looking angle	32°	Antenna looking angle	62.8°	Transmitted bandwidth	150 MHz
Velocity	7000 m/s	Velocity	200 m/s	Sample rate	165 MHz
Beam velocity on ground	6477.2 m/s	Beam velocity on ground	200 m/s	PRF	3000 Hz
Equivalent squint angle	0.10°	Squint angle	0°	—	
Antenna size	4.78 m (az) 0.704 m (rg)	Antenna size	1 m (az) 0.25 m (rg)		

Note: In this table "az" means azimuth and "rg" means range.

[29–31] and good images have been obtained. In this book, the bistatic echo of a spaceborne–airborne bistatic SAR is simulated and processed by NuSAR, introduced here. The simulated parameters are listed in Table 6.10. The satellite and the aircraft are assumed to be flying along the y axis in this simulation.

Using the space–time diagrams introduced in Chapter 3, the calculated co-illuminating time is about 0.8 s and the co-illuminating area is about 1.2 km in the azimuth and about 16 km in the range. The Doppler bandwidth is about 1600 Hz. According to the results of the co-illuminating area, the positions of the simulated targets are chosen and are shown in Figure 6.37. An antenna pattern generated by the "sinc" function according to the antenna size is added to the echo, and the simulated echo is shown in Figure 6.38. Because the co-illuminating area in the azimuth is narrower than that in the range direction, the echo is rather narrow and long. In this simulation, about 3000 pulses are received by the airborne receiver.

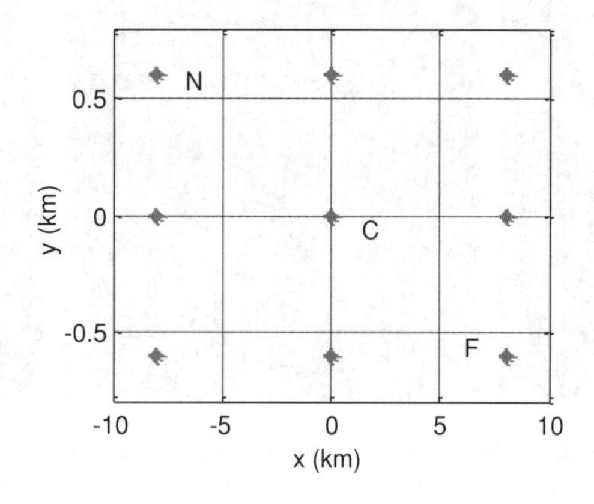

Figure 6.37 Simulated targets of the spaceborne–airborne configuration

Figure 6.38 Bistatic SAR echo of the spaceborne–airborne configuration

The differential phases for those targets with the same range gate and different azimuth positions are calculated and the maximal differential phase without the linear term is 37.20°, which is less than $\pi/4$, so the azimuth perturbation step is omitted here. A −18 dB Taylor window is added to the azimuth spectrum and a −23 dB Taylor window is added to the range spectrum while imaging using NuSAR. The imaging results are shown in Figures 6.39 and 6.40 and Table 6.11. It can be seen that

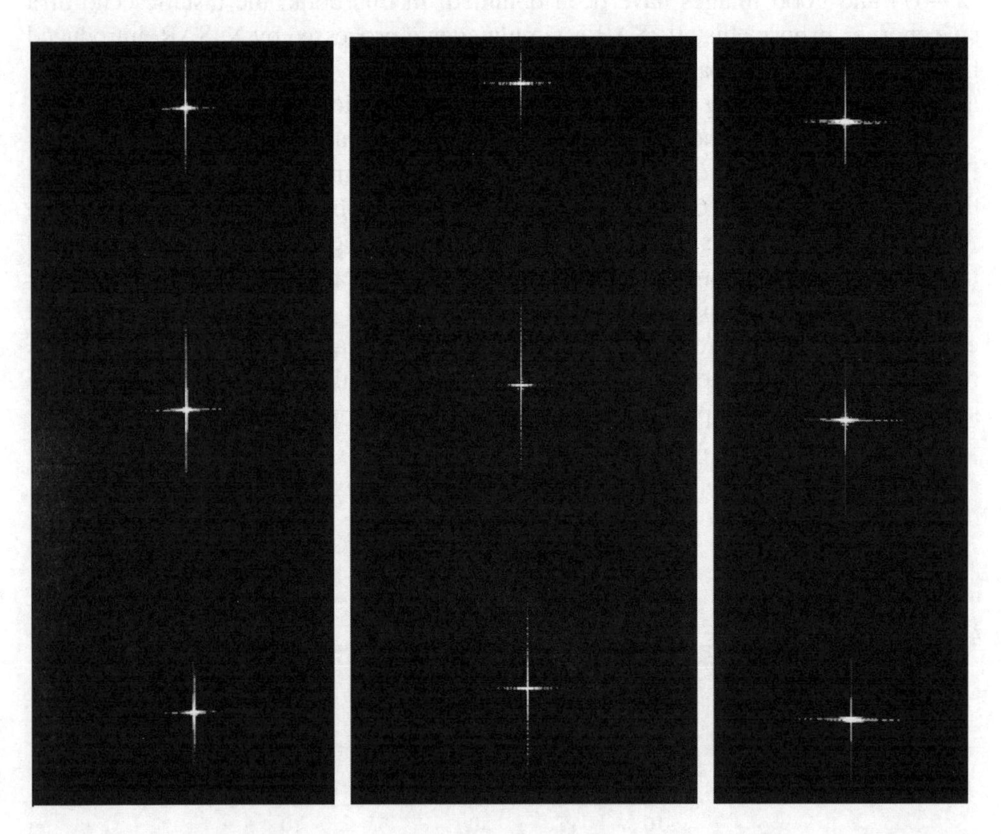

Figure 6.39 Imaging results of the simulated targets (notethat the horizontal is the range direction and the vertical is the azimuth direction, and to show the targets more clearly, three patches cut from the whole image are shown here)

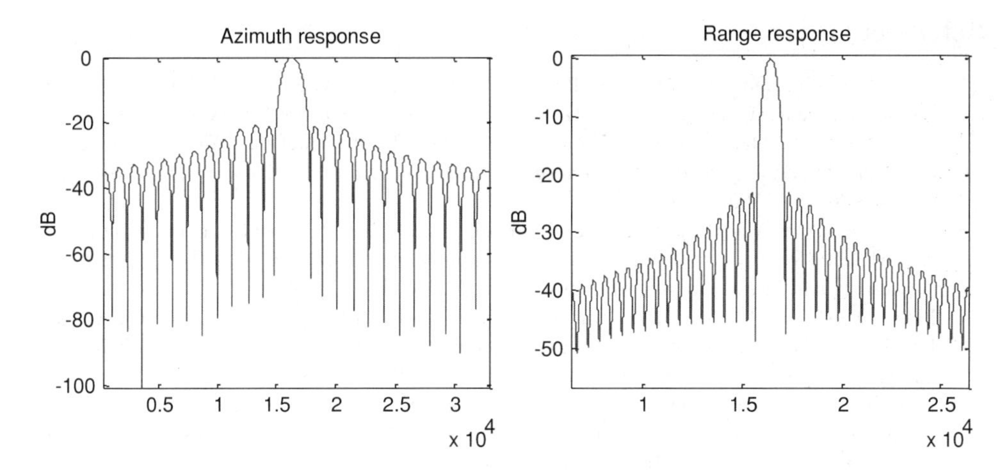

Figure 6.40 The azimuth and range responses of target N

all the simulated targets are well focused, which means that the NuSAR processor introduced here can handle this configuration of bistatic SAR.

6.4 Summary

In this chapter, the efficient algorithms taking advantage of the fast Fourier transform for the bistatic SAR of different translational variant configurations are described. The tomography sensitivity is not taken into consideration in this chapter and the scenario is assumed to be quite flat. The two-dimensional variation is the main difficulty talked about in this chapter. The azimuth perturbation and perturbation error compensation method combined with subswath processing can deal with the azimuth variation of the signal with an acceptable efficiency. However, if the resolution is enhanced and the scenario is heavily undulated, then the universal but inefficient BP algorithm may have to be used. As computers are becoming faster and faster, the BP algorithm may be acceptable in practice some day not far away. Then the BP algorithm will become the best choice for the translational variant configuration with severe variations.

Table 6.11 Imaging qualities of the targets

	Range				Azimuth			
Target	Resolution (m)	Widen ratio	PSLR (dB)	ISLR (dB)	Resolution (m)	Widen ratio	PSLR (dB)	ISLR (dB)
N	1.521	1.035	−23.178	−17.520	6.295	0.980	−20.669	−16.057
C	1.448	1.035	−23.155	−17.505	6.299	0.981	−20.679	−16.081
F	1.401	1.033	−23.153	−17.505	6.299	0.981	−20.621	−16.068

Note: The resolution listed in this table is the ground range resolution.

References

[1] Sanz-Marcos, J., Lopez-Dekker, P., Mallorqui, J.J., *et al.* (2007) SABRINA: a SAR bistatic receiver for interferometric applications. *IEEE Geoscience and Remote Sensing Letters*, **4** (2), 307–311

[2] Sanz-Marcos, J., Mallorqui, J.J., Broquetas, A. (2004) Bistatic parasitic SAR processor evaluation. In: *Proceedings of the IEEE International Geoscience and Remote Sensing Symposium (IGARSS'04)* pp. 3666–3669.

[3] Lopez-Dekker, P., Mallorqui, J.J., Serra-Morales, P., *et al.* (2007) Phase and temporal synchronization in bistatic SAR systems using sources of opportunity. In: *Proceedings of the IEEE International Geoscience and Remote Sensing Symposium (IGARSS'07)*, pp. 97–100

[4] Lopez-Dekker, P., Merlano, J.C., Duque, S., *et al.* (2007) Bistatic SAR interferometry using ENVISAT and a ground based receiver: experimental results. In: *Proceedings of the IEEE International Geoscience and Remote Sensing Symposium (IGARSS'07)*, pp. 107–110.

[5] Lopez-Dekker, P., Mallorqui, J.J., Serra-Morales, P., *et al.* (2008) Phase synchronization and Doppler centroid estimation in fixed receiver bistatic SAR systems. *IEEE Transactions on Geoscience and Remote Sensing*, **46** (11), 3459–3471.

[6] Duque, S., Lopez-Dekker, P., Merlano, J.C., *et al.* (2008) Bistatic interferometry using fixed receiver configurations. EUSAR Conference, pp. 1–4.

[7] Reuter, S., Behner, F., Nies, H., *et al.* (2010) Development and experiments of a passive SAR receiver system in a bistatic spaceborne/stationary configuration. IEEE International Conference on Geoscience and Remote Sensing Symposium, pp. 118–121.

[8] Nies, H., Behner, F., Reuter, S., *et al.* (2010) Polarimetric and interferometric applications in a bistatic hybrid SAR mode using TerraSAR-X. IEEE International Conference on Geoscience and Remote Sensing Symposium, pp. 110–113.

[9] Qiu, X.L., Behner, F., Reuter, S., *et al.* (2012) An imaging algorithm based on Keystone transform for one-stationary bistatic SAR of spotlight mode. *EURASIP Journal on Advances in Signal Processing*, **2012** (221) (provisional PDF available in http://asp.eurasipjournals.com/content/2012/1/221/abstract, formatted PDF is in production).

[10] Sanz-Marcos, J., Mallorqui, J.J., Aguasca, A. and Prats, P. (2005) Bistatic fixed-receiver parasitic SAR processor based on the back-propagation algorithm. In: *Proceedings of the IEEE International Geoscience and Remote Sensing Symposium (IGARSS'05)*, pp. 1056–1059.

[11] Sanz-Marcos, J., Mallorqui, J.J., Aguasca, A., *et al.* (2006) First envisat and ERS-2 parasitic bistatic fixed receiver SAR images processed with the subaperture range-Doppler algorithm. IEEE International Conference on Geoscience and Remote Sensing Symposium, pp. 1840–1843.

[12] Wong, F.H. and Yeo, N.L. (2001) New applications of nonlinear chirp scaling in SAR data processing. *IEEE Transactions on Geoscience and Remote Sensing*, **39** (5), 946–953.

[13] Qiu, X.L., Hu, D.H. and Ding, C.B. (2007) Non-linear chirp scaling algorithm for one-stationary bistatic SAR. 1st Asian and Pacific Conference on Synthetic Aperture Radar (APSAR), pp. 111–114.

[14] Qiu, X.L., Hu, D.H., and Ding, C.B. (2008). An improved NLCS algorithm with capability analysis for one-stationary BiSAR. *IEEE Transactions on Geoscience and Remote Sensing*, **46** (10), 3179–3186.

[15] Klare, J., Walterscheid, I., Brenner, A.R., *et al.* (2006) Evaluation and optimisation of configurations of a hybrid bistatic SAR experiment between TerraSAR-X and PAMIR. IEEE International Conference on Geoscience and Remote Sensing Symposium, pp. 1208–1211.

[16] Qiu, X.L., Hu, D.H. and Ding, C.B. (2008) Some reflections on bistatic SAR of forward-looking configuration. *IEEE Geoscience and Remote Sensing Letters*, **5** (4), 735–739.

[17] Balke, J. (2005) Field test of bistatic forward-looking synthetic aperture radar. IEEE International Radar Conference, pp. 424–429.

[18] Balke, J., Matthes, D. and Mathy, T. (2008) Illumination constraints for forward-looking radar receivers in bistatic SAR geometries. European Radar Conference, EuRAD, pp. 25–28.

[19] Espeter, T., Walterscheid, I., Klare, J., *et al.* (2011) Bistatic forward-looking SAR experiments using an airborne receiver. International Radar Symposium (IRS), pp. 41–44.
[20] Chang, C.Y., Jin, M. and Curlander, J.C. (1989) Squint mode SAR processing algorithms. International Geoscience and Remote Sensing Symposium (IGARSS'89), 12th Canadian Symposium on Remote Sensing, vol. 3, pp. 1702–1706.
[21] Tan, N.L., Yeo, T.S., Lu, Y.H., *et al.* (1999) SAR data processing at high squint. In: *Proceedings of the IEEE International Geoscience and Remote Sensing Symposium (IGARSS'99)*, vol. 1, pp. 561–563.
[22] Wang, R., Loffeld, O., Nies, H., *et al.* (2009). Focusing spaceborne/airborne hybrid bistatic SAR data using wavenumber-domain algorithm. *IEEE Transactions on Geoscience and Remote Sensing*, **47** (7), 2275–2283.
[23] Wang, R., Deng, Y.K., Loffeld, O., *et al.* (2011) Processing the azimuth-variant bistatic SAR data by using monostatic imaging algorithms based on two-dimensional principle of stationary phase. *IEEE Transactions on Geoscience and Remote Sensing*, **49** (10), Part 1, 3504–3520.
[24] Wang, R., Loffeld, O., Neo, Y.L., *et al.* (2010) Extending Loffeld's bistatic formula for the general bistatic SAR configuration. *IET Radar, Sonar and Navigation*, **4** (1), 74–84.
[25] Wong, F.H., Cumming, I.G. and Neo, Y.L. (2008) Focusing bistatic SAR data using the non-linear chirp scaling algorithm. *IEEE Transactions on Geoscience and Remote Sensing*, **46** (9), 2493–2505.
[26] Bamler, R. and Boerner, E. (2005) On the use of numerically computed transfer functions for processing of data from bistatic SARs and high squint orbital SARs. In: *Proceedings of IEEE International Geoscience and Remote Sensing Symposium (IGARSS'05)*, vol 2, pp. 1051–1055.
[27] Bamler, R., Meyer, F. and Liebhart, W. (2006) No math: bistatic SAR processing using numerically computed transfer functions. IEEE International Conference on Geoscience and Remote Sensing Symposium, pp. 1844–1847.
[28] Qiu, X.L., Hu, D.H. and Ding, C.B. (2009) A new calculation method of NuSAR for translational variant bistatic SAR. IEEE International Geoscience and Remote Sensing Symposium (IGARSS'09), vol. 2, pp. II-45–II-48.
[29] Baumgartner, S.V., Rodriguez-Cassola, M., Nottensteiner, A., *et al.* (2008) Bistatic experiment using TerraSAR-X and DLR's new F-SAR system. EUSAR Conference, pp. 1–4.
[30] Espeter, T., Walterscheid, I., Klare, J., *et al.* (2008). Progress of hybrid bistatic SAR: synchronization experiments and first imaging results. EUSAR Conference, pp 1–4.
[31] Walterscheid, I., Espeter, T., Brenner, A.R., *et al.* (2010) Bistatic SAR experiments with PAMIR and TerraSAR-X – setup, processing, and image results. *IEEE Transactions on Geoscience and Remote Sensing*, **48** (8), 3268–3279.

7

Bistatic SAR Parameter Estimation and Motion Compensation

7.1 Introduction

Previous discussions are based on the assumption that the transmitter and receiver of the bistatic SAR system is fully synchronized, the platform trajectories are known precisely and the antenna beam directions are under precise control. However, the actual condition is not so perfect. Even if the problem of synchronization error is not taken into account, the problem of motion error always exists in airborne SAR due to the effect of airflow and other factors. In this case, the movement of the aircraft will deviate from the uniform linear motion, which can be described as: (1) the acceleration along the nominal track is not zero and (2) the velocity perpendicular to the nominal track is not zero. In addition, the platform usually cannot be moving translationally, but will swing in three directions, so there are angular motions in three directions, named yaw, pitch and roll. Motion errors and attitude errors make the antenna phase center (APC) deviate from the uniform linear motion and the direction of beam pointing changes all the time. As a result both the phase and the amplitude of the SAR echo are affected by these errors. Not only does the Doppler frequency centroid drift and the Doppler frequency modulate rate change, but there are also high-order or other types of phase errors adding to the Doppler signal. Moreover, the signal amplitude modulation varies due to motion and attitude errors. Therefore, the Doppler parameter estimation and motion compensation are indispensable to airborne SAR in order to achieve high image quality, and consequently they should also be done in the airborne bistatic SAR.

Compared with monostatic SAR, the motion error analysis and motion compensation of bistatic SAR will face greater challenges due to the existence of two platforms. The challenge mainly lies in the following two facts: (1) the sources of motion error increase due to the existence of two platforms, so the motion error analysis is more complex, and (2) the effects of motion error on different target echoes are also

Bistatic SAR Data Processing Algorithms, First Edition. Xiaolan Qiu, Chibiao Ding and Donghui Hu.

complex due to the complexity of the bistatic SAR geometry, so motion compensation is more difficult than that in monostatic SAR. However, the motion error analysis and compensation in bistatic SAR are consistent with those in monostatic SAR in principle, which is to analyze and correct the phase and amplitude distortion. Therefore, some of the motion compensation methods and ideas for monostatic SAR can also be adopted for bistatic SAR.

This chapter first analyzes the effects on SAR imaging caused by motion errors and then considers the change of Doppler parameter caused by attitude errors. On this basis, the Doppler parameter estimation method is introduced and the SAR motion compensation method is highlighted. After a brief introduction to the basic principle of SAR motion compensation, the motion compensation strategy of bistatic SAR is introduced in view of its special characteristic, which further improves the bistatic SAR imaging process.

7.2 Analyzing the Effects of Motion Errors

As described in Section 7.1, SAR motion errors include not only the errors caused by the deviation from uniform linear motion of the platforms but also the errors caused by the changeable attitude of the platforms. The effects of these two errors are analyzed as follows.

7.2.1 Attitude Error Analysis

The variance in the platform attitude mainly causes error in the antenna beam direction. The variance of the antenna attitude can be expressed by yaw angle θ_y, pitch angle θ_p and roll angle θ_r. The two-dimensional beam direction errors $\Delta\delta_a$ and $\Delta\delta_r$ [1] in the azimuth and range directions, respectively, can be derived from these three attitude angles. As shown in Figure 7.1, the platform moves along the y axis, the z axis is

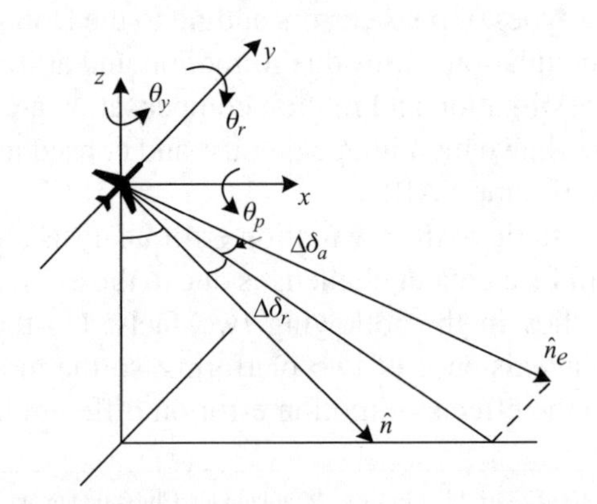

Figure 7.1 Geometry of the antenna beam direction error

perpendicular to the ground in an upward direction and the x axis forms a Cartesian coordinate system with the y and z axes. Then the yaw angle, pitch angle and roll angle are defined as the rotation angle around the z axis, x axis and y axis, respectively. In Figure 7.1, $\hat{n}$ is the beam direction without attitude error, ϕ is the side-looking angle of the antenna without attitude error and $\hat{n}_e$ is the beam direction with attitude errors. It can be seen from the geometry in Figure 7.1 that

$$\hat{n} = [\sin\phi \quad 0 - \cos\phi]^{\mathrm{T}} \tag{7.1}$$

In the Cartesian coordinate system, the yaw, pitch and roll rotation matrixes are as follows:

$$\begin{aligned} M_y &= \begin{pmatrix} \cos\theta_y & \sin\theta_y & 0 \\ -\sin\theta_y & \cos\theta_y & 0 \\ 0 & 0 & 1 \end{pmatrix}, \quad M_p = \begin{pmatrix} 1 & 0 & 0 \\ 0 & \cos\theta_p & \sin\theta_p \\ 0 & -\sin\theta_p & \cos\theta_p \end{pmatrix}, \\ M_r &= \begin{pmatrix} \cos\theta_r & 0 & -\sin\theta_r \\ 0 & 1 & 0 \\ \sin\theta_r & 0 & \cos\theta_r \end{pmatrix} \end{aligned} \tag{7.2}$$

Rotated by M_y, M_p and M_r in order, the beam direction turns out to be

$$\begin{aligned} \hat{n}_e &= \begin{pmatrix} x \\ y \\ z \end{pmatrix} = M_r M_p M_y \hat{n} \\ &= \begin{pmatrix} \cos\theta_r\cos\theta_y\sin\phi - \sin\theta_r\sin\theta_p\sin\theta_y\sin\phi + \sin\theta_r\cos\theta_p\cos\phi \\ -\cos\theta_p\sin\theta_y\sin\phi - \sin\theta_p\cos\phi \\ \sin\theta_r\cos\theta_y\sin\phi + \cos\theta_r\sin\theta_p\sin\theta_y\sin\phi - \cos\theta_r\cos\theta_p\cos\phi \end{pmatrix} \end{aligned} \tag{7.3}$$

so

$$\Delta\delta_r = \arctan\left(-\frac{x}{z}\right) - \phi \approx -\theta_r \tag{7.4}$$

$$\Delta\delta_a = \arctan\left(\frac{y}{\sqrt{x^2 + z^2}}\right) \approx \theta_y\sin\phi + \theta_p\cos\phi \tag{7.5}$$

It can be found from the above equations that the roll angle mainly causes the beam direction error in the range direction, while the yaw and pitch angles mainly cause the beam direction error in the azimuth direction. The antenna beam direction error has no effect on the phase of the SAR echo; it only causes modulation on the echo

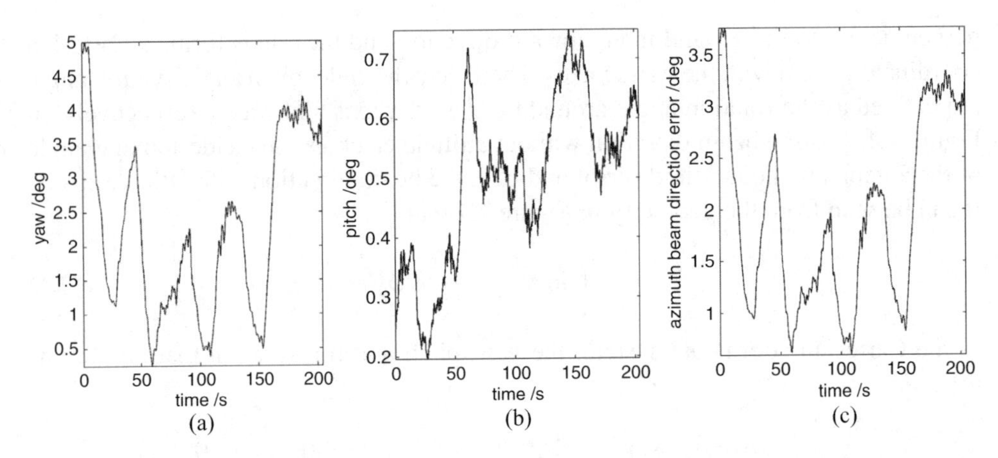

Figure 7.2 The measured aircraft attitude and the calculated azimuth beam direction error

amplitude. Commonly, the antenna beam width in the elevation direction (i.e., the range direction) is relatively quite broad, so the beam direction error in the range direction only causes a drop of echo power at the imaging swath margin, which causes few effects. However, the azimuth beam width is relatively narrow; thus the effects of the azimuth beam direction error are usually noticeable. The effects will be analyzed in the following text.

Figure 7.2 is a set of real measured yaw angle and pitch angle of aircraft attitude and the calculated beam direction errors in the azimuth. Usually, the synthetic aperture time is used as the threshold that divides the azimuth direction error into a low-frequency component and a high-frequency component. A high-frequency change indicates the fluctuation within a synthetic aperture time, while a low-frequency change indicates the fluctuation in the scale of multiple synthetic aperture times. Typically, the low-frequency component is caused by a relatively stable yaw angle, and causes the offset and a slow drift of Doppler centroid. Meanwhile, the high-frequency component is caused by the high-frequency jitter of attitude, and makes the echo have a high-frequency amplitude modulation in the azimuth direction. This modulation, according to the analysis in [2], will cause artificial peaks (often referred to as paired echoes) on both sides of the target's peak after azimuth compression. Thus the SAR image quality indexes such as the PSLR and the ISLR will deteriorate, and the false alarm probability in image interpretation will increase.

In the bistatic SAR, if the attitude error exists in both platforms, they will cause the beam direction error in each antenna. Thus the synthetic beam direction drifts, and will even cause lack of synchronization between the transmit and receive antenna coverage when the attitude error is large. In addition,as is the case in monostatic SAR, the attitude error in bistatic SAR also mainly causes the Doppler centroid frequency offset and the echo amplitude modulation, which affect the imaging quality, but it usually will not destroy the signal phase.

7.2.2 *Motion Error Analysis*

Different from the attitude error, the platform motion error will cause changes in the phase of SAR echo. Thus it will directly affect the SAR imaging quality. Serious motion error even makes it fail to form an image. Hence, the effects caused by motion error are analyzed in detail in the following text.

7.2.2.1 Monostatic SAR Motion Error Analysis

First take the monostatic SAR as an example. In the ideal condition, SAR moves along a straight path and each position where the pulse transmitted/received is of an identical space interval. However, these cannot be guaranteed in practice:

1. The acceleration along the nominal trajectory of the platform cannot always be controlled to zero, so the space interval between pulses will be nonuniform when the pulses are transmitted/received at a fixed pulse repetition frequency.
2. The velocity vector in the plane perpendicular to the nominal trajectory cannot always be controlled to be zero, which makes the real path deviate from a straight track.

The effects on SAR imaging caused by the above facts are analyzed separately as follows:

1. The effects caused by nonuniform sampling along the trajectory. When the aircraft moves along the nominal trajectory in nonuniform velocity and transmitting/receiving pulses are in a fixed PRF, the echo pulses will be nonuniformly aligned along the azimuth direction (as shown in Figure 7.3). From the imaging point of view, this makes the range history different from the ideal one. Suppose that the difference between the real position and ideal position along the trajectory

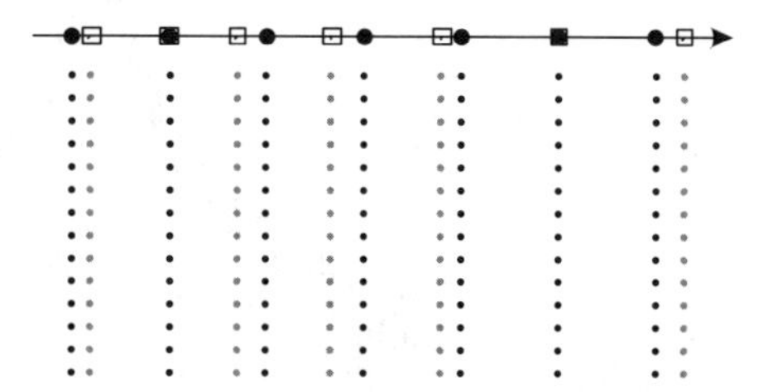

Figure 7.3 Illustration of uniform/nonuniform spatial intervals of pulses

when azimuth time is η is $\delta_y(\eta)$ and $\delta_y(\eta) = \int_0^{\eta} \delta_V(\eta)d\eta$, where $\delta_V(\eta)$ is the variation of the azimuth velocity caused by the azimuth nonzero acceleration; then the range history of the target at (R_0, y_0) in the slant-range plane will be

$$R'(\eta) = \sqrt{R_0^2 + \left[V\eta + \int_0^{\eta} \delta_V(\eta)d\eta - y_0\right]^2}$$
$$\approx R(\eta) + \frac{[V\eta - y_0]\int_0^{\eta} \delta_V(\eta)d\eta + \frac{1}{2}\left(\int_0^{\eta} \delta_V(\eta)d\eta\right)^2}{R(\eta)} \tag{7.6}$$

where $R(\eta) = \sqrt{R_0^2 + (V\eta - y_0)^2}$ is the range history under uniform motion. It can be seen that the existence of $\delta_V(\eta)$ will cause an additional error in the range history. Usually, $\delta_V(\eta)$ is very small compared to V; thus the range difference $|R(\eta) - R'(\eta)|$ caused by $\delta_V(\eta)$ is often much smaller than the range sampling interval. For this reason, $\delta_V(\eta)$ mainly affects the phase. It adds a phase error to the Doppler signal, thereby worsening the resolution, PSLR and ISLR. Figure 7.4 shows a curve of a real measured azimuth velocity error and the azimuth compression results with/without the existence of this error. It can be seen that the degradation of focusing quality caused by the nonuniform sampling is somewhat obvious, so generally this error needs compensation. In addition, it should be noted according to Equation (7.6) that the range history difference caused by $\delta_V(\eta)$ is a function of

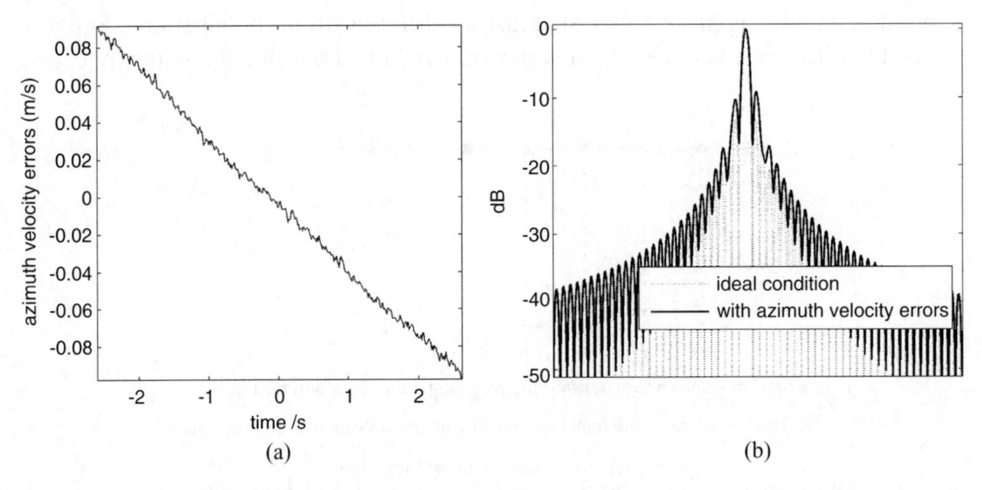

Figure 7.4 Azimuth velocity error and comparison of the compression results: (a) real measured azimuth velocity error, (b) the azimuth compression results with/without azimuth velocity error

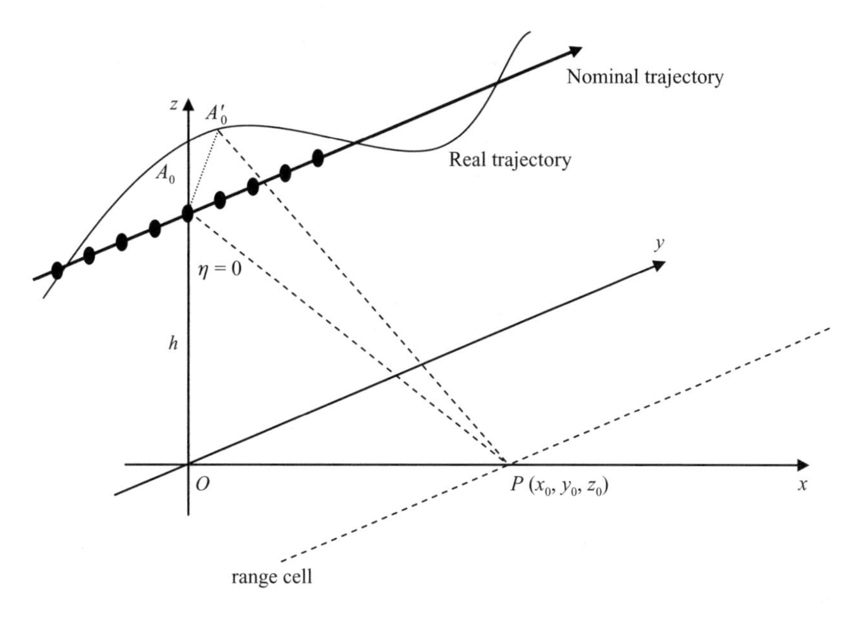

Figure 7.5 Illustration of the range history difference between the ideal trajectory and the real trajectory

R_0, y_0 and η. This means that the range history error caused by $\delta_V(\eta)$ is spatially variant, to which attention must be paid when compensating.

2. The effects caused by motion errors in the track perpendicular plane. When the platform velocity in the track perpendicular plane is not zero, as shown in Figure 7.5, the range history also deviates from the ideal one. Supposing that the nominal trajectory is $[0, V\eta, h]$, the real trajectory is $[x_d(\eta), V\eta, h + z_d(\eta)]$, where $x_d(\eta)$,$z_d(\eta)$ are two components that the motion error projected on to the xOz plane at η; then the range history from target P to the real trajectory is

$$R'(\eta) = \sqrt{[x_d(\eta) - x_0]^2 + (V\eta - y_0)^2 + [h + z_d(\eta) - z_0]^2} \tag{7.7}$$

The range history difference for target P between the nominal and the real trajectory is

$$\Delta R_d(\eta) = R'(\eta) - R(\eta) \approx -\frac{x_0}{R(\eta)} x_d(\eta) + \frac{h - z_0}{R(\eta)} z_d(\eta) \tag{7.8}$$

where $R(\eta) = \sqrt{x_0^2 + (V\eta - y_0)^2 + (h - z_0)^2}$ is the range history under the nominal trajectory. This range history difference results in disorder of the echo samples in the range direction, as shown in Figure 7.6. It can be seen that the RCM curve of a target is no longer a hyperbola as in the ideal case, but has additional fluctuation associated with the motion errors. Therefore, if RCMC is done according to the nominal trajectory without motion error correction, there may be a residual range

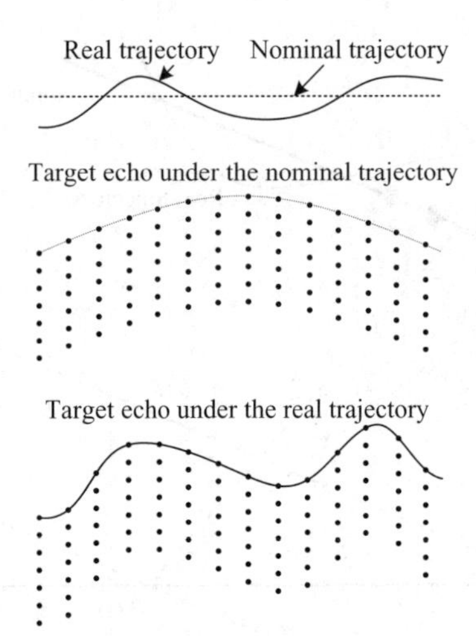

Figure 7.6 The effect on SAR data storage caused by motion errors in the track perpendicular plane

migration that exceeds one range cell in the echo data, and this will lead to defocusing in the final image. Moreover, the range history error of Equation (7.8) will seriously affect the Doppler phase of the SAR signal, leading to a sharp deterioration in the imaging result. For example, suppose the range between the target and the nominal trajectory is 15000 m, the platform velocity is 150 m/s, the pulse repetition frequency is 1500 Hz, the synthetic aperture time is 10 s, the carrier wavelength λ is 0.03 m and the pulse sampling frequency is 500 MHz, the range history error of the target caused by the platform motion error is shown in Figure 7.7(a). It is noticed that the range

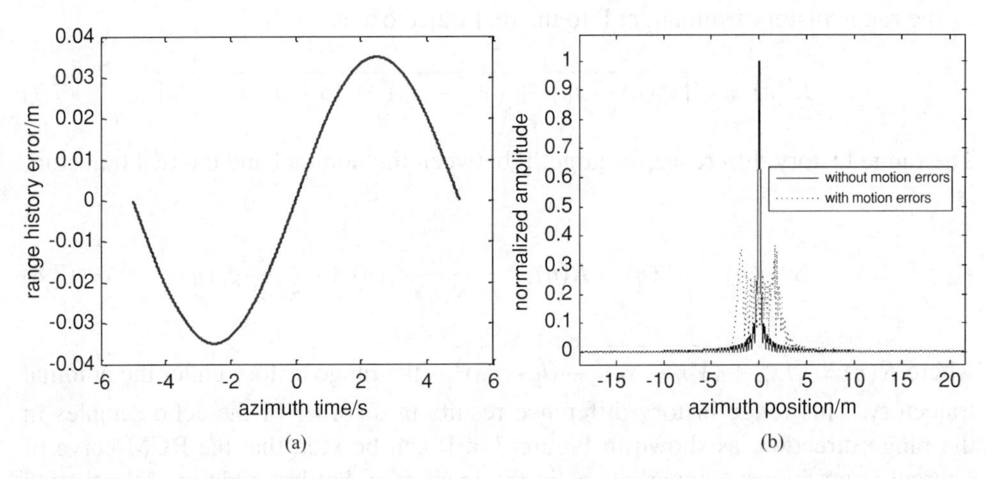

Figure 7.7 The effect on azimuth compression caused by motion errors in the track perpendicular plane: (a) range history error caused by motion errors in the track perpendicular plane, (b) azimuth compression results with/without motion errors

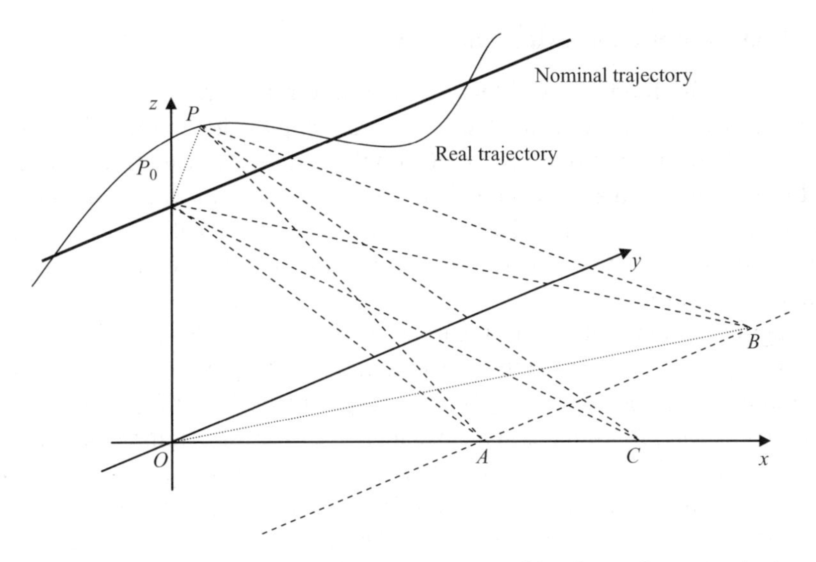

Figure 7.8 Variance of the effects on different targets caused by the motion errors in the trajectory perpendicular plane

history error is less than 0.04 m while the range sampling interval is 0.3 m, so the effect of disorder in the echo storage location could be ignored. However, the azimuth compression result in Figure 7.7(b) shows that even a small motion error like the one shown in Figure 7.7(a) causes the azimuth compression quality to suffer from serious deterioration. Therefore, the motion error in the track perpendicular plane must be appropriately compensated.

In addition, it should be noted that the motion errors in the plane perpendicular to the nominal trajectory cause different effects on different targets in the scenario. As shown in Figure 7.8, suppose the real position of SAR is P, corresponding to the ideal position P_0, and PP_0 is in the xOz plane to the three targets A, B and C in the imaging area (A and B have the same range position and A and C have the same azimuth position); the range errors caused by the deviation of P from P_0 are as follows:

$$r_{eA} = |PA| - |P_0A|, \quad r_{eB} = |PB| - |P_0B|, \quad r_{eC} = |PC| - |P_0C| \tag{7.9}$$

Generally, $|PA| - |P_0A| \neq |PB| - |P_0B| \neq |PC| - |P_0C|$; therefore, the same motion error in the trajectory perpendicular plane causes different range history errors for different targets in the scene. Usually, the difference between r_{eA} and r_{eC} is called "the range space-variant error" and the difference between r_{eA} and r_{eB} is called "the azimuth space-variant error". In addition, if two targets are of different heights, their range history errors will also be different. Therefore, it can be seen that the three-dimensional space-variant characteristic of motion errors further increases the complexity of motion compensation.

7.2.2.2 Analysis of Bistatic SAR Motion Errors

Compared with the monostatic SAR, the motion error sources of the bistatic SAR increase and the error analysis is relatively more complex due to the existence of two independent platforms. Currently, research on the bistatic SAR motion compensation has not been adequate, most of which [3–10] being based on the translational invariant configuration. Therefore, in this chapter, the analysis of the bistatic SAR motion errors and the introduction of the compensation method are both limited to the translational invariant configuration.

The imaging geometry of the bistatic translational invariant SAR with motion errors is shown in Figure 7.9. Suppose the transmit and receive platforms both move along the y axis and their velocities are V_R and V_T, respectively. Take the position where the synthetic beam center illuminates at the instant $\eta = 0$ to be the origin of the coordinates. Suppose the transmitter and receiver are respectively located at (X_T, Y_T, H_T) and (X_R, Y_R, H_R) at the moment $\eta = 0$. Here in Figure 7.9, the straight dashed lines parallel to the y axis are the nominal trajectories of the transmitter and receiver, while the solid curves indicate the real trajectories and the position errors of the transmitter and receiver are represented by $[X_T + \Delta X_T(\eta), Y_T + \Delta Y_T(\eta), H_T + \Delta Z_T(\eta)]$ and $[X_R + \Delta X_E(\eta), Y_R + \Delta Y_R(\eta), H_R + \Delta Z_R(\eta)]$, respectively. Supposing the scene is flat, the range history of target $P(x, y, 0)$ with motion errors is

$$\begin{aligned}\tilde{R}(\eta; x, y) = &\sqrt{[X_T + \Delta X_T(\eta) - x]^2 + [Y_T + \Delta Y_T(\eta) + V_T\eta - y]^2 + [H_T + \Delta Z_T(\eta)]^2}\\ &+\sqrt{[X_R + \Delta X_R(\eta) - x]^2 + [Y_R + \Delta Y_R(\eta) + V_R\eta - y]^2 + [H_R + \Delta Z_R(\eta)]^2}\end{aligned} \tag{7.10}$$

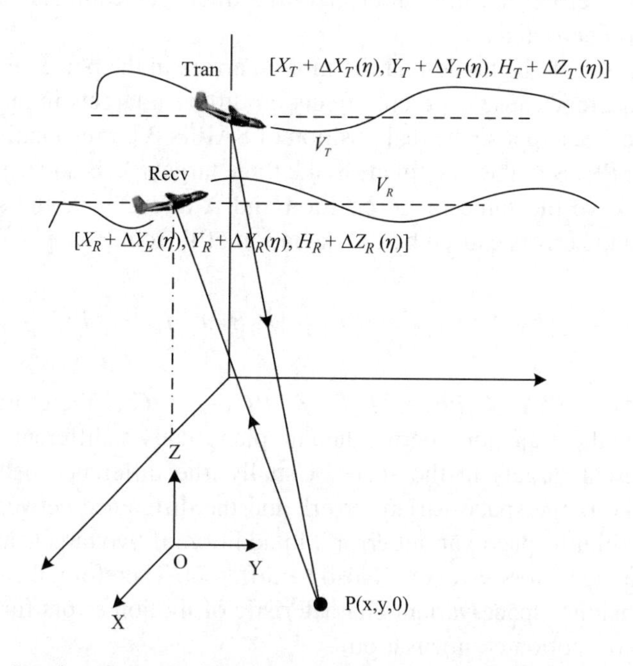

Figure 7.9 Imaging geometry of the bistatic translational invariant SAR with motion errors

while the bistatic range history of target P in the ideal condition is

$$R(\eta;x,y) = \sqrt{[X_T - x]^2 + [Y_T + V_T\eta - y]^2 + H_T^2} + \sqrt{[X_R - x]^2 + [Y_R + V_R\eta - y]^2 + H_R^2} \tag{7.11}$$

Thus the motion error is

$$\Delta R(\eta;x,y) = \tilde{R}(\eta;x,y) - R(\eta;x,y) \tag{7.12}$$

Similar to the monostatic case, it can be decomposed into two components, which are the motion error in the trajectory perpendicular plane and the motion error along the azimuth direction, written as

$$\Delta R(\eta;x,y) = \Delta R_{XZ}(\eta;x,y) + \Delta R_Y(\eta;x,y) \tag{7.13}$$

where

$$\Delta R_{XZ}(\eta;x,y) = \tilde{R}(\eta;x,y) - \tilde{R}_Y(\eta;x,y) \tag{7.14}$$

is the error in the trajectory perpendicular plane. Similar to that in monostatic SAR, this error will be different for different targets of the scenario and

$$\Delta R_Y(\eta;x,y) = \tilde{R}_Y(\eta;x,y) - R(\eta;x,y) \tag{7.15}$$

is the range difference caused by the motion errors along the azimuth direction. In Equations (7.14) and (7.15),

$$\tilde{R}_Y(\eta;x,y) = \sqrt{[X_T - x]^2 + [Y_T + \Delta Y_T(\eta) + V_T\eta - y]^2 + H_T^2} + \sqrt{[X_R - x]^2 + [Y_R + \Delta Y_R(\eta) + V_R\eta - y]^2 + H_R^2} \tag{7.16}$$

is the bistatic range history when only the azimuth error exists. This is different from monostatic SAR, since the transmitter and receiver move along two different trajectories and transmitting and receiving are not at the same position. Because the azimuth motion errors of the transmitter and the receiver are usually different from each other, the space intervals between the transmitting places and between the receiving places for the pulses are both nonuniform. Therefore the azimuth motion error compensation method is different from that of monostatic SAR. Section 7.4 will further explain the compensation method of bistatic SAR.

7.3 Estimation of Doppler Parameters

Section 7.2 has analyzed the effects of attitude error and motion error on SAR imaging. Of these, attitude error does not result in a change of the signal phase, but can induce the offset of the Doppler centroid. Thus, when the attitude is not accurately measured, it is often necessary to estimate the Doppler centroid through echo data. Besides, if the azimuth velocity and/or the time delay of the inner system are not accurately measured, the calculated result of the Doppler FM rate will deviate from its true value, and consequently the image quality will be degraded. As a result, it is often necessary to estimate the Doppler FM rate or aircraft velocity [11] based on echo data. As for SAR imaging, the Doppler centroid and the Doppler FM rate are two extremely important parameters whether in the monostatic system or in the bistatic system. Since the meanings of these two parameters in bistatic SAR are the same as in monostatic SAR, the estimation methods of the Doppler parameter in monostatic SAR can be adopted in bistatic SAR. There has been a lot of research on Doppler centroid and Doppler FM rate estimation in the existing literature [12–19]; hence they are not intended to be introduced in this book. Instead, a Doppler centroid estimation method, namely, the Doppler centroid estimation method based on time–frequency domain prefiltering, which is from the authors' own work, will be introduced in this section. It can be applicable to both monostatic and bistatic SAR cases and can make an improvement on estimation accuracy in nonhomogeneous scenes.

The existence of the Doppler centroid error will result in a loss of image SNR, an increase in the azimuth ambiguity level and degradation in spatial resolution. Thus, an accurate and robust algorithm for Doppler centroid estimation is important for obtaining high-quality SAR images. For bistatic SAR, because of the separation of the receiver and transmitter, error sources of beam direction control increase. Consequently, the Doppler centroid calculated from the orbit (or trajectory) and attitude information may be seriously lacking in accuracy. Hence, the Doppler centroid estimation based on echo data may be even more important for bistatic SAR than for monostatic SAR. In addition, an accurate estimated Doppler centroid value can be used to validate the antenna pointing precision. Therefore, an accurate and robust Doppler centroid estimation algorithm is of great significance.

The existing Doppler centroid estimation algorithms can be roughly divided into two kinds. One is based on the dominant scatters [15, 16], which estimates the Doppler centroid using methods like the Radon transform to extract the slope of the range migration curve. The advantage of this kind of method is that the unambiguous value of the Doppler centroid can be obtained directly, while the disadvantage is that the extremely strong scatters must exist in the scenario and the computation load is usually quite large. The other kind of estimation algorithm is based on the correspondence between the azimuth spectrum shape of the echo and the antenna beam pattern [13, 14, 17]. This kind of method is more practical for its simplicity, but it can easily be affected by the echoes of those partially exposed targets at the border of the scenario. Especially when the scenario has high-contrast borders (for example, the scenario includes the conjunction of land and sea), the estimated Doppler centroid based on the second kind of method will deviate from the true value by a large amount. Although azimuth compressed data can be used for estimating the release of the effect of partially exposed targets, it is not the best choice in practice due to the fact that one has to iterate many times during the imaging and so the efficiency of the processing system declines.

Besides, it is possible that the transmitting and receiving antenna patterns are quite different in shape and deviate somewhat in the pointing direction. Consequently, the synthetic beam pattern, which modulates the targets' echoes, may not be symmetrical. Therefore, the definition of the Doppler centroid in such cases is clarified firstly in this section and then a Doppler centroid estimation method based on time–frequency domain pre-filtering is introduced.

7.3.1 Definition of the Doppler Centroid

In monostatic SAR, the center or mean value of Doppler instant frequency is usually referred to as the Doppler centroid [20]. As the receiver and the transmitter usually share the same antenna in monostatic SAR, and the antenna beam pattern usually has good symmetry, the expressions that are not so strict, such as the center, mean or the frequency according to the peak of Doppler frequency spectrum, usually indicate the same frequency, and thus there will be no confusion. However, in the bistatic SAR case, the beam direction deviation caused by separation and the shape difference between the receiving and transmitting antennas can possibly result in an asymmetrical synthetic beam pattern, which will further induce an asymmetrical azimuth spectrum of the SAR echo. Figure 7.10 shows a simulation result of the bistatic SAR synthetic antenna beam pattern when there is a beam direction deviation. As can be seen from the projection of the azimuth beam on the ground, the asymmetry in the

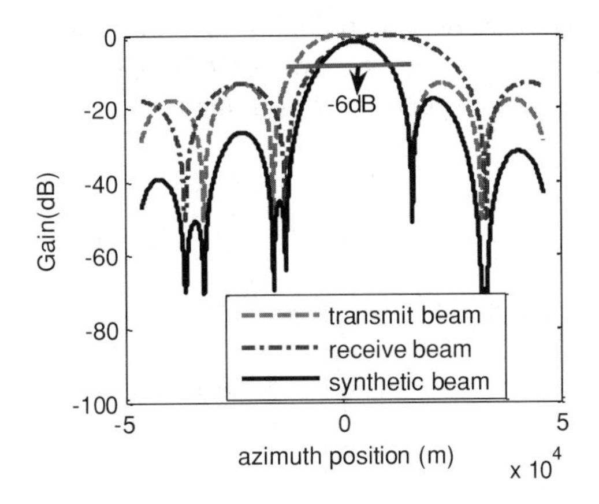

Figure 7.10 Diagram of the asymmetrical bistatic synthetic beam pattern

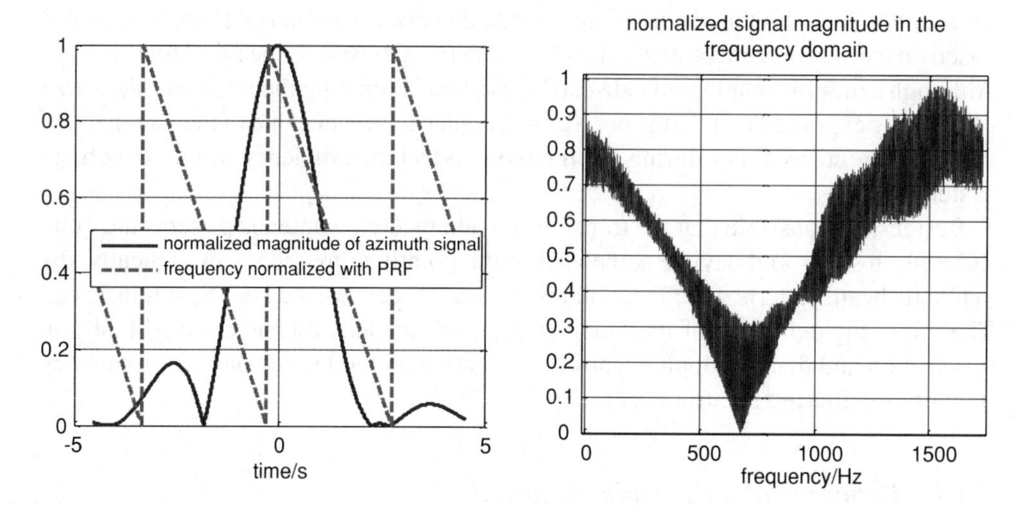

Figure 7.11 The asymmetrical azimuth signal modulated by the beam pattern shown in Figure 7.10

synthetic beam pattern between the left part and the right part is quite obvious. The left figure of Figure 7.11 is the normalized magnitude of the time domain azimuth echo modulated by the antenna pattern considering the first pair of sidelobes; the right figure is the normalized magnitude of the azimuth spectrum in the frequency domain within one PRF. Obviously, due to the asymmetry of the mainlobe and the sidelobes of the synthetic beam pattern, the signal spectrum shows asymmetry. Hence, it is necessary to clarify the definition of the Doppler centroid in the bistatic SAR case.

In imaging processing, generally not all the spectrum within the whole PRF band are processed. Instead, the matched filter usually has a limited bandwidth of $B_{a,pro}$ ($B_{a,pro}$ is called the azimuth processing bandwidth) and the limit band matched filter is applied to the signal spectrum with centres at a certain frequency and $\pm B_{a,pro}/2$ bandwidth on each side. Signals outside this processing band are set to zero for the purpose of improving SNR and reducing azimuth ambiguity. Therefore, it is reasonable to give a definition of the bistatic SAR Doppler centroid in terms of SNR in the case where the azimuth spectrum is asymmetric:

$$f_{dc} = f \text{ to maximize} \frac{\displaystyle\int_{f_{dc}-B_{a,pro}/2}^{f_{dc}+B_{a,pro}/2} |S(f)|^2 df}{\displaystyle\int_{f_{dc}-B_{a,pro}/2}^{f_{dc}+B_{a,pro}/2} |N_R(f)|^2 + |N_{amb}(f)|^2 df} \tag{7.17}$$

where $S(f)$ is the spectrum of the signal modulated by the mainlobe, $N_R(f)$ is the spectrum of the receiver noise and $N_{amb}(f)$ is the spectrum of the azimuth ambiguity.

The Doppler centroid under this definition can lead to a maximum SNR (which is the ratio of useful signal to receiver noise and azimuth ambiguity noise) within the processing bandwidth. In the case of unknown receiver noise and azimuth ambiguity, the Doppler centroid in Equation (7.17) is difficult to calculate. Thus, it is reasonable to adopt the definition of the following equation as an approximate definition in practice:

$$f_{dc} = f \text{ to maximize} \int_{f_{dc}-B_{a,pro}/2}^{f_{dc}+B_{a,pro}/2} |\tilde{S}(f)|^2 df \tag{7.18}$$

Here $\tilde{S}(f)$ represents the spectrum of the entire signal modulated by the entire beam pattern.

7.3.2 Doppler Centroid Estimation Method Based on a Time–Frequency Domain Pre-Filter

The precondition for estimating the Doppler centroid based on the above definition is that the azimuth spectrum of the echo signal can well represent modulation of the beam pattern. However, if the partially exposed targets exist and make the shape of the azimuth spectrum deviate from the shape of the beam pattern, the estimation error will increase. In the following, a Doppler centroid estimation method based on eliminating the partially exposed target signal using a time–frequency domain pre-filter is introduced.

7.3.2.1 Basic Principle

Among the many algorithms of Doppler centroid estimation, the correlation Doppler estimator (CDE) [13] proposed by Madsen is simple to implement without iterations and has been commonly used. This algorithm exploits the fact that the Fourier transformation of the autocorrelation function of a signal represents the power spectrum of the signal. Thus, the Doppler centroid can be estimated by calculating the phase of the correlation coefficients. The formula is expressed as follows:

$$f_{DC} = \frac{1}{2\pi kT} \arg\{r(k)\} \tag{7.19}$$

where $r(k)$ is the autocorrelation function of the signal, T is the pulse repetition interval, f_{DC} is the baseband Doppler centroid and $k = 1$ is usually preferred. Since it is only necessary to calculate the time-domain correlation, the algorithm is quite fast.

Usually, the azimuth echo signal can be approximated to the LFM signal for the purpose of illustrating some concepts, because the components higher than a quadratic

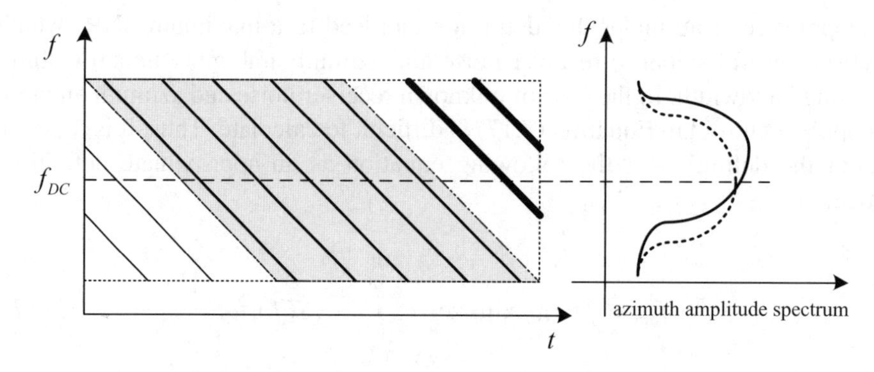

Figure 7.12 The distribution of the azimuth echo signal in the time–frequency domain

one in target phase history are relatively small. Hence, the distribution of the signal in the time–frequency domain can be illustrated by Figure 7.12.

In Figure 7.12, the signals within the shaded quadrilateral have complete Doppler history. By contrast, the Doppler histories of the signals outside the quadrilateral are incomplete. If the scene is relatively homogeneous and the signal intensity on both sides of the scene is comparable, the azimuth spectrum of the signal can well represent the shape of antenna beam pattern, with the peak value at the centroid frequency and good symmetry on both sides (as is shown by the dotted line on the right side). Thereby, an accurate estimate is possible using the CDE algorithm. On the contrary, if the signal intensity on the two sides of the scenario contrasts sharply (in the figure, the thick and thin lines represent strong and weak signals, respectively), the power spectrum of the azimuth signal will lean to one side (as is shown by the solid line on the right side of the figure). In that case, there will be large errors in the estimation of CDE. If the incomplete signal is eliminated in the time–frequency domain while the complete Doppler signal is reserved, the phenomenon of spectrum offset can be avoided and the estimation precision of the CDE algorithm can be improved.

As for the time–frequency analysis, there are quite a lot of methods, such as the Winger–Ville (WV) distribution, the short-time Fourier transform (STFT) and the wavelet analysis. Among them, the definition of the WV time–frequency distribution is

$$WVD_x(t, f) = \int_{-\infty}^{+\infty} x\left(t + \frac{\tau}{2}\right) x^*\left(t - \frac{\tau}{2}\right) \exp\{-j2\pi f\tau\} d\tau \tag{7.20}$$

It can be seen that the spectrum of

$$h(\tau|t) = x\left(t + \frac{\tau}{2}\right) x^*\left(t - \frac{\tau}{2}\right)$$

at a certain time t is observed by the WV time–frequency distribution. Because the time width of τ in $h(\tau|t)$ can be very wide and the interval of the observation time t can be as short as required, this method can offer a good time–frequency resolution. However, as the integral kernel function is in quadratic form, there will be cross terms for multicomponent signals. Thus, this method is usually applied to a scene containing dominant scatters, where the Doppler centroid and the Doppler FM rate can be estimated by analyzing the WV time–frequency distribution of a certain dominant scatter; however, it is not suitable for the elimination of partially exposed targets. On the other hand, the wavelet transform projects the signal into a set of wavelet bases, and thus the signal can be analyzed in the time–frequency domain for different scales. However, the calculation is a little complex and not as efficient as STFT. Therefore, the STFT method, which is simple and direct, is chosen here for the time–frequency analysis. As shown by the experiments, the STFT method is efficient and appropriate.

The idea of STFT is very simple: in order to make the frequency spectrum reveal time characteristics, the spectrum is observed only over a certain period of the time-domain signal for each time; that is, the signal is multiplied by a time-domain window. Here, to insure the uniqueness of the segmental signal and to simplify the processing, there is no intersection between the time segments. The STFT adopted here can be expressed as

$$\begin{cases} F(nT, f) = \displaystyle\int_{-T/2}^{T/2} x(nT - \tau)\exp\{-j2\pi f\tau\}d\tau \\ x(nT + \tau) = \displaystyle\int_{-T/2}^{T/2} F(nT, f)\exp\{j2\pi f\tau\}df \end{cases}, n = \ldots, -1, 0, 1, \ldots \quad (7.21)$$

where T is the length of each time segment. As shown in the equation above, owing to the division of time into several segments, in the time–frequency domain, the time resolution of the signal is T, whereas the frequency resolution is $1/T$. As a consequence, by STFT, a good time resolution and a good frequency resolution cannot be gained at the same time. If T is too long, the time resolution is so low that it is impossible to obtain the incomplete signal region accurately during time–frequency filtering and will result in poor filtering accuracy. In the same way, if T is too short, the frequency resolution becomes too low. Besides, as the azimuth echo signal is approximately a linear FM signal, T that is too short can make the time bandwidth product of the signal too small, which can lead to a spread spectrum. This will also cause poor filtering accuracy. Figure 7.13 explains the above phenomenon. In the figure, a rectangular grid is a time–frequency resolution cell and the full-aperture signal corresponds to the dot-shaded area, while the hatched area represents the reserved area during filtering. Obviously, if T is too long or too short, the reserved

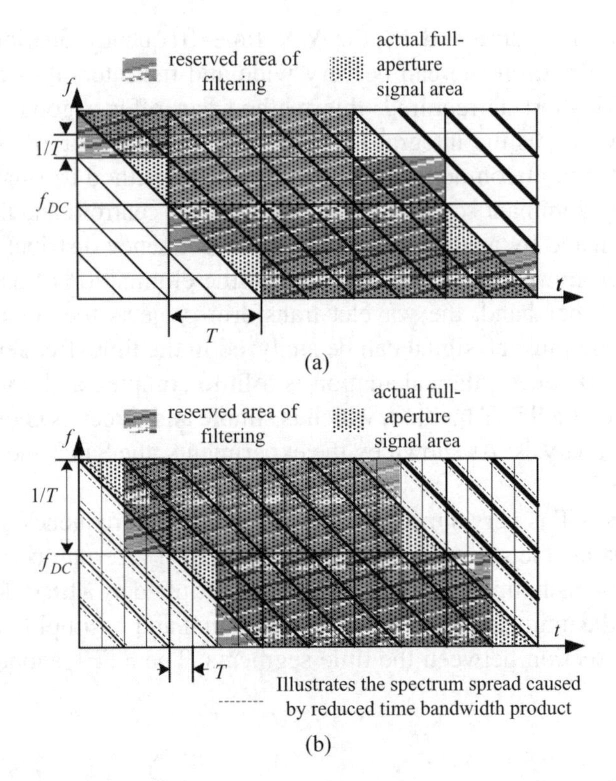

Figure 7.13 Diagram of signal filtering in the time–frequency domain: (a) T is too long, (b) T is too short

area of filtering cannot fit the full-aperture signal area very well, which will increase the filtering error (shown as the "sawtooth" area that contains only a dot-shaded area or only a hatched area in Figure 7.13).

As it is a discrete signal that is processed in digital signal processing, here M represents the number of samples contained in each time segment of length T and N_{syn} is the number of samples during the synthetic aperture time. Simulation experiments show that the mean value and standard deviation of estimation errors (which indicate the accuracy and stability of the estimation) keep a relatively steady level within a quite large range of M. Besides, when the time bandwidth product is larger than 10 and the time and frequency resolution are well balanced (i.e., M is close to $\sqrt{N_{syn}}$), the mean value and standard deviation of estimation error tend to the minimum (as illustrated by Figure 7.14).

Therefore, with regard to the choice of segmental sample number M, the following experiential method can be adopted. Firstly, the value of M should ensure that the time bandwidth product is larger than 10, that is, $|f_r M^2/f_s^2| > 10$. Secondly, there should be a balanced tradeoff between time and the frequency resolution, that is, M is close to $\sqrt{N_{syn}}$.

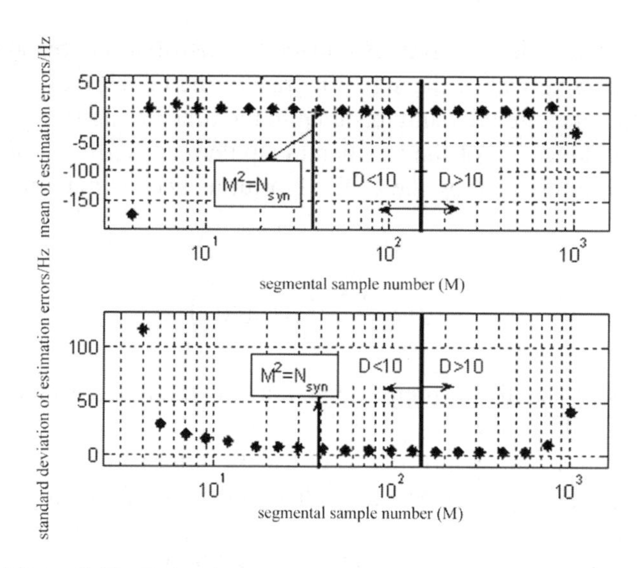

Figure 7.14 Estimation error versus the segmental sample number

7.3.2.2 Algorithm Flow

1. The initial Doppler parameters (including the Doppler centroid f_{DCO} and the Doppler FM rate f_{r0}) are obtained through calculations based on the space-borne/airborne SAR geometry and the information of beam direction, or through traditional estimation methods.
2. The echo data (either before or after range compression) are divided into different blocks along the range direction. Some or all of the blocks are chosen as estimation blocks. Meanwhile the processed signal length in the azimuth direction is larger than one (better if larger than two) synthetic aperture time.
3. Every azimuth line within each estimated block is processed by STFT; that is, the azimuth signal is divided into time segments of length M and each time segment is transformed by Fourier transform, thus forming a two-dimensional time–frequency domain, with one axis denoting the sequence number of each time segment and the other denoting frequency. The length of time segment M is chosen according to the experiential method mentioned above.
4. Based on the initial Doppler centroid and the FM rate, the distribution area of the full-aperture signal in the time–frequency domain is calculated. The time range of this area at each f is determined by the range of t_a, which satisfies

$$0 \le t = \frac{f - f_{DCO}}{f_{r0}} + t_a \le T_a, \ \forall f \in [f_{DCO} - f_P/2, f_{DCO} - f_P/2]$$

where f_P is the pulse repetition frequency and T_a represents the time length of the azimuth signal.

5. The time–frequency domain signals outside the distribution area above are set to zero. Then each segment of the signal is transformed back into the time domain, resulting in the pre-filtered azimuth signal in the time domain.
6. The estimated value $\hat{f}_{DC}$ of the baseband Doppler centroid is estimated from the pre-filtered azimuth signal by the CDE algorithm. The estimation method can be expressed as

$$\hat{f}_{DC} = \frac{f_p}{2\pi k} \arg\{r(k)\}$$

   where $k = 1$ will usually be preferred and $r(k)$ is the autocorrelation function of the azimuth signal.
7. If the estimation accuracy still does not satisfy, return to step 3 to estimate the parameters again, using the estimated values above as the initial parameters.
8. Given the estimates from each estimation block, a polynomial fit is made to obtain the Doppler centroid estimates of each range cell in the echo data.

As can be seen from the steps above, if there are errors in the initial Doppler parameters, there will be errors in the calculation of full-aperture signal distribution, resulting in imperfect elimination of partially exposed signals. This will affect the accuracy of the algorithm. However, as the time length of each azimuth segment is far shorter than that of the original signal, the frequency resolution of the time–frequency domain is relatively low. Thus, this method is not sensitive to the errors of the initial Doppler parameters. This will be verified by the experimental results below.

7.3.2.3 Experimental Results

This algorithm was applied to the Doppler centroid estimation of both simulated echo data and real echo data of RadarSAT-1 (obtained from the CD included in [20]). The results are as follows:

1. Simulated data. To observe the estimation accuracy of the algorithm in the case where the scenario has high contrast on two sides, the simulation is based on a scenario of the land and sea border shown in Figure 7.15. The X-band spaceborne SAR echo data are simulated by considering the first sidelobe in the azimuth. The synthetic aperture length of the simulated data is 1509 pulses; 4096 pulses are used in the estimation while the range block depth is 256 range gates and f_p is 3460 Hz.

   The azimuth spectrum before and after time–frequency domain filtering is shown in Figure 7.16. It can be seen that the azimuth spectrum centroid deviates from the real Doppler centroid before filtering, while the azimuth spectrum centroid becomes very close to the real Doppler centroid after filtering. The estimation errors of 16 range estimation blocks are shown in Figure 7.17. The length of each

Figure 7.15 The scenario for echo data simulation

subsegment used in the time–frequency analysis is $M = 146$. As illustrated in the figure, when there is no time–frequency domain pre-filtering, the estimates deviate seriously from the real values, with estimation errors up to 300 Hz ($10\% f_p$). After time–frequency domain pre-filtering using these inaccurate estimations as initial Doppler parameters, the estimation errors decrease to less than 10 Hz. Furthermore, as illustrated in Figure 7.17, even if the initial FM rate error is up to 5%, the Doppler centroid estimation error is still less than 30 Hz, which indicates that this method is not sensitive to errors of the initial Doppler parameters. Table 7.1 shows the statistics of estimation errors of the 16 estimation blocks. It can be seen that the time–frequency domain pre-filtering does improve the estimation accuracy of the Doppler centroid in scenarios where there is a high contrast on the two sides.

2. Real echo data of RadarSAT-1. In the following, this method is applied to real echo data of the RadarSAT-1 in the Vancouver scenario (as shown in Figure 7.18). There are 704 pulses within a synthetic aperture time. Here 16 384 azimuth pulses are processed and estimated separately by dividing the pulses into eight different azimuth blocks. While estimating, the range block depth is 256 range gates and the conventional CDE algorithm where there is no time–frequency pre-filtering is used for the first estimation. The results of this first estimation are used as the initial values of the second estimation (i.e., the first iteration) and the second estimation results are used as the initial values of the third estimation (i.e., the second iteration). A quadratic polynomial fit is then made to the estimation results (as shown by Figure 7.19) and the root-mean-square (RMS) values of the deviations between estimated values and the fitted curve are calculated and shown in Table 7.2.

It can be seen from Figure 7.19 that, without time–frequency domain pre-filtering, the estimation results fluctuate violently with the variation of the scene contents, which can be especially noticed in block 2 where the scene has high contrast on the two sides

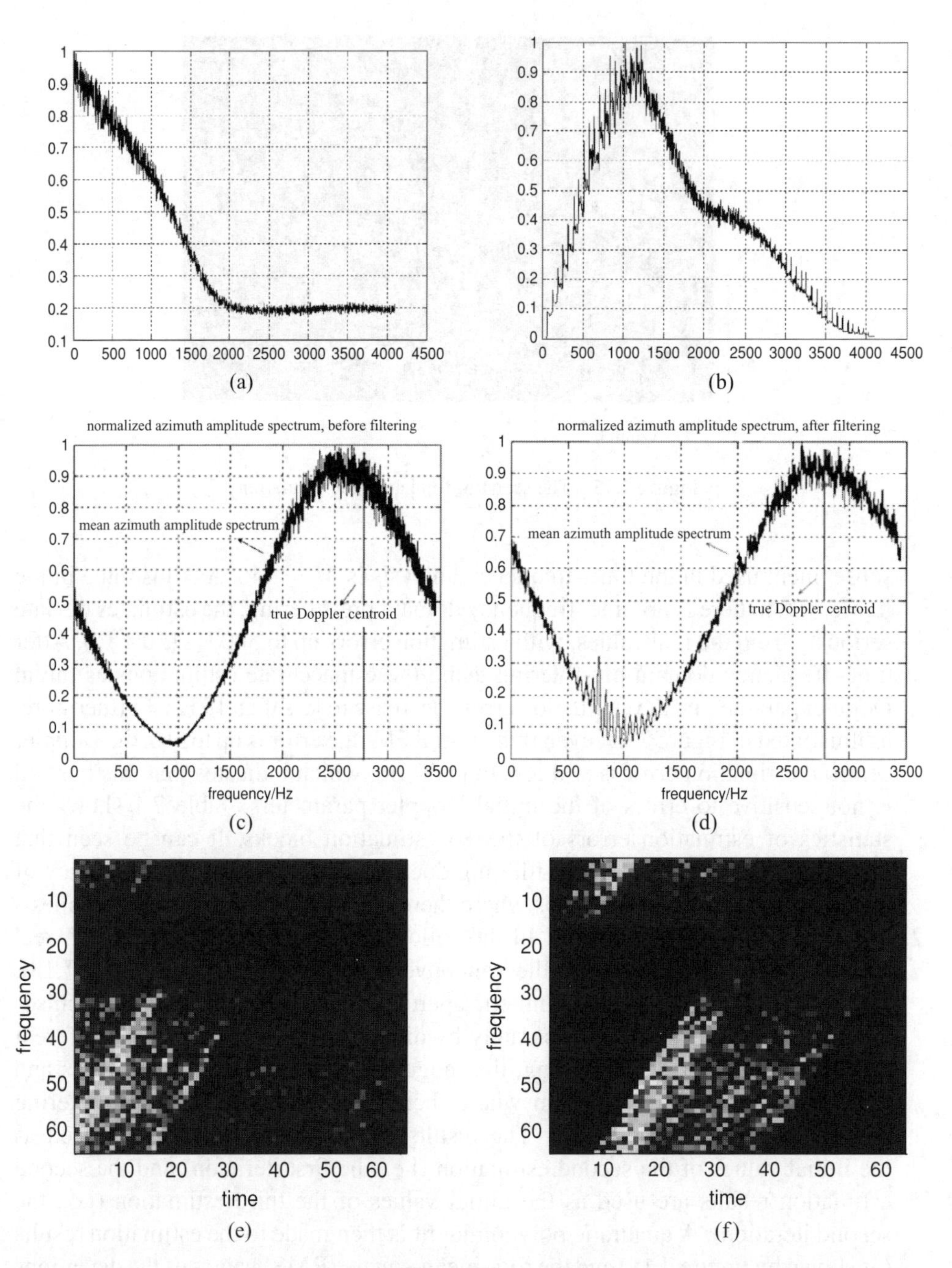

Figure 7.16 Comparison of simulated data before and after time–frequency pre-filtering: (a) the magnitude of the time domain signal before filtering, (b) the magnitude of the time domain signal after filtering, (c) the azimuth spectrum before filtering, (d) the azimuth spectrum after filtering, (e) the magnitude in the time–frequency domain before filtering, (f) the magnitude in the time–frequency domain after filtering

Table 7.1 Estimation results of simulated data

	Without time–frequency filtering	Time–frequency filtering with FM rate error of 5%	Time–frequency filtering with no FM rate error
Mean value (Hz)	−114.91	−2.99	3.18
Standard deviation (Hz)	74.06	8.05	3.26

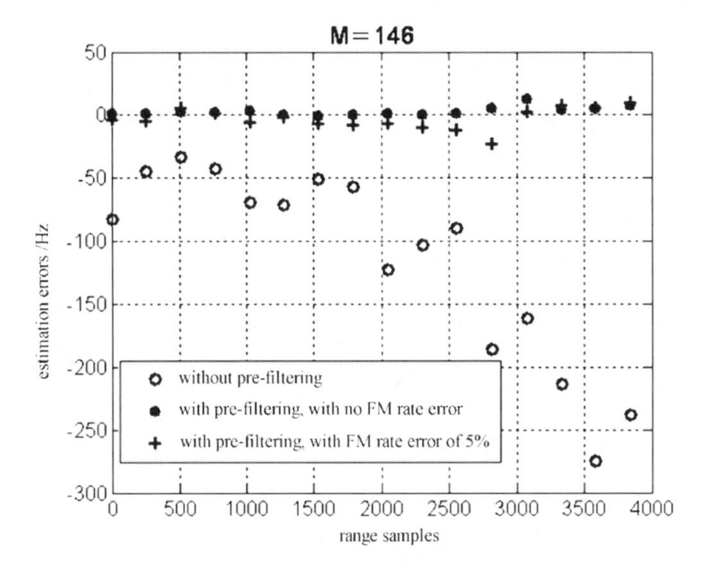

Figure 7.17 The estimation results of simulated data

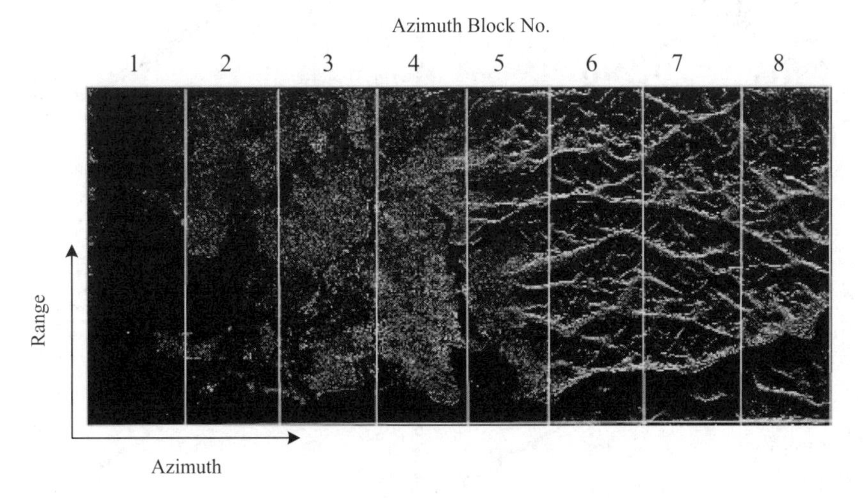

Figure 7.18 Vancouver scenario of RardarSAT-1

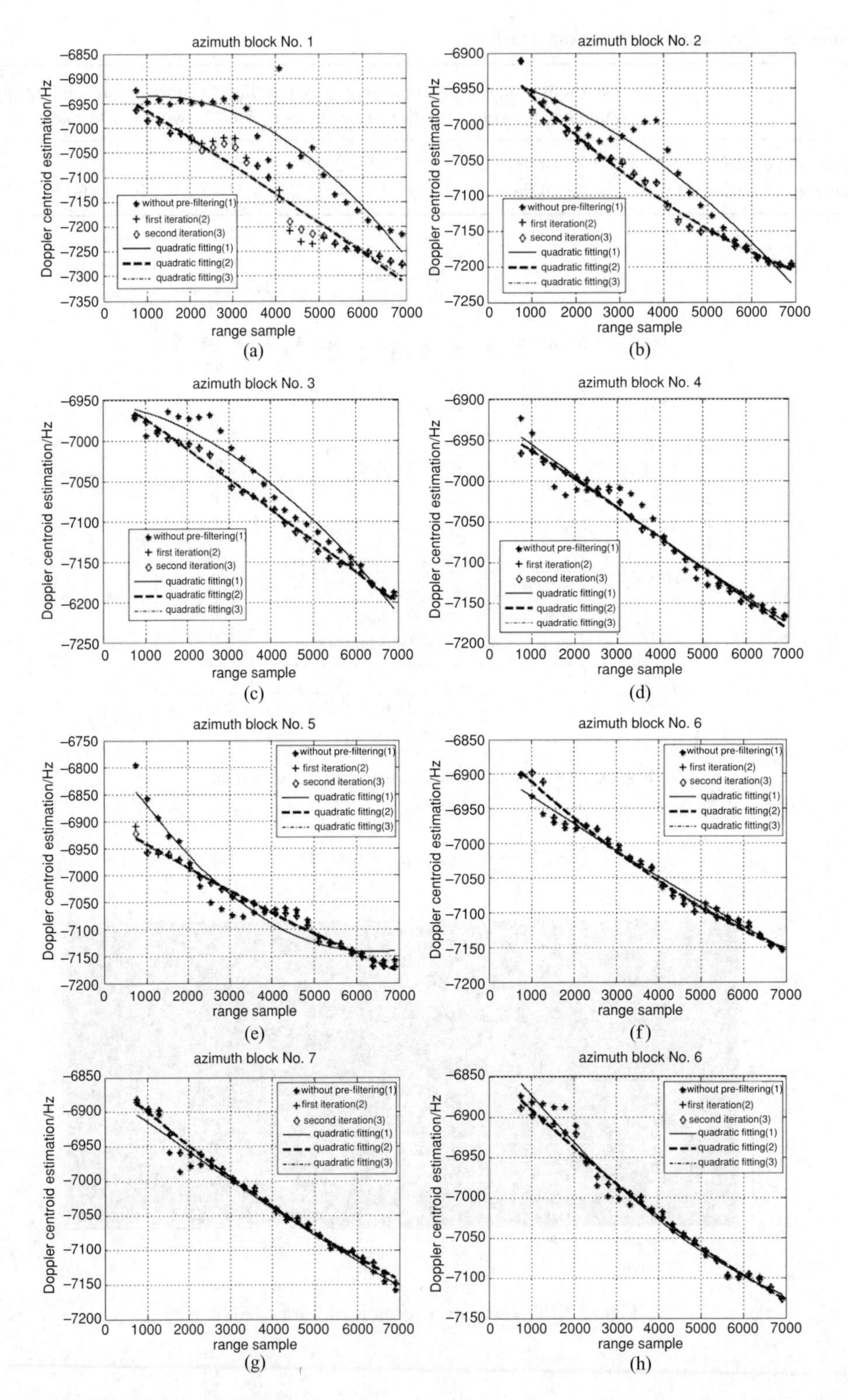

Figure 7.19 The Doppler centroid estimation values and the fitted curve of the Vancouver scenario: (a) azimuth block 1, (b) azimuth block 2, (c) azimuth block 3, (d) azimuth block 4, (e) azimuth block 5, (f) azimuth block 6, (g) azimuth block 7, (h) azimuth block 8

Table 7.2 The RMS deviations between the Doppler centroid estimation values and the fitted quadratic curve (Hz)

Block	1	2	3	4	5	6	7	8
Without time–frequency filtering	35.96	21.91	14.77	16.68	26.31	9.59	12.67	13.53
The first iteration	29.68	11.43	6.07	6.52	8.30	9.28	7.07	7.13
The second iteration	22.27	11.99	5.94	6.54	7.17	9.40	6.98	7.12

in many range blocks (as shown by Figure 7.19(b)). By contrast, after time–frequency domain pre-filtering, the fluctuation of the estimates decreases. Besides, when the scene is relatively homogeneous, the estimation results after time–frequency domain pre-filtering agree very well with the estimations without pre-filtering (such as in azimuth block 6). It can be seen from Table 7.2 that, with time–frequency domain pre-filtering, the RMS deviations of most azimuth blocks are less than 10 Hz, except blocks 1 and 2, where most are ocean areas. This means that the scene contents have little effect on Doppler centroid estimation with the pre-filtering method, and thus it is robust. Besides, compared with the second estimation, the third estimation results improve only a little bit, which implies that this method is not sensitive to the initial Doppler centroid errors. Thus, in most situations, only one iteration is enough.

In addition, it should be noted that as long as the estimation of the Doppler centroid is based on the correspondence between the azimuth signal spectrum and the azimuth beam pattern modulation, the method of time–frequency domain pre-filtering will work. Accordingly, though the experiments here are of the monostatic case, the method can also be applicable for Doppler centroid estimation in the bistatic SAR.

7.4 Principle and Methods of SAR Motion Compensation

7.4.1 Principle of Motion Compensation

The analysis in Section 7.2 shows that motion errors have a serious negative effect on SAR imaging; thus they should be appropriately compensated to get well-focused images. Motion compensation involves measurement or extraction of the motion errors of the antenna phase center as precisely as possible by motion sensors and/or signal processing methods and then removal of them from the SAR echo data. The motion compensation algorithm is a modification of the SAR imaging algorithm according to the nonideal aircraft motion conditions. Motion compensation makes the imaging algorithm match the SAR data and thus gets well-focused SAR images. Theoretically, the motion errors affect different targets of the scenario differently; therefore, only through motion compensation applied pixel by pixel can we get precisely focused images. However, pixel-by-pixel processing is a kind of ideal but is an unrealistic processing approach. As for fast imaging algorithms, the motion compensation algorithm should also be taken in the batch mode to guarantee the efficiency of SAR imaging.

Batch mode motion compensation cannot avoid introducing some approximation during the processing, but it could be applied as long as the approximation does not affect the SAR imaging quality to a certain degree.

Generally speaking, motion compensation can be divided into sensor-based motion compensation and SAR data-based motion compensation. Sensor-based motion compensation uses the strapdown inertial navigation system (SINS), the global positioning system (GPS) and some other instruments to measure the aircraft trajectory, and then carry out compensation for SAR data during an imaging process based on the measurements.

Early, the inertial navigation system (INS) is generally used as the motion sensor on the airborne platform and sensor-based motion compensation is carried out according to the aircraft motion information output from INS. However, as the main purpose of INS is navigation, it is designed specially for aircraft navigation; therefore, the point of emphasis is different from that for motion error measurements. At the same time, the radar antenna phase center is somewhat distant from the aircraft INS in general. When the aircraft attitude changes, the motion information measured by the aircraft INS cannot represent the motion condition of the antenna phase center and leverage correction must be done. As the radar antenna phase center is quite distant from the INS, the variance in the lever is too complex to accurately simulate through a simple model; therefore, motion compensation based on the information gathering by INS has great limitations.

In order to measure the motion errors as accurately as possible, a modern airborne SAR usually installs the strapdown inertia measuring unit (SIMU) – which is close to the antenna phase center as much as possible – to be a motion sensor. SIMU consists of accelerators and gyros, both in three directions. During the aircraft movement, SIMU measures the angular velocity and acceleration of the antenna platform relative to the inertial space and then the accurate motion information of the SIMU mount point can be obtained through combining the SINS mechanical equations. As the SIMU mount point is quite close to the antenna phase center, the structure between them can be considered as rigid, so accurate motion information of the antenna phase center can be obtained through a short leverage correction.

However, SIMU is an inertial measurement device, which has its unavoidable shortcoming, that is, the drift phenomenon, which means the positioning errors will increase rapidly with time; thus it is unsuitable for being able to work independently over a long time. Therefore, it is important to study methods that use external information to control the inertial device drift. In the 1980s, Difilippo in Canada combined the Doppler velocity sensor and atmospheric pressure altimeter with the aircraft INS and used the Kalman filter to provide the signal combination. Field experiments demonstrated that the combined system could obtain the desired motion compensation result.

With the development of GPS technology, people found that although the short-term precision of the GPS signal is not high enough, it had the feature of no divergence and good long-term accuracy, which is coincidently the complementary characteristic of the inertial navigation system. Therefore, the combination of SINS and GPS becomes the most accurate integrated navigation system used today. Lynx SAR, which weas developed by American Sandia Nation Laboratory, is equipped with a high-accuracy motion compensation system called Momeans to achieve the ultrahigh resolution of 0.1 m. Its motion measuring device SIMU uses gyros with drift less than 0.01° and accelerometers with an accuracy of 50 μg. In order to solve the drift problem of the inertial device, it combines the carrier-phase differential GPS with SIMU and uses the Kalman filter to fulfill integrated filtering. The final measurement accuracy of Momeans could achieve less than 0.025 m in high-frequency positioning and less than 1 cm/s in velocity measuring.

As mentioned above, the trajectory recorded by the sensor is usually not the trajectory of the antenna phase center, but the trajectory of the sensor itself. So some transformations between the trajectory of the sensor and the trajectory of the antenna phase center are needed based on the geometric relationship. Since here only the principle of motion compensation methods is introduced, the "aircraft trajectory" referred to in the following indicates the trajectory of the antenna phase center. The precision of the sensor-based motion compensation method greatly depends on the measurement accuracy of the sensor and more analysis about the sensor-based motion compensation method can be seen in [21].

If the accuracy of the motion sensor is high enough, the motion errors can be well corrected by the sensor-based motion compensation, and thus the desired image quality may be obtained. However, when the sensor has limited measurement accuracy about the trajectory, after the sensor-based motion compensation there will still be some motion errors in the SAR data. These can be further extracted and compensated based on the SAR data itself; this is the SAR data-based motion compensation. The SAR data-based motion compensation does not depend on any external information, but only uses the SAR data itself and the provided initial imaging parameters. The SAR data-based motion compensation corrects the imaging parameters or compensation motion errors based on the knowledge of the essential characteristics of the SAR data; therefore, it is usually called "auto focusing". In the following, the bistatic SAR motion compensation methods are introduced from these two points of view, combined with the bistatic characteristic.

7.4.2 Sensor-Based Motion Compensation

Generally, the sensor-based motion compensation is the first choice for SAR motion compensation, because it can compensate the various effects on SAR data caused by motion errors, as it uses the aircraft trajectory information as a basis. The sensor-based motion compensation usually includes the following steps: range resampling, azimuth resampling and the azimuth phase error compensation. Among these steps, range resampling is used to eliminate disorder in the echo storage location caused by the motion errors, as long as the range history difference between errors with and without motion is known; there is no difference in this step for bistatic SAR and for monostatic SAR. Azimuth resampling is used to solve the problem of nonuniform space intervals between azimuth samples. Some monostatic SAR maintains the uniform azimuth space intervals in nonuniform flight by real-time adjusting of PRF, which is done according to the information of ground speed tracking. However, the accuracy of ground speed tracking is usually limited, so this approach may not guarantee a satisfying precision. In bistatic SAR, because of the separation of the transmitter and receiver, the PRF real-time adjustment can only guarantee the uniform space intervals between the places where pulses are transmitted, but it cannot simultaneous guarantee a set of uniform space intervals between where pulses are received. Therefore, this problem can be solved only through the post-processing method. Finally, the azimuth phase error compensation is used to eliminate the effects of motion errors on the phase of the SAR echo. The principle of bistatic SAR azimuth phase error compensation is consistent with that in monostatic SAR.

From the aspect of realization, the sensor-based motion compensation algorithm is generally divided into two steps, which are referred to as the first-order motion compensation and the second-order motion compensation. Here, the first-order motion compensation is the bulk compensation that ignores the range space variance and the second-order motion compensation handles the range-variant errors. If the motion errors are large and the desired resolution is high, then this two-step motion compensation may still not meet the requirement of focusing accuracy. In this case, further compensation of the azimuth space-variant errors is needed. Readers interested can refer to [22] to [24] for more detail, but it will not be introduced here. Besides, the sampling frequency of the aircraft motion trajectory data recorded by the sensor is usually less than PRF and the aircraft trajectory recorded by the sensor is typically indicated by longitude, latitude, altitude and sampling time, which cannot be directly used for motion compensation. Therefore, it often needs preprocessing of the motion data before applying the motion compensation. According to the realization process of motion compensation, the following steps will be introduced one after another, which are the motion data pre-processing, the first-order motion compensation and the second-order motion compensation.

7.4.2.1 Motion Data Pre-processing

The motion data pre-processing usually includes the following steps:

1. Conversion from the latitude–longitude–altitude coordinates into the north–east–height coordinates. The aircraft trajectory data in the latitude–longitude coordinates is firstly converted into the north–east coordinates (e.g., the x axis indicates the east direction and the y axis indicates the north direction). Then, the origin of the north–east coordinates is set, for example, at the position where the first pulse is transmitted (that is, the first pulse corresponds to $x = 0$, $y = 0$). Besides, the height value given by the sensor is usually the altitude of the aircraft, so it usually should subtract the mean altitude of the scenario when converted into the north–east–height coordinates for the convenience of further processing.

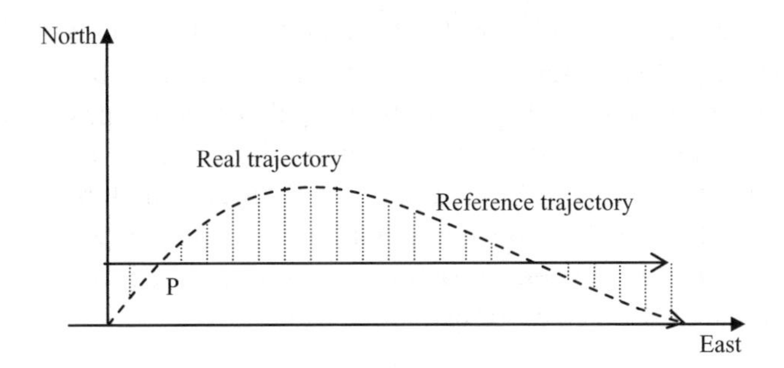

Figure 7.20 Diagram of the reference trajectory

2. Reference trajectory estimation. Choose an appropriate horizontal straight line to be the reference trajectory. Consequently, the purpose of motion compensation is to compensate the errors caused by deviation from this reference trajectory. As shown in Figure 7.20, the reference trajectory estimation includes the following two aspects:

   (a) Choose the appropriate reference path angle $\hat{\theta}$. The path angle is the one between the horizontal projection of the aircraft flight and the east direction (counterclockwise). In Figure 7.20, the reference path angle is 0°. (b) Choose an appropriate reference point $P(x_{ref}, y_{ref}, z_{ref})$. The point $P(x_{ref}, y_{ref}, z_{ref})$ is chosen to be the position through which the reference trajectory must pass. It determines the position of the reference trajectory in the three-dimensional space and the height of the reference trajectory ($h_{ref} = z_{ref}$). The reference path angle $\hat{\theta}$ and the value of x_{ref}, y_{ref} of the reference point P can be obtained by linear fitting to the horizontal projection of the trajectory within the entire azimuth span; z_{ref} (i.e., h_{ref}) can be chosen to be the mean value of the trajectory height within the entire azimuth span. Thus, the reference trajectory is the horizontal straight line that passes through the reference point P and intersects the east direction with angle $\hat{\theta}$ (counterclockwise).
3. Conversion from the north–east–height coordinates into the path-headed coordinates. The path-headed coordinate is a right-hand coordinate taking the reference trajectory as the x axis, the vertical upward direction is the z axis and the x position related to the first pulse transmitted time is zero. If expressed by functions, then for (x, y) in the north–east coordinates, the corresponding (x', y') in the path-headed coordinates is

$$x' = x \cos \hat{\theta} + y \sin \hat{\theta} \tag{7.22}$$

$$y' = -x \sin \hat{\theta} + y \cos \hat{\theta} \tag{7.23}$$

with further translation to make the first pulse transmitted to be at $x' = 0$ and the y'_{ref} of reference point P equal to zero. Besides, there is no processing for the height data during this coordinates conversion.

4. Motion data interpolation. Generally, the sampling rate of the sensor is far lower than PRF of the SAR system; thus interpolation is usually needed to obtain the aircraft trajectory data with a high sampling rate that corresponds to each SAR echo pulse. The interpolation needs to be done in all three axes in path-headed coordinates according to the time information of the SAR echo and the time information of the motion data. As for the interpolation method, piecewise cubic Hermite interpolation is recommended. In order to ensure the interpolation accuracy, the sampling frequency of the sensor cannot be too low.

In the bistatic SAR, there are two trajectories because of the existence of two different antennas, so each trajectory needs to be processed separately. In the translational invariant configuration, the path angles of the transmitter and receiver trajectories are almost the same; therefore, the path angle and reference point of any trajectory can be used as a reference when establishing the path-headed coordinates.

7.4.2.2 First-Order Motion Compensation

The first-order motion compensation is to compensate the bulk effects on the SAR echo. The bulk effects are the motion error effects on the echo of targets that are at the reference range (usually the range of the swath center is taken as the reference range). As shown in Figure 7.21, y_η indicates the intersection of the bistatic synthetic beam center and the swath center (i.e., where $x = x_{ref}$) at the instant η. According

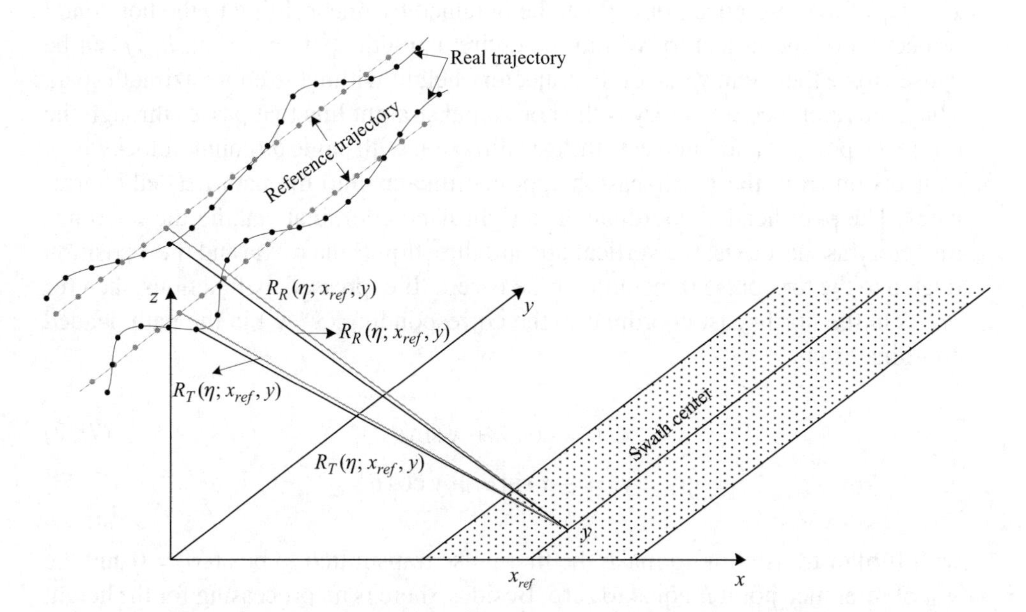

Figure 7.21 Diagram of first-order motion compensation of the bistatic SAR

to Equations (7.10), (7.11) and (7.14), the motion error of the swath center can be calculated, which is

$$\Delta R_{XZ}(\eta; x_{ref}, y_\eta) = \tilde{R}(\eta; x_{ref}, y_\eta) - \tilde{R}_y(\eta; x_{ref}, y_\eta) \tag{7.24}$$

The above range difference is used for the first-order phase compensation and range resampling. The first-order phase compensation performs the following multiplication for each pulse of $s(t, \eta)$, which is

$$s_1(t, \eta) = s(t, \eta) \exp\left\{ j \frac{2\pi}{\lambda} \Delta R_{XZ}(\eta; x_{ref}, y_\eta) \right\} \tag{7.25}$$

The range resampling is usually conducted in the range frequency domain; that is, $s_1(t, \eta)$ is transformed to the range frequency domain by Fourier transform to get $S_1(f, \eta)$ and then the following process is conducted:

$$s_2(t, \eta) = IFT\left\{ S_1(f, \eta) \exp\left\{ j \frac{2\pi f}{c} \Delta R_{XZ}(\eta; x_{ref}, y_\eta) \right\} \right\} \tag{7.26}$$

where IFT indicates the inverse Fourier transform.

The above compensation considers the motion errors in the trajectory perpendicular plane; the compensation of along trajectory errors will be introduced in the following. As mentioned previously, because of the separation of the transmitter and receiver in bistatic SAR, the places where a pulse is transmitted and received are different, and the motion errors of the transmitter and receiver are also different. Therefore, the space intervals between the places where pulses are transmitted and received are both nonuniform. Then how can the azimuth motion errors for the bistatic SAR be compensated? Here a bistatic SAR echo azimuth resampling method based on an equivalent average sampling time is introduced. The resampling method is as follows.

From Figure 7.21, the bistatic range of the n pulse when there is no motion error can be written as follows:

$$\begin{aligned} R_Y(n; x, y) = & \sqrt{[X_T - x]^2 + [Y_T + V_T n\Delta\eta - y]^2 + H_T^2} \\ & + \sqrt{[X_R - x]^2 + [Y_R + V_R n\Delta\eta - y]^2 + H_R^2} \end{aligned} \tag{7.27}$$

where $\Delta\eta = 1/PRF$. Similarly, when there are azimuth motion errors, the range history of the n pulse is

$$\begin{aligned} \tilde{R}_Y(n; x, y) = & \sqrt{[X_T - x]^2 + [Y_T + V_T(n\Delta\eta + \Delta Y_T(n\Delta\eta)/V_T) - y]^2 + H_T^2} \\ & + \sqrt{[X_R - x]^2 + [Y_R + V_R(n\Delta\eta + \Delta Y_R(n\Delta\eta)/V_R) - y]^2 + H_R^2} \end{aligned} \tag{7.28}$$

From Equation (7.28), it can be perceived that the transmitter and the receiver are supposed to be moving at constant speeds V_T and V_R, respectively, while the n pulse is supposed to be transmitted and received at $n\Delta\eta + \Delta Y_T(n\Delta\eta)/V_T$ and $n\Delta\eta + \Delta Y_R(n\Delta\eta)/V_R$, respectively. As the transmitting and receiving of the same pulse are considered as simultaneous under the stop–go assumption, here we define $\tilde{\eta}_n$ as the equivalent time at which the n pulse is transmitted/received under the condition of azimuth motion errors:

$$\tilde{\eta}_n = n\Delta\eta + \frac{1}{2}\left[\frac{\Delta Y_T(n\Delta\eta)}{V_T} + \frac{\Delta Y_R(n\Delta\eta)}{V_R}\right] \tag{7.29}$$

Based on the equivalent time above, azimuth resampling of the bistatic SAR echo can be performed to bistatic SAR raw data from nonuniform $\tilde{\eta}_n$ to uniform η_n by interpolation (such as the Lagrange interpolation method) and the SAR echo, which is hypothetically received under the uniform motion condition, can be obtained approximately.

The azimuth equivalent sample time in Equation (7.29) is a hypothetical equivalent time, so the resampling based on this equivalent time will induce errors. Apart from the error introduced by the resampling itself (this error also exists in the monostatic SAR resampling), the error caused by approximation of the equivalent sample time is the unique error in bistatic SAR. Thus, the second error is analyzed in the following.

If the transmitter and receiver are moving with constant velocities V_T and V_R, respectively, the bistatic range at $\tilde{\eta}_n$ is

$$\begin{aligned}\tilde{R}'_Y(n;x,y) = &\sqrt{[X_T - x]^2 + \left[Y_T + \frac{1}{2}\Delta Y_T(n\Delta\eta) + \frac{V_T}{2V_R}\Delta Y_R(n\Delta\eta) + V_T n\Delta\eta - y\right]^2 + H_T^2}\\ &+\sqrt{[X_R - x]^2 + \left[Y_R + \frac{1}{2}\Delta Y_R(n\Delta\eta) + \frac{V_R}{2V_T}\Delta Y_T(n\Delta\eta) + V_R n\Delta\eta - y\right]^2 + H_R^2}\end{aligned} \tag{7.30}$$

Therefore the error introduced by the equivalent time is

$$\begin{aligned}&|\tilde{R}'_Y(n;x,y) - \tilde{R}_Y(n;x,y)|\\ &\approx \left|\frac{V_T}{2}\sin\theta_T(n\Delta\eta;x,y) - \frac{V_R}{2}\sin\theta_R(n\Delta\eta;x,y)\right|\\ &\times|\Delta Y_T(n\Delta\eta)/V_T - \Delta Y_R(n\Delta\eta)/V_R|\end{aligned} \tag{7.31}$$

Typically, in the translational invariant configuration, V_T and V_R are the same (in practice they may not be exactly the same, but are controlled to be very close). If θ_T and θ_R are nearly the same, the first factor on the right side of Equation (7.31) may be close to zero. Thus, the azimuth resampling based on above method can effectively compensate the azimuth motion errors in the translational invariant case, especially when θ_T and θ_R are nearly the same.

7.4.2.3 Second-Order Motion Compensation

The second-order motion compensation is to compensate the residual error of the SAR data after the first-order motion compensation. Comparing Equation (7.14) with Equation (7.24), it can be found that the residual error after the first-order motion compensation is as follows:

$$\Delta R'_{XZ}(\eta; x, y) = \Delta R_{XZ}(\eta; x, y) - \Delta R_{XZ}(\eta; x_{ref}, y_\eta) \tag{7.32}$$

When ignoring the azimuth space-variance character of the error, the error that should be compensated in the second-order motion compensation is

$$\Delta R''_{XZ}(\eta; x, y_\eta) = \Delta R_{XZ}(\eta; x, y_\eta) - \Delta R_{XZ}(\eta; x_{ref}, y_\eta) \tag{7.33}$$

Typically, $\Delta R''_{XZ}(\eta; x, y_\eta)$ is often smaller than the range sampling interval. In this case, only the effect of this error on the echo phase needs to be compensated; that is, for each range cell of the time-domain SAR data after range compression and RCMC (denoted as $s_3(t, \eta)$), the compensation step is as follows:

$$s_4(t, \eta) = s_3(t, \eta) \exp\left\{ j\frac{2\pi}{\lambda} \Delta R''_{XZ}(\eta; x, y_\eta) \right\} \tag{7.34}$$

It should be noted that the imaging processing usually requires to process pulses more than one synthetic aperture length at one time and, in addition, it would usually like to process as many pulses as possible to generate a continuous long strip image without mosaic. However, as for the motion compensation, the more pulses it processes, the greater the motion errors come to deviate from the reference trajectory. As shown in Figure 7.22, for the aircraft trajectory C, within the short processing time T_A, the corresponding reference trajectory is A and it can be seen that the motion errors are

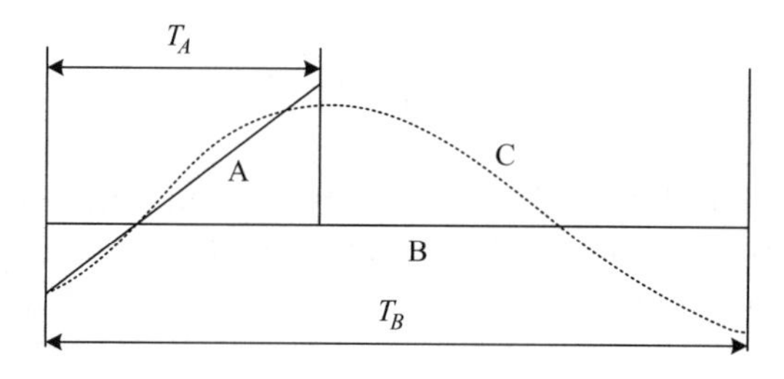

Figure 7.22 The number of processing pulses versus the amount of motion error

relatively small. However, if the processing time is the longer T_B, the corresponding reference trajectory is B and then the motion errors are bigger. Therefore, on the requirement of certain compensation accuracy, an appropriate number of processing pulses are needed.

7.4.3 SAR Data-Based Motion Compensation

After the sensor-based motion compensation, the SAR image quality is usually greatly improved, sometimes even already meeting the quality requirement. However, with the improvement of the SAR resolution, the accuracy of the motion sensor is required to be higher and higher. If the motion sensor is not accurate enough, then after the sensor-based motion compensation, there will usually be some azimuth phase errors left in the SAR data that affect the imaging quality. In this case, the SAR data-based motion compensation, which is also called auto focusing, is needed to estimate and compensate the azimuth phase error. With regard to auto focusing, there is no difference between the bistatic SAR and the monostatic SAR; therefore, the data-based phase estimation and compensation methods of the monostatic SAR can also be applied to the bistatic SAR.

There are various auto focusing methods that are commonly used, such as the Mapdrift method [25], the phase difference (PD) method [26], the phase gradient autofocus (PGA) algorithm [27], the rank one phase estimation (ROPE) [28] and so on. Among them, methods like Mapdrift and PD are based on the assumption that only low-order polynomial term phase errors exist, so they can only effectively remove the low-order components of the phase error but have no effect on the high-order phase components. The PGA algorithm, on the contrary, is not limited to the finite-order phase error model. It can estimate the phase error of any order and therefore is one of the most popular auto focusing algorithms, and has been studied in depth [27, 29–31]. The basic ideals of this auto focusing method are: looking for the strong targets of the scenario, estimating the azimuth phase error of each strong target and then gradually approaching the real phase error through iterations. Here the PGA algorithm is simply introduced in the following.

7.4.3.1 Principle of the PGA Algorithm

For a discrete chirp signal

$$s(n) = \exp\{j\pi k(nT_s)^2\} \tag{7.35}$$

where k is the linear FM rate, T_s is the sampling time interval, $n = -N/2, \ldots, N/2 - 1$ is the sampling count and N is the total sampling number. The matched filter of $s(n)$ in the frequency domain is

$$R(k) = FFT^*\{s(n)\} \tag{7.36}$$

where "*" means conjugate. Filtering $s(n)$ by this matched filter, this becomes

$$i(n) = IFFT\{FFT[s(n)]R(k)\} \tag{7.37}$$

where $i(n)$ is the compressed signal, whose peak appears when $n = 0$.

If there is a phase error $\varphi(n)$ in $s(n)$, that is,

$$s'(n) = s(n)\exp\{j\varphi(n)\} \tag{7.38}$$

then the corresponding compressed signal is

$$i'(n) = IFFT\{FFT[s'(n)]R(k)\} \tag{7.39}$$

If $i'(n)$ is known, the following method can be used to extract the phase error $\varphi(n)$, which is

$$s'(n) = IFFT\{FFT[i'(n)]R^*(k)\} \tag{7.40}$$

$$\varphi(n) = \angle\{s'(n)s^*(n)\} \tag{7.41}$$

where the operation of Equation (7.41) is often called "dechirp processing".

However, the angle-seeking operation of Equation (7.41) can only obtain the angle value within $[-\pi, \pi]$ for a complex signal and thus it cannot represent the original phase error curve. To solve this problem, the gradient domain angle-seeking method is used to rebuild $\varphi(n)$. Suppose that $f(n) = s'(n)s^*(n) = \exp\{j\varphi(n)\}$; then

$$f(n)f^*(n-1) = \exp\{j(\varphi(n) - \varphi(n-1))\} = \exp\{j\Delta\varphi(n)\} \tag{7.42}$$

$$\Delta\varphi(n) = \angle\{f(n)f^*(n-1)\} \tag{7.43}$$

$$\varphi(n) = \sum_{k=0}^{n} \Delta\varphi(k) \tag{7.44}$$

Because $\Delta\varphi(n)$, which is the gradient of $\varphi(n)$, is generally within $[-\pi, \pi]$, the angle obtained by Equation (7.43) is generally unambiguous. Therefore, the phase error curve $\varphi(n)$ can be rebuilt by accumulation (i.e., Equation (7.44)).

7.4.3.2 Implementation of PGA Algorithm

Based on the principle above, the implementation idea of the PGA algorithm is as follows. Firstly, find several isolated strong targets in the SAR complex image. Then add appropriate windows to the strong targets and estimate its phase error. Thirdly, average all the estimation results to get the estimated phase error curve. Then compensate the phase error according to the estimated phase error curve to obtain a better focused SAR complex image. Finally, iterate the above operations several times to obtain the final estimated phase error curve for the scenario and make compensation to the azimuth signal to produce the final quality improved image. The detailed steps of implementation of the classic PGA algorithm are as follows:

1. Target selection. Firstly, select a strong target in each range cell. The purpose of selecting a strong target is to guarantee a high SNR and a high SCR (signal-to-clutter ratio).
2. Circular shifting. Shift the signal of each strong target circularly to locate the peak of the signal at the azimuth left endpoint. Circular shifting is intended to remove the linear phase term in the phase spectrum caused by the position of each target.
3. Windowing. An appropriate unified window is added to the shifted signals of all the targets, thus further improving the SCR of each strong target.
4. Phase error gradient estimation. Apply the FFT to the windowed target signal $g(x)$ (x corresponds to the azimuth time domain coordinate) to obtain the spectrum, which is

$$G(u) = |G(u)| \exp(j[\phi_\varepsilon(u) + \theta(u)]) \tag{7.45}$$

where u corresponds to the azimuth frequency domain coordinate, $\phi_\varepsilon(u)$ is the frequency domain azimuth phase error and $\theta(u)$ is the residual phase term. After the circular shifting step, $\theta(u)$ only contains the constant component. Then the gradient phase can be calculated as follows:

$$\Delta\phi(u) = \Delta\phi_\varepsilon(u) + \Delta\theta(u) = \angle\{G(u)G^*(u-1)\} \tag{7.46}$$

where

$$\Delta\phi_\varepsilon(u) = \phi_\varepsilon(u) - \phi_\varepsilon(u-1) \tag{7.47}$$

$$\Delta\theta(u) = \theta(u) - \theta(u-1) \tag{7.48}$$

As $\theta(u)$ is constant after circular shifting, we have $\Delta\theta(u) = 0$, and then the gradient of the phase error $\phi_\varepsilon(u)$ is

$$\Delta\phi_\varepsilon(u) = \angle\{G(u)G^*(u-1)\} \tag{7.49}$$

5. Averaging. As $\phi_\varepsilon(u)$ in all the range cells are considered to be identical, the average of the estimated $\Delta\phi_\varepsilon(u)$ for each range cell is considered to be the estimation result of the phase error gradient in this iteration, which is still denoted as $\Delta\phi_\varepsilon(u)$.
6. Phase error rebuilding. Accumulate the phase gradient $\Delta\phi_\varepsilon(u)$ as

$$\phi_\varepsilon(u) = \sum_{m=1}^{u} \Delta\phi_\varepsilon(m) \tag{7.50}$$

to obtain the phase error curve $\phi_\varepsilon(u)$. Then eliminate the constant component of the phase error curve.

7. Phase error correction. Compensate the phase error $\phi_\varepsilon(u)$ for $G(u)$ and then apply IFFT to get the SAR image after phase error correction.
8. Algorithm iteration. Iterate the above seven steps several times to obtain the optimal focused result.

The PGA algorithm above is proposed for spotlight SAR and applies the phase estimation, averaging and correction in the frequency domain. It is based on an

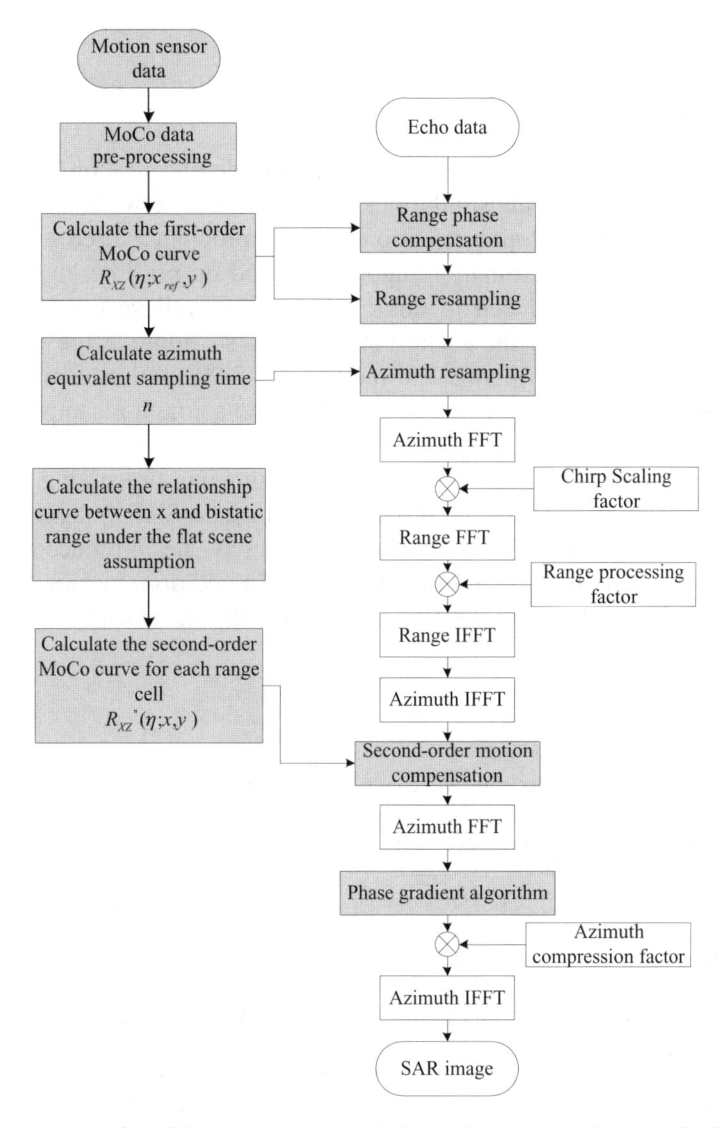

Figure 7.23 Diagram of the CS algorithm combined with motion compensation (MoCo) for the bistatic SAR

implied assumption, which is that all the targets in the scenario have experienced the same phase error in the time domain and they are all illuminated by the SAR antenna within the entire azimuth time span. However, in stripmap SAR, the azimuth time span of each target is different and the experienced phase error is different as well. Therefore, the application of the PGA algorithm in stripmap SAR needs further consideration. Readers interested in this could refer to [24].

In summary, combined with the CS algorithm based on the hyperbolic approximation, which was introduced in Chapter 5, the diagram of motion compensation and imaging processing for the bistatic translational invariant SAR can be found, as shown in Figure 7.23.

7.5 Summary

This chapter first analyzes the effects that the attitude and motion errors have on SAR imaging and point out the similarities and differences between monostatic SAR and bistatic SAR. For theDoppler centroid drifting problem caused by the attitude error, a Doppler centroid estimation algorithm based on pre-filtering in the time–frequency domain is introduced in this chapter. It can relieve the dependence of the estimation accuracy on the scene contents and so improve the robustness of the estimation. Then, the principle and methods of SAR motion compensation are introduced and the special issue of bistatic SAR echo azimuth resampling is briefly described.

This chapter is a complement for bistatic SAR imaging processing and the purpose of the chapter is to make the discussion of bistatic SAR imaging processing more complete. Therefore, only some classic algorithms are introduced here, which might provide a preliminary guidance for readers. In addition, bistatic SAR motion compensation and parameter estimation are both worth further study.

References

[1] Tang, X. (2009) Study on modeling and compensation for the motion errors in airborne interferometric SAR. PhD thesis, Institute of Electronics, Chinese Academy of Sciences, Beijing.

[2] Zhang, C. (1989) *Synthetic Aperture Radar: Principles, Analysis and Utilization*, Science Press, Beijing.

[3] Rigling, B.D. and Moses, R.L. (2006) Motion measurement errors and autofocus in bistatic SAR. *IEEE Transactions on Image Processing*, **15** (4), 1008–1016.

[4] Li, Y., Zhang, Z., Xing, M., *et al.* (2008) Motion compensation for the airborne bistatic SAR. *Acta Electronica Sinica*, **36** (3), 421–427.

[5] Zhu, Z., Tang, Z. and Jiang, X. (2006) Research on bistatic SAR motion compensation. International Conference on Radar, pp. 1–4.

[6] Cantalloube, H.M.J. and Krieger, G. (2007) Elevation-dependent motion compensation for frequency-domain bistatic SAR image synthesis. IEEE International Geoscience and Remote Sensing Symposium (IGARSS'07), pp. 2148–2151.

[7] Nies, H., Loffeld, O., Natroshvili, K., *et al.* (2006) A solution for bistatic motion compensation. IEEE International Geoscience and Remote Sensing Symposium (IGARSS'06), pp. 1204–1207.
[8] Nies, H., Loffeld, O. and Natroshvili, K. (2007) Analysis and focusing of bistatic airborne SAR data. *IEEE Transactions on Geoscience and Remote Sensing*, **45** (11), 3342–3349.
[9] Yang, Y. and Pi, Y. (2008) Motion error of spaceborne–airborne hybrid bistatic SAR. *Acta Geodaetica et Cartographica Sinica*, **2**, 163–16.
[10] Yang, Y. and Pi, Y. (2008) Azimuth space-variant properties of spaceborne–airborne hybrid bistatic SAR. *Modern Radar*, **30** (3), 47–49.
[11] Peng, H. and Hu, D. (1993) Effect of satellite attitude parameters on SAR imaging processing. *Journal of Electronics (China)*, **15** (6), 567–574.
[12] Maeda, A., Tsuboi, A. and Komura, F. (1986) A fast and accurate algorithm for SAR Doppler center frequency estimation. IEEE International Conference on Acoustics, Speech and Signal Processing (ICASSP'86), vol. 11, pp. 1913–1916.
[13] Madsen, S.N. (1989) Estimating the Doppler centroid of SAR data. *IEEE Transactions on Aerospace and Electronic Systems*, **25** (2), 134–140.
[14] Bamler, R. (1991) Doppler frequency estimation and the Cramer–Rao bound. *IEEE Transactions on Geoscience and Remote Sensing*, **29** (3), 385–390.
[15] Zhu, Z., Tang, Z., Zhang, Y., *et al.* (2008) The estimation of doppler parameter of bistatic SAR based on radon translation. *Journal of Electronics and Information Technology*, **30** (6), 1331–1335.
[16] Ma, D., Li, W. and Le, Z. (1999) A new method to estimate Doppler parameters for synthetic aperture radar in target echo signals. In: *Proceedings of the IEEE International Geoscience and Remote Sensing Symposium (IGARSS'99)*, vol.1, pp. 521–523.
[17] Wei, Z. (2001) *Synthetic Aperpture Radar Satellite*, Science Press, Beijing.
[18] Chang, C.Y. and Curlander, J.C. (1992) Application of the multiple PRF technique to resolve Doppler centroid estimation ambiguity for spaceborne SAR. *IEEE Transactions on Geoscience and Remote Sensing*, **30** (5), 941–949.
[19] Jin, M.Y. (1996) Optimal range and Doppler centroid estimation for a ScanSAR system. *IEEE Transactions on Geoscience and Remote Sensing*, **34** (2), 479–488.
[20] Cumming, I.G. and Wong, F.H. (2004) *Digital Processing of Synthetic Aperture Radar Data: Algorithms and Implementation*, Artech House.
[21] Moreira, J.R. (1989) A new method of aircraft motion error extraction from radar raw data for real time SAR motion compensation. *IGARSS*, **4**, 2217–2220.
[22] Kirk, D.R., Maloney, R.P. and Davis, M.E. (1999) Impact of platform motion on wide angle synthetic aperture radar (SAR) image quality. In: *The Record of the 1999 IEEE Radar Conference*, pp. 41–46.
[23] Carrara, W., Tummala, S. and Goodman, R. (1995) Motion compensation algorithm for widebeam stripmap SAR. *Proceedings of the SPIE*, **2487**, 13–23.
[24] Meng, D. (2006) Research on Motion Compensation Algorithms for Airborne SAR. PhD thesis, Institute of Electronics, Chinese Academy of Sciences, Beijing.
[25] Ding, C. (2000) Research Report on Motion Compensation Technology. Research Report on Research of Spotlight SAR and Signal Processing. Institute of Electronics, Chinese Academy of Sciences, Beijing.
[26] Liu, X. (1998) Research on Spotlight SAR Signal Processing. PhD thesis, Institute of Electronics, Chinese Academy of Sciences, Beijing.
[27] Wahl, D.E., Eichel, P.H. and Ghiglia, D.C. (1994) Phase gradient autofocus – a robust tool for high resolution SAR phase correction. *IEEE Transactions on Aerospace and Electronic Systems*, **30** (3), 827–835.
[28] Li, L. (1998) Research on Problems of Motion Compensation for High Resolution Airborne SAR. PhD thesis, Beijing University of Aeronautics and Astronautics, Beijing.

[29] Guo, Z. (2003) Research and System Implementation on Motion Compensation for High Resolution Airborne SAR. PhD thesis, Institute of Electronics, Chinese Academy of Sciences, Beijing.
[30] Wahl, D.E., Jakowatz, C.V. and Thompson, P.A. (1994) New approach to StripMap SAR autofocus. In: *Proceedings of the 1994 6th IEEE Digital Signal Processing Workshop*, pp. 53–56.
[31] Thompson, D.G., Bates, J.S., Arnold, D.V. and Long, D.G. (1999) Extending the phase gradient autofocus algorithm for low-altitude stripmap mode SAR. In: *Proceedings of the IEEE International Geoscience and Remote Sensing Symposium (IGARSS'99)*, vol. 1), pp. 564–566.

Index

Bistatic SAR Data Processing Algorithms, First Edition. Xiaolan Qiu, Chibiao Ding and Donghui Hu.